The Scientific Renaissance
1450–1630

harper ✦ torchbooks

A reference-list of Harper Torchbooks, classified by subjects, is printed at the end of this volume.

THE RISE OF MODERN SCIENCE

GENERAL EDITOR: A. RUPERT HALL

MARIE BOAS

The Scientific
Renaissance
1450-1630

HARPER TORCHBOOKS ❧ The Science Library
Harper & Row, Publishers, New York

In Memoriam R. P. B.

CONTENTS

GENERAL INTRODUCTION *page* vii

PREFACE xi

I The Triumph of Our New Age 17

II The Pleasure and Delight of Nature 50

III The Copernican Revolution 68

IV The Great Debate 90

V The Frame of Man and its Ills 129

VI Ravished by Magic 166

VII The Uses of Mathematics 197

VIII The Organisation and Reorganisation of
Science 238

IX Circles Appear in Physiology 265

X Circles Vanish from Astronomy 287

XI Debate among the Stars 313

Epilogue 344

BIBLIOGRAPHY AND NOTES 351

INDEX 367

ILLUSTRATIONS

*The following illustrations will be found
in a group following page 96:*

Map from a sixteenth-century portolano atlas
A Ptolemaic map
The pea
The aurochs
The bishop fish
Francis Bacon, studio of P. Van Somer
The Portuguese man-of-war
Tycho Brahe and his mural quadrant
Mathematical instruments, detail from Holbein's Ambassadors
An anatomical demonstration
Vesalius demonstrating the muscles of the arm
A human skeleton from Vesalius
An alchemical laboratory
Renaissance engineering: a pump
Renaissance engineering: a crane
Kepler's House of Astronomy
Galileo in old age

GENERAL INTRODUCTION

The endeavour to understand events in nature is as old as civilisation. In each of its three great seminal areas—the Chinese, the Indian and the West Asian-European—men tried to find a logic in the mysterious and an order in the chaotic. They made many attempts, sometimes revealing strange similarities in these totally different societies, to express general truths from which particular events would follow as rational, comprehensible consequences. They tried to describe and analyse in order to understand, for men could not live in the world without seeking to assign causes to the things that happen in it.

This series of volumes on *The Rise of Modern Science* describes the fruition in Europe of one of these attempts to describe and analyse nature. Modern science is not merely European ; even before it had entered upon its triumphant age its establishment in North America and China had begun, and the origins of the intellectual tradition from which it sprang must be sought in Egypt and Western Asia. But the revolution in ideas which alone made modern scientific achievements possible occurred in Europe, and there alone, creating an intellectual instrument so universal and so powerful that it has by now entirely displaced the native scientific traditions of non-European societies.

The present volume, *The Scientific Renaissance*, describes the early stages of this Scientific Revolution, beginning with what is traditionally (but somewhat inaccurately) known as the Renaissance of Learning in the fifteenth century. The Scientific Revolution was the effect of a unique series of innovations in scientific ideas and methods ; it gave the key to the understanding of the structure and relations of things. It was (and still remains) the greatest intellectual achievement of man since the first stirrings of abstract thought, in that it opened the whole physical universe—and ultimately human nature and behaviour—to cumulative exploration. Of its practical and moral implications we only now begin to have an inkling. For this colossal accomplishment Europe owed much to the Oriental world of which it then knew little. The vehicles of modern science, paper and printing, derived from China ; the language of science is still expressed in numerals devised in India ; Europe drew likewise on the East for its first knowledge of some phenomena (such as those of the magnetic compass), of some substances (such as saltpetre), and of some industrial techniques that relate to experimental science. But Europe did not borrow scientific ideas from the East, and in any case the borrowings had ceased before the rise of modern science began.

For this reason these volumes will make only incidental allusion to science outside the Europeanised world. Europe took nothing from the East without which modern science could not have been created ; on the other hand, what it borrowed was valuable only because it was incorporated in the European intellectual tradition. And this, of course, was founded in Greece. The Greek philosophers, imposing no bounds on intelligence but those of the universe itself, set at the very root of the European tradition of science the ideal of an interlocking system of ideas sufficient to explain all the variety of nature. They were, above all, theoretical scientists but at the same time they discussed critically the relationship between theories and the actual perception of events in nature. They began both observational biology and mathematical physics. Through most of two thousand years Europe continued to see nature through Greek eyes. Although the Scientific Revolution ultimately came as a reaction against the dogmatism inherent in the emulation of antiquity, it too drew its inspiration in part from neglected aspects of the Greek legacy. As Galileo admired Archimedes no less than Harvey did Aristotle, so the " mechanical philosophy " that flourished in the seventeenth century looked back to Epicuros and Lucretius. The Scientific Revolution did not reject Greek science ; it transformed it. Therefore the first volume in this series will be devoted to the scientific attitude of the Greeks, and its relation to the modern achievements of science. For it is impossible to understand fully what kind of changes in ideas were required to bring modern science into being, without considering the strengths and limitations of the Greek outlook.

This outlook reached the Europe of early modern times in complex ways, partly directly, partly through the Romans, partly through the Arabic-speaking peoples, partly through the immediately antecedent philosophy and mathematics of the Middle Ages. In a history of modern science it is unnecessary to describe the slow and devious process by which, after the fall of the Roman Empire, Greek science (with some accretions) was partially recovered and assimilated in Europe. On the other hand, it is very important to analyse the effect that the fresh exploration of Greek sources had on the fifteenth and sixteenth centuries, when mediaeval science seemed to have become sterile. Such an analysis is a major interest of this volume. At the same time—and this also falls into place here—more justice must be done than the Renaissance allowed to the permanent merits of mediaeval scientific thought, especially in the study of motion. It possessed, as we can see, a certain richness which mediaeval philosophers themselves scarcely knew how to employ, but which gained its true expression in the hands of Galileo and his contemporaries.

The firm foundations that Galileo laid are treated in another volume of *The Rise of Modern Science*. Here, the sixteenth century is poised between the old thought and the new, between authority and originality, between common sense and wild extravagance. To some, at this moment, anything seems possible when mathematics shades into mysticism and experiment promises a key to esoteric marvels. Yet the logic of science grows stronger, creating while it destroys. The descriptive method in biology, turned against the ideas of its Greek founders, makes fresh conquests ; Copernicus is vindicated by the mathematical analysis of the very authority he overturns. If the universe, no longer finite, no longer comfortably spinning round the earth, seemed a strange and terrifying place ; if a new scientific metaphysic was reducing everything to the play of matter and motion, nevertheless reason still offered, as in the past, the only road to reality. In the last resort the universe is to man what he sees in it. The sixteenth century effected a profound change in the point of view ; it was for later generations to see what that would disclose.

A. RUPERT HALL

PREFACE

This book will, I hope, show that the period from 1450 to 1630 constitutes a definite stage in the history of science. It was an era of profound change ; but the change was curiously consistent. Equally, this era marks a break with the past. I do not wish to deny the importance or validity of the mediaeval contribution to science, especially to mathematical physics ; but however much sixteenth-century scientists drew from the science of the fourteenth century, they were separated from it by three generations' passionate attempt to revive Graeco-Roman antiquity in fifteenth-century Europe. The attempt to re-discover and re-learn what the Greeks had known dominated men's minds in 1450 ; the brilliant innovations of the sixteenth century showed that this knowledge, once assimilated, had surprising implications. The revolutionary theories and methods of the 1540's were fully realised by 1630. Harvey's work on the circulation of the blood, published in 1628, and Galileo's brilliant *Dialogue on the Two Chief Systems of the World*, completed in 1630, both mark at once the culmination of the work of a preceding century and the beginning of a new age. Both were admired by two quite different generations, for different but equally valid reasons.

Evidence of my debt to many scholars is recorded in the Bibliography and Notes. I am particularly grateful to those who have eased my path by providing English translations of sixteenth-century authors, though I have compared the translations with the originals where these were available to me and have not hesitated to make my own translations where this seemed preferable. Mr. Stillman Drake kindly made available to me two of his Galileo translations in advance of publication.

Indiana University   MARIE BOAS

CHAPTER I

THE TRIUMPH OF OUR NEW AGE

The world sailed round, the largest of Earth's continents discovered, the compass invented, the printing-press sowing knowledge, gun-powder revolutionising the art of war, ancient manuscripts rescued and the restoration of scholarship, all witness to the triumph of our New Age.[1]

These words of a French physician writing in 1545 might have been those of any renaissance intellectual trying to characterise his age. Happily unaware of our modern consciousness that history is a continuous process, and that each new development has its roots in the past, men in the fifteenth century claimed complete emancipation from their mediaeval ancestors, proud to believe that they were founding a new stage in history which would rival that of classical antiquity in brilliance, learning and glory. As a sign and symbol of their success they could point proudly to two areas of discovery : the exploration of the intellectual world of the ancients by scholars, and the exploration of the terrestrial world by seamen. Two technical inventions aided men in their search for new worlds : the printing-press and the magnetic compass. The first was a product of the fifteenth century, the second had been introduced into Europe nearly two centuries before ; neither was devised by scientists, yet science somehow participated in both, and gained in importance as scholarship and practical geography each flourished in their different ways.

Nothing is more paradoxical than the relation of science and

17

scholarship in the fifteenth century. This was the time when a man could become famous in wide intellectual circles for his profound pursuit of the more arid reaches of philological scholarship, or for the rediscovery of a forgotten minor work of a Greek or Roman author. Humanism had already stolen from theology the foremost place in intellectual esteem. The term humanism is ambiguous ; it meant in its own day both a concern with the classics of antiquity and a preoccupation with man in relation to human society rather than to God. Most humanists were primarily concerned with the recovery, restoration, editing and appraisal of Greek and Latin literature (theological literature not being entirely excluded) ; they regarded themselves as in rebellion against scholasticism, the intellectual discipline of the mediaeval schools, which they saw as concerned with logic and theology rather than with literature and secular studies. Far from rebelling in turn against this literary and philological emphasis, which seems superficially more remote from science than the scholastic curriculum with its all-embracing interest in the works of God, the fifteenth-century scientist cheerfully submitted to the rigidity of an intellectual approach which was rooted in the worship of the remote past, and thereby strangely prepared the way for a genuinely novel form of thought about nature in the generation to follow.

Scientists were ready to adopt the methods of humanism for a variety of reasons. As men of their age, it seemed to them as to their literary contemporaries that the work of the immediate past was inferior to that of natural philosophers of Graeco-Roman antiquity, and that the last few centuries were indeed a " middle age," an unfortunate break between the glorious achievements of the past and the glorious potentialities of the present. Humanists were anxious to recover obscure or lost texts, and to make fresh translations to replace those current in the Middle Ages, sure that a translation into correct (that is, classical) Latin direct from a carefully edited Greek text would mean more than a twelfth or thirteenth century version in barbarous (that is, Church) Latin, made from an Arabic translation of the Greek original, and full of strange words reflecting its devious origin. Scientists agreed that to understand an author one needed correct texts and translations ; and that there were many interesting and important scientific texts little known or not at all understood in the Middle Ages. Scientists were very ready to learn Greek and the methods of classical scholarship, and to enroll themselves in the humanist camp. So the English physicians Thomas Linacre (c. 1460–1524) and John Caius (1510–73) saw the restoration and retranslation of Greek medical texts as an end in itself, a proper part of medicine, for the Greeks had been better physicians than themselves. So, too, the German astronomers George Peurbach (1423–69) and Johann Regiomontanus (1436–76) happily lectured at the University of Vienna on Vergil and Cicero, drawing larger audiences and more pay than they could hope for as professors of any scientific subject ; they were nevertheless able and influential professional astronomers. Scientists of the fifteenth century saw nothing " unscientific " about an interest or competence in essentially linguistic matters, and in editing Greek scientific texts they saw themselves aiding both science and humanism.

Indeed, science was not, as yet, a recognised independent branch of learning. Scientists were mostly scholars, physicians or magicians. The practising physician had always been in demand ; with the increase in epidemic disease which had begun with the Black Death in the fourteenth century and continued with the appearance of syphilis and typhus in the late fifteenth, there was more need for him than ever. A physician, especially one with a fashionable practice, was often a very wealthy man, and the professor of medicine held the best paid chair in most universities, to the envy of his colleagues. The success of a physician had nothing to do with his knowledge of anatomy or physiology, for the art was still almost entirely empirical; but the practising physician had, if he chose, abundant opportunity for medical research and discovery, either literary or practical.

A slightly less respectable scientific profession—but one which was sometimes very lucrative—was that of the astrologer. For many reasons—as complex and diverse as the psychological shocks of the great plagues of the fourteenth century, the shattered prestige of the Church consequent on schism and heresy, the increased tempo of war, the wider attention paid to observational astronomy, the popularisation of knowledge through increased education and the role of the printing-press—belief in the occult flourished exceedingly in the fifteenth century and showed little sign of decrease in the sixteenth. This was the height of the witchcraft delusion, especially in Germany. It was a great age of magic and demonolatry : the age of Faust. Astrology, previously almost the private domain of princes (especially in the Iberian peninsula, where every court had its official astrologer) was made available to the masses, again partly through the medium of the printing-press. (It also transferred its centre to Germany.) There was soon an enormous demand for ephemerides (tables of planetary positions), the essential tool of proper astrology : Regiomontanus, after he ceased lecturing on classical literature, devoted himself to their production. And every striking celestial occurrence—the conjunction of planets, the appearance of comets (especially plentiful in this period), eclipses and new stars (novae)—called forth a flood of fugitive literature scattered far and wide by the printing-press, prognosticating not merely for princes but for the masses as well. Even the illiterate enjoyed the advantages of being assured by astrologers that the future held as certain doom as the past, that famine, pestilence, war and rebellion would continue to dominate the Earth; for crude but vivid woodcuts portrayed both the heavenly bodies which presaged disaster and the inevitable and all-too-familiar disaster itself. Amid the calamities of the fifteenth and sixteenth centuries, astrologers could hardly fail in their prognostications, as long as they made them dire enough.

Mystic science was, in this period, the most widely known : astrology catered for the masses by whom it was so readily understood that in the popular mind astrologer and astronomer were one. The alchemist's dream, too, was widely known. Almost unheard of in Western Europe before the thirteenth century, alchemy became the preoccupation of more and more learned and semi-learned men in the Renaissance ; yet, rather curiously, it was often viewed with scepticism, as it had been by Chaucer's pilgrims. And now nascent experimental science was popularised as natural magic, properly the study of the seemingly inexplicable forces of nature (like magnetism, the magnification of objects by lenses, the use of air- and water-power in moving toys), more generally the wonders of nature and the tricks of mountebanks. Mathematics contributed its share to magic in the form of number mysticism, useful for prognostication.

Non-mystical aspects of science were also increasingly popularised, and turned to useful ends. Scholars were beginning to be proud to boast that they had mastered the secrets of a craft, believing that knowledge would thereby be acquired such as was not to be found in books. They repaid the debt by spreading knowledge of applied science. As in the Middle Ages, all literate men now knew something of astronomy, if only in its humbler aspects : the astronomy of time-keeping and the calendar. In the fifteenth and sixteenth centuries astronomers developed a further interest in practical applications and began to make attempts, ultimately successful, to introduce astronomical methods of navigation to reluctant and conservative seamen. Mathematical practitioners, half applied scientist, half instrument-maker, became common, and provided a new profession for the scientist. New maps and new exploration made geography an ever more popular subject. Map-makers flourished more on the proceeds of the beautiful and colourful maps sold to the well-to-do than on the profits from the manufacture of seamen's charts, but both were produced in quantity. Algorism, reckoning with pen and paper and Arabic numerals (modern arithmetic), instead of the older practice of using an abacus and Roman numerals, had been known to scholars since the introduction (in the twelfth century) of the Hindu-Arabic numerals ; but it was the sixteenth century which saw the production of a spate of simple and practical books on elementary arithmetic. These, mainly in the various vernaculars, were the contribution of mathematicians to merchants, artisans and sailors.

Much of the rediscovered Greek theoretical learning was also soon made available to the non-learned, as a process of translation from Latin to the vernacular succeeded the first stage of translation from Greek to Latin, by which the learned had been made free of the new literature by the more learned. Indeed, one aspect of humanism was the popularisation of ancient learning. To be sure, the humanist theory of education, designed to produce gentlemen, was an aristocratic ideal (though in fact it aimed at *creating* gentlemen, not merely at training gentlemen born). But humanism battered its way into scholastic strongholds only by adroit and clever propaganda which won sympathy from powerful forces outside the learned world of the university. To secure support from public opinion necessitated the creation of a limited but ever widening audience ; and as this audience increased, it began to demand the enjoyment of humanism without its tediums. Hence the flood of translations, making science (and literature) available in a language the layman could read.

Soon, following the example of his humanist predecessors, the scientist tried to make his learning easily available to the ordinary man. To this end, the sixteenth-century scientist burned with the (somewhat premature) desire to teach the ignorant artisan how to improve his craft through better theory or more knowledge. For this purpose an increasing number of simplified manuals were written, like those of the English mathematician Robert Recorde : *The Grounde of Arts* (1542, on arithmetic), *The Pathway to Knowledge* (1551, on geometry), and *The Castle of Knowledge* (1556, on astronomy) ; in the process, vernacular prose was much improved. Scientists were, in this period, very ready to learn from craftsmen ; having learned what the craftsman could teach them, they naturally became convinced that they had much to teach him in turn. They were constantly disappointed to find this more difficult, when the craftsman failed to show himself eager to be taught.

The heroic stage of humanism belongs to the period before 1450 : it was in 1397 that the Greek diplomat Manuel Chrysoloras (*c.* 1355–1415) began those lectures on Greek language and literature which had seduced clever young Florentines from their proper university studies and made them vehemently enthusiastic for Greek letters. The early fifteenth century had seen an avid international search for manuscripts of Greek and Latin authors previously forgotten, neglected or unknown. Though the major interest of the humanists was naturally in the literary classics, they took all ancient learning as their province, and scientific works were cherished equally with literary ones, always providing that they had not been studied in earlier centuries. In 1417 the Italian humanist Poggio Bracciolini was as pleased with his discovery in a "distant monastery" of a manuscript of Lucretius (little read in the Middle Ages, but to become immensely popular in the Renaissance) as he was with the manuscripts of Cicero that he found at the monastery of St. Gall. Guarino of Verona, hot in pursuit of Latin literature, was happy in finding the medical work of Celsus (in 1426) unknown for over 500 years. When Jacopo Angelo returned from Constantinople with manuscripts for baggage, only to be shipwrecked off Naples, one of the treasures he managed to pull to shore was Ptolemy's *Geography*, mysteriously unknown to the Christian West that had revered Ptolemy's work on astronomy for three centuries ; he had already translated it into Latin (1406) so that it was ready for the public.

By the mid-fifteenth century this great and exciting work of collection and discovery was, of necessity, ended : the monasteries of Europe had been thoroughly pillaged, and the fall of Constantinople to the Turks in 1453 meant, as the humanists lamented, the end of the richest source of supply for Greek texts. One of the strangely persistent myths of history is that the humanist study of Greek works began with the arrival in Italy in 1453 of learned refugees from Constantinople, who are supposed to have fled the city in all haste, laden with rare manuscripts. Aside from the essential improbability of their doing any such thing, and the well-established fact that the opening years of the fifteenth century had seen intense activity in the collection of Greek manuscripts in Constantinople, there is the testimony of the humanists themselves that the fall of Constantinople represented a tragedy to them. Characteristic is the cry of the humanist Cardinal Aeneas Sylvius Piccolomini (later Pope Pius II), who wrote despairingly to Pope Nicholas in July, 1453, "How many names of mighty men will perish ! It is a second death to Homer and to Plato. The fount of the Muses is dried up for evermore." [2]

Cut off from the possibility of finding new manuscripts, humanists now turned from physical to intellectual discovery, from finding manuscripts to editing and translating them in ever more thorough, critical and scholarly a fashion, establishing the canons of grammar and restoring corrupt and difficult manuscripts to what was hopefully believed to be the state in which the author had left them. Here again the humanists showed a surprising impartiality, to the advantage of science. No one could be considered to have finished his apprenticeship to humanism unless, as his masterpiece, he produced a creditable Latin translation of a Greek original : the author chosen might be a medical or scientific one, especially in the sixteenth century when the supply was running short. Thus Giorgio Valla (d. 1499), a perfectly ordinary literary humanist, counted among his treasures two of the three most important manuscripts of Archimedes ; he also owned manuscripts of Apollonios and of Hero of Alexandria, and made partial translations of these and other scientific texts which appeared in 1501 as part of his encyclopedic work, *On Things to be Sought and Avoided* (*De Expetendis et Fugiendis Rebus*). Guarino, discoverer of Celsus, translated Strabo's *Geography* into Latin, along with purely literary texts. Linacre was long better remembered for his share in introducing Greek studies into England than for his encouragement of medical learning through new translations of Galen and the foundation of the Royal College of Physicians (1518), but contemporaries found this mixture of activities quite natural.

It is important to realise that it was primarily the humanists who made the work of the "new" Greek science available. Although much Greek science had been widely known in Latin versions to the Middle Ages, this was chiefly either early science (fifth and fourth century B.C.) or late (second century A.D.). The works of the best period of Greek science, of the Hellenistic scientists of c. 300–150 B.C., was little known in the Middle Ages, partly because it was often highly mathematical and always complex and difficult. The humanists' role had important consequences both for what was available and how it was studied. Humanism, by nature, was intensely concerned with the establishment of the exact words of the author, with the correction of scribal errors and the restoration of doubtful passages. Consequently, humanists inevitably looked with both scorn and distrust on translations of Greek works made in the twelfth and thirteenth centuries indirectly through Arabic : these translations, whose Latin words were often separated from the Greek by four or more other languages, so tortuous had been the path of translation, were necessarily far from exact and often included what, to fifteenth-century ears, were horrible Arabicisms and neologisms, though the sense of the original was doubtless more or less preserved. (The Roman medical writer Celsus was above all at this time valued because he provided pure and proper Latin equivalents of Greek anatomical terms to replace the Latin forms of Arabicised Greek terms.) This preoccupation with exact rendering of an author's words mattered far less for scientific purposes, of course, than it did for literary ones, but no distinction was made, which explains what now seems an excessive preoccupation with "pure" texts.

The fifteenth- and sixteenth-century scientist was in complete sympathy with these ideas, imbued as he was with humanist ideals: hence his concern with "returning to" Galen or Ptolemy (*tout pur*, purged from Islamic or mediaeval commentary) and hence the time spent on the study of purely verbal aspects of ancient scientific texts. No doubt much of this was time wasted ; on the other hand, it did force a return to original sources which was beneficial : it was certainly more useful to read Galen and Euclid direct than to read what a commentator thought an Arabic paraphrase of Galen or Euclid meant. Many ambiguities were undoubtedly cleared up. Above all, return to the original enforced a more serious consideration of what Aristotle, Hippocrates, Galen and Ptolemy had actually said, and this in turn involved recognition of the truth, error, fruitfulness or uselessness of the contributions the great scientists of the past had made. This constituted a first step towards scientific advance. Greek science had by no means exhausted its inspiration in the fifteenth century ; it could still, as it was to do for at least two centuries, suggest different topics of exploration to each succeeding age and above all it provided authority for departing from orthodox thought. Humanism did, therefore, have much to offer science.

How is it that, nevertheless, humanists like Erasmus often seem to have attacked science ? When they did so, they were attacking the science of the universities, which they regarded as part of the sterility of scholasticism. An age determined to be new must of necessity repudiate the ideas of the immediate past ; so the humanists turned the much-praised "subtle doctor" of the late thirteenth century (Duns Scotus) into the nursery dunce of the sixteenth. Modern historians, admiring the ingenuity of fourteenth-century mathematics and physics, deplore this anti-pathy and regard the humanist worship of antiquity as having been harmful to the smooth advance of science. But however high the achievements of the fourteenth-century philosophers in certain directions, some other ingredient was needed to stimulate the development of modern science. *Il faut reculer pour mieux sauter* is often true in intellectual matters : the mediaeval inspiration was at a low ebb by the beginning of the fifteenth century, and the Greek inspiration had, at the moment, more to offer. When the humanist attacked mediaeval science, he was attacking an intellectual attitude that seemed to him over-subtle, and sterile ; he was emphatically not attacking science as such. He admired equally Aristotle the literary critic and Aristotle the biologist, while attacking Aristotle the cosmologer and semantic philosopher. He praised both Plato the Socratic rejecter of the material world, and Plato the cosmologer, who had insisted on the study of geometry as a preliminary to the study of higher things. Indeed, Plato's precepts were followed ; for wherever humanist schools were set up, mathematics, pure and applied, was always associated with the purely literary study of Latin and Greek. Infatuated as the humanist was with all that had, in his eyes, constituted the glory of the Greek past, he was eager to impart the image of that past as a whole, and to show that the

Greeks had contributed to all areas of secular knowledge.* The humanist emphasis on Greek learning may have cast mediaeval learning temporarily into the shade, but it brought to light much that was fruitful and useful for a contemporary scientist to know, and which he would otherwise not have considered.

That science was not solely a scholar's concern, but was truly a part of the popular learning of the age, even if not the main point of emphasis, appears from the list of books printed before 1500, the incunabula which modern collectors have so lovingly collected and catalogued, and have made so expensive. The earliest surviving book printed in Western Europe is dated 1447 ; by 1500 at least 30,000 individual editions had been published in all the countries of Western Europe (the Iberian peninsula seeing the establishment of presses only at the very end of the century). Of these titles the bulk was, naturally, religious, from the Bible to theology ; other books, equally reflecting the demand, followed, for no more then than now did printers wish to publish what they thought the public would not want to buy.

Yet perhaps ten per cent of the incunabula deal with scientific subjects, not at all a bad proportion : there is a mixture of popular science, scientific encyclopedias, Greek and Latin classics, mediaeval and contemporary textbooks and elementary treatises, especially on medicine, arithmetic and astronomy. There were, as yet, relatively few Greek editions of scientists—reasonably enough since the Latin translations were bound to be more popular—and few very difficult or advanced works. Thus instead of Ptolemy's *Almagest* entire (perhaps the most influential treatise on astronomy ever written, but a work of interest only to competent mathematical astronomers) one finds the up-to-date *Epitome* of that great work by Regiomontanus. On the other hand,

* Thus Erasmus edited the first Greek edition of Ptolemy's *Geography* (Froben Press, 1533).

Ptolemy's *Geography* was printed before 1500 (in Latin, in a number of editions), reflecting the wide contemporary interest in cartography. There is nothing surprising in all this : specialist works, of interest only to a limited scientific audience were not printed as books ; they remained, for the time being, in manuscript, much as specialist work to-day remains in learned journals ; then, as now, the wider demand was for semi-popular expositions. That so many scientific works appeared early in the sixteenth century shows how effective was the popularisation of learning and science.

The printing-press undoubtedly had a twofold influence on science : first, by making texts more readily available, it " sowed knowledge," providing a wider audience than could ever have been the case without printing, while serving as well to emphasise the authority of the written word. Secondly, it peculiarly influenced the development of the biological sciences, by making possible the dissemination of identical illustrations. Much fifteenth- and sixteenth-century work in anatomy, zoology, botany and natural history depended for its effect primarily on illustrations, which enormously aided identification (as well as standardisation of technical terms) ; accurate illustrations could only be produced in quantity through printing. (What happened when manuscript illustrations were copied by scribes is obvious in the woeful degeneration that overtook the originally fine illustrations to Dioscorides' botanical work in the course of centuries : miniaturists could draw flower illustrations accurately, but they had no notion of scientific exactitude.) With the co-operation of contemporary artists, books of astonishing beauty, as well as technical competence and importance, poured from the presses, and again increased the popularity of science. The printing-press also made easier the progress of science : it became increasingly normal to publish one's discoveries, thus assuring that new ideas were not lost, but were available to provide a basis for the work of others.

Scientific advance was not dependent on the printed word : indeed, many scientists, like Copernicus (1473–1543), withheld their work from the press for many years, or like Maurolyco (1494–1577) and Eustachius (1520–74) failed to publish important works in their lifetimes ; but this attitude was increasingly rare. Publication enormously facilitated dissemination, and it is generally true that scientific work not printed had very little chance of influencing others—the work of Leonardo da Vinci being the most notable case in point. In this, as in so much else, the situation was only stabilised in the course of the sixteenth century, but the late fifteenth century prepared the way.

Just as the invention of the printing-press immensely furthered the spread of past knowledge, so too it furthered an interest in the new knowledge which resulted from the great age of exploration and discovery. The great discoveries were made with few new techniques ; but the result of the discoveries stimulated major advances in mathematical geography and in astronomical methods of navigation, advances disseminated by printed books. The early fifteenth-century seamen, venturing blindly and hopefully into the Southern Atlantic, provoked learned men at home into a frenzy of dismay over the inadequacy of their methods. This was ultimately to the advantage of the seamen ; for it was nearly always landsmen, not sailors, who introduced improvements, showing how to find one's way in unknown seas, and how accurately to portray the Earth's surface on a flat map or chart. The first was something new, totally unknown to antiquity ; the second was a revival of knowledge lost in the centuries when Ptolemy's *Geography* was not known nor read.

Seamen's methods and landsmen's knowledge joined hands only slowly, and, throughout the fifteenth century, sea charts and land maps were constructed on entirely different principles. At first the seaman was in the better position, for his charts were more accurate by far than scholar's maps. The humanist re-discovery of Ptolemy's *Geography* changed the picture by intro-ducing a knowledge of mathematical cartography which applied mathematicians were quick to adopt. Meanwhile, the seaman continued to use portolan charts, developed after the introduction of the compass in the late thirteenth century. The earliest such charts are of the Mediterranean ;. later, especially in the fifteenth century, they were drawn for the Atlantic coast of Europe ; finally, both shores of the North Atlantic were mapped. The principle of the portolans was simple : as sea charts, they gave careful outlines of the coast, with distances between landmarks very precisely determined ; land features were marked only if they were of interest to the seaman, as ports or as navigational aids. All lettering was on the landward side, leaving the sea area clear for the characteristic feature of the portolan : the network of fine lines (rhumbs) radiating from a series of compass roses, giving the points of the compass as they were gradually being standardised. This permitted the sailor to work out his approxi-mate course from one place to another by tracing appropriate rhumbs (indicating constant compass heading) from one compass rose to the next. Precise though these charts were in regard to distance and direction, they were not intended to represent large areas of the Earth's surface, and indeed were inadequate to do so. They are the marine equivalent of the mediaeval road map, a kind of schematic representation of a route or pilot-book.

The scholar had always wanted something different, namely a pictorial representation of the whole spherical surface of the Earth, or at least of its inhabited parts, traditionally a hemisphere.*

---

* It should be noted that the idea of a spherical earth had never been completely lost, and after the revival of Ptolemy's astronomy and Aristotle's cosmology in the late twelfth century every educated man had at his fingertips the classical arguments supporting this position. Columbus, in spite of the familiar schoolbook version, never had to convince anyone that the Earth

For many centuries he had been content with a purely symbolic method of representation : he drew a circle, and with two strokes like a T divided it into three separate areas of a roughly triangular shape ; these areas represented the three continents, Europe, Asia and Africa, while the spaces between them and around the edge of the circle represented the seas and the all-encircling ocean. In these so-called " T.O." maps the biggest area is assigned to Asia, and it is put at the top, reflecting thereby the importance of the Holy Land. (Ancient maps had South at the top ; the portolans initiated the modern Northern orientation.) Gradually sections were taken out and expanded, and distances and directions began to approximate to accuracy as geographical knowledge of the Earth's surface increased, but mathematical geography remained an unknown science until the rediscovery of Ptolemy's *Geography*, which summarised all Greek learning on the subject to the second century A.D. It was made available by the Latin translation of Jacopo Angelo, and after 1410 there are numerous manuscript copies ; with its first printing, in 1475, it became even more popular, and there were seven editions in the fifteenth century. So successful was the work, and so busy were the printers kept supplying new editions, that there was no demand for new collections of maps ; new maps, and maps of new areas, were merely added on to Ptolemy.

Ptolemy's *Geography* had originally included maps ; though the text survived the centuries intact, the maps were lost, and there were none given in the first manuscript. One of the first tasks facing geographers was to re-draw them according to Ptolemy's careful directions. He had left tables of distance and

---

was a sphere ; what he had to do was to convince people that the Earth's circumference was as small as he (wrongly) thought it to be, and that the land distance from Spain to the Indies was proportionally as great as he (wrongly) supposed it to be. The scholars were quite right ; he could not have sailed from Spain to Japan.

direction of the principal points of the inhabited world of Graeco-Roman antiquity, and rules for drawing maps ; but the construction of such maps was not at first an easy task, since it involved mathematical factors with which no fifteenth-century mathematician was familiar. It is impossible to open up a hollow sphere and lay it flat ; equally it is impossible to wrap a flat sheet of paper around a sphere and have it fit smoothly. Hence the spherical surface of the Earth cannot be represented properly by direct transference to a continuous two-dimensional surface : even a hemisphere cannot be accurately represented on a plane.* It is true that the makers of the portolans proceeded as if one could ; but sea charts usually involved only relatively small areas ; when they did not, they contained dangerous errors, as sixteenth-century writers on navigation endlessly pointed out to recalcitrant seamen who would not believe that the plane chart which they were accustomed to use was faulty. The map-maker must be resigned to error ; his worry is how to reduce that error to a minimum and to represent distance, direction, size and shape as accurately as may be.

Until they read Ptolemy, European map-makers had been unaware of the problem. Ptolemy had discussed two methods of projection : the first, and that he mostly used, involved imagining the spherical surface of the Earth to be the lower portion of a cone, since cones can be represented as plane surfaces. This gave straight lines for equator and meridians, and gave good results for regions not very far from the equator, though increasing distortion in the northern regions of interest to a fifteenth-century European.

---

* Astronomers were, of course, familiar with the stereographic projection of the astrolabe ; but aside from the fact that it is not obvious that one can represent the Earth in the same way as the heavens, the astrolabe gave position only ; it was not an attempt to map the relative direction and distance of the stars, and distortion was of no importance. A map was supposed to be as exact a representation as possible, not a conventional table from which positions of natural objects and cities could be determined.

The second projection (often called the Donis projection after the German Benedictine who drew maps in this way to illustrate a fifteenth-century edition of the *Geography*) was a modification of the first in which meridians and parallels were curved ; the geometry involved is more complex. Once the methods described by Ptolemy had been mastered, cartographers were ready to invent new and better projections, but this lay in the future, as did the re-estimation of a length of a terrestrial degree at various latitudes.

Only one new method of representation was utilised in the fifteenth century : the terrestrial globe. Celestial globes, showing the constellations in fanciful detail and the positions of the major stars, had been known for some time, but the first surviving terrestrial globes appear only at the end of the century. These were hand drawn ; the most famous is one which Martin Behaim (1459–1507) made at the request of the town fathers of Nuremberg. Behaim was a native of that city, so famous for its astronomical instrument-makers ; but he was interested in the navigational aspects of astronomy, rather than in its astrological aspects and therefore chose to spend most of his life in Portugal, returning only briefly to Nuremberg, where he finished his globe in 1492. This showed the notable Portuguese discoveries along the west coast of Africa, but not, of course, the more recent discoveries of Vasco da Gama to the East nor those of Columbus to the West. As long as globes had to be drawn by hand they were uncommon ; by the turn of the century, however, methods of printing strips (gores) to be pasted on the globe itself had been invented, though various problems remained to be solved in accurate representation. To be useful in sailing, globes had to be inconveniently large ; in spite of sixteenth-century attempts to persuade sailors to use them, they tended to remain, as they have done ever since, mainly decorative objects.

Preoccupied as they were with the problems inherent in

learning to make accurate land maps, the cartographers of the fifteenth century had no time to spare to inquire what use these could be to sailors ; indeed, it was all they could do to record the new discoveries the voyages of exploration were producing. The

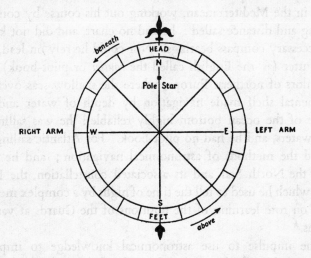

FIG. I. THE SAILOR'S GUIDE TO CELESTIAL DIRECTION

When the Guards pointed to " ten o'clock " fifteenth-century sailors said they pointed to " two hours beneath the head ", while " five o'clock " was " one hour above the feet ". Seventeenth-century sailors said " northwest by north " and " south by east "

see footnote on page 36

application of scientific cartography to sea charts was to wait until well into the next century. Astronomers and mathematicians had other possible ways of associating themselves with the new discoveries, however, ways in which they could and did help the explorers : namely, in developing new navigational methods. These were sorely needed. They necessarily came from the scholar, who understood the problem in theoretical terms, rather than from the practical navigator, highly skilled in traditional methods but at a loss when these methods failed, as they were bound to do as ships sailed out of the well-known waters of the Mediterranean and northern Europe. No longer could the seaman rely on compass, chart and table, as he had been accustomed to do in the Mediterranean, working out his course by compass bearing and distance sailed : he had no chart, and did not know the necessary compass bearings. Nor could he rely on lead, line and rutter (as the English called the *routier* or pilot-book), like the sailors of northern Europe, where the shallow seas over the continental shelf made navigation by depth of water and the nature of the ocean bottom highly reliable : he was sailing in deep waters, and he had no pilot-book. For Atlantic sailing, he needed the methods of astronomical navigation ; and he only knew the North Star and its associated constellation, the Little Bear, which he used to tell the time of night by a complex method based on rote learning of the positions of the Guards at various seasons.*

The impulse to use astronomical knowledge to improve navigational methods came at first, like the impulse to Atlantic exploration, from the Portuguese prince, Henry the Navigator. He himself went on only one expedition, and that in his early youth : across the straits of Gibraltar to assist in the capture of Ceuta in 1415. But until his death in 1460 he was the chief European exponent of the value of Atlantic exploration along the coast of Africa and of the value of astronomical aids to such navigation. To this joint end he set up a veritable research

* The Guards of the Little Bear are those two stars which are not on the line ending in Polaris ; the sailor memorised their relative positions at each hour, for each season of the year. The primitive nature of his astronomical outlook is indicated by the fact that he had to imagine a human figure whose head was North, whose feet were South and whose arms pointed East and West ; these four basic positions were further subdivided, anatomically or by fractions, to give eight or sixteen positions. (See figure 1.)

institute at Sagres, on the south-west tip of Portugal ; though this lapsed with his death, the impetus to improve methods of navigation, like the impulse to explore, continued strong. Portugal became so famous for its navigational interests that in the late fifteenth century foreigners like Abraham Zacuto of Salamanca and Martin Behaim of Nuremberg found Lisbon the most active and interesting centre for practical astronomers.

To teach the sailor how to navigate unknown oceans in the fifteenth century required two developments. One was to devise simplified methods by which he might determine his position on the globe ; in practice this meant methods of determining latitude from the altitude of the Sun or a star. The second was to devise instruments which would work on the unstable deck of a small and lively ship, in the hands of an untrained seaman. For an astronomer's instruments were no more suited to life at sea than were his observational methods to a sailor. The astronomer's favourite instrument was, as it had long been, the astrolabe, a complicated device designed to supply astrologically useful information about the motion and position of the Sun and stars ; it was only incidentally provided with a sighting device (the alidade) for measuring the angular distance between two objects. A simplified form, the mariner's astrolabe (essentially a heavy ring carrying a scale, with the alidade as a movable pointer) was not developed until the next century. The second most common device was the quadrant ; this also carried an immense amount of information, as well as two sighting holes and a plumb-line ; but it had no moving parts (as did the astrolabe) and could be simplified readily by replacing the astronomical tables engraved on the back with navigational ones ; in this form it was the instrument most used in the fifteenth century. There was also the cross-staff, named by analogy with the crossbow which it resembled, and the seaman could use it like a crossbow to shoot a star ; it was little else than a long narrow piece of wood or bone, fitted with a short movable crosspiece ; to use it, one held the end of the staff to the eye, and slid the crosspiece to and fro until its ends just covered the two objects whose angular distance apart was sought. The staff could be calibrated in various ways to give the information desired.

The latitude of a point on the Earth's surface can be determined by measuring the height of the Sun above the horizon and adding (or subtracting) the angle of declination of the Sun for that day. This method was not much used in the fifteenth century, because it was easier to determine the altitude of the Pole Star and compare it with the altitude at some known point ; but in actual practice neither astronomer nor seaman spoke in terms of latitude, still a concept known only to the scientific cosmographer. Instead, the seaman was taught to take an *altitude* by means of his quadrant, and to compare it with the altitude of his home port by means of a simple rule (regiment was the more usual contemporary term) devised by the astronomers. As a Portuguese explorer, sailing to Guinea in 1462, commented, " I had a quadrant when I went to those parts, and I marked on the table of the quadrant the altitude of the arctic pole." [3] From such data as this, accumulated slowly through the century, astronomers worked out tables of celestial altitudes, which became an essential part of the rules for the use of seamen devised in the later fifteenth century. The seaman was taught to observe the Pole Star and to correct his observation according to the position of the Guards in the Little Bear, the same Guards he already watched to tell the time at night. In addition he was given rules to help him to work out, from an altitude determination, how to sail by wind and compass in various directions so as to arrive at the altitude of his destination, along which he could then sail East or West until he arrived, no strange currents interfering : " running down the latitude " it was called in the sixteenth century. When manuals attempted to teach the seaman to find his latitude by shooting the

Sun, it was a laborious business, involving many tables and much calculation, as the manuals which used Abraham Zacuto's *Almanach Perpetuum* (written about 1473) amply demonstrate.

The Portuguese were in this period as far in advance in astronomical navigation as they were in exploration, though in both cases the Spanish were about to rival them. Even the Portuguese astronomers could not help the seaman much in determining longitude ; for the only method known involved comparison of the time at which there occurred some celestial event—an eclipse or a planetary conjunction—and the time at which it was predicted to take place at some known city in Europe. This was so obviously an unsatisfactory method to rely on, depending as it did on relatively rare events, difficult to observe and involving inherent errors of timing and prediction, that it is hardly surprising that little attempt was made to determine longitude in this period except by dead reckoning and guesswork. The seaman was quite content to use his new knowledge of winds, currents, speeds and distances, unconsciously acquired skills which distinguished navigators like Columbus who, in his later voyages, arrived at his destinations with almost uncanny precision.

The fifteenth-century astronomer had good reason to be proud of his achievement in solving navigational problems : he had developed methods of using the stars unknown to either antiquity or the Middle Ages, had translated them into simple terms which seamen could use to good purpose, and had plenty of ideas on further new methods. He had less reason to be happy about his ability to solve the problems of astronomical theory which confronted him. Outwardly serene in possession of a cosmological system which had endured for over thirteen centuries, he was well aware that this system contained serious flaws which threatened its existence. This was a cause of widespread scientific unease, one of the recurring crises to which the theoretical superstructure of the sciences has always been, and still is, periodically liable.

Traditional cosmology involved a comfortable and tidy universe, supported by Aristotelian philosophy, made scientifically effective by the mathematical synthesis of Ptolemy, and Christianised by the scholars of the thirteenth century. Imagination had contrived an orderly series of spheres, one within the other, moving according to divine law, here and there adorned with shining heavenly bodies. At the centre was the Earth, lowly in position and nature, yet dignified as the centre of all and the abode of man ; fixed immovably in its place, it was subject to the influences of the ever-widening spheres which surrounded it. First in the terrestrial region, below the moon, came the spheres of the four elements, earth, water, air and fire, the region of generation and corruption and change. Then, in the celestial region, the eternal and unchanging heavens, came the crystalline spheres of the Moon, Mercury, Venus, the Sun, Mars, Jupiter, Saturn ; hollow spheres nesting one within the other, so that although their radii were large, the outer surface of one touched the inner surface of the next bigger one. Beyond the planetary spheres lay the sphere of the fixed stars ; and beyond that again, the ninth sphere of the Primum Mobile. The size of these spheres was presumed to be arranged in some sort of harmonious proportion, and indeed it was commonly held that in turning, these spheres produced a heavenly harmony, in which, perhaps, all the stars joined :

> Look how the floor of heaven
> Is thick inlaid with patines of bright gold :
> There's not the smallest orb which thou behold'st
> But in his motion like an angel sings,
> Still quiring to the young-ey'd cherubim.[4]

So far, all was well, and men were pleasantly aware of being at the centre of a neat cosmos, designed and ordered for the

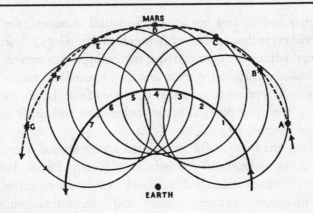

FIG. 2. THE MOTIONS OF MARS ACCORDING TO PTOLEMY

1-7 are successive positions of the centre of the epicycle (thin line)
at intervals of one month as it is borne round by the deferent (thick
line); A-G, corresponding positions of the planet on the epicycle
as the epicycle rotates. The dotted line is the apparent path of
the planet

benefit of man. But astronomically speaking, the universe was
less tidy. The mathematical devices of Greek astronomy were
necessary to give an accurate representation of the exact motions
of the planets, and they were abstractions of a totally different
order from the solid realities of the crystalline spheres.* For
thousands of years astronomer and layman had agreed, however,
that the evidence of the senses was a reliable guide even to the
heavens ; and the evidence of the senses confirmed what mathematical and philosophical argument deduced, that all heavenly
motion was truly circular. One had but to watch the sun rise
and set, or the moon travel nightly through the skies, to see that
circular motion was natural to the heavens. So the mathematicians' epicycle and eccentric circles (figure 2) had to be considered only mathematical fictions ; reality knew only, as even

* See below, p. 43.

Ptolemy had held in his less mathematical moments, the solid, material crystalline spheres of Aristotelian cosmology. Fifteenth-century astronomers concurred ; increasingly astronomers were to demand reality, in the sense that they wanted astronomy to be a science of representation rather than one of calculation, a science dealing with real physical bodies rather than with mere mathematical magnitudes.

Even mathematically considered, conventional astronomy was in an unsatisfactory condition. Nearly fifteen hundred years of observation had revealed many discrepancies, real and imaginary, between theory and observation. Some of these discrepancies (like the imaginary trepidation "discovered" by Thabit ibn Qurra in the ninth century, a " shaking " of the spheres) could be handled by adding on more spheres, or more mathematical devices. But discrepancies between the predicted location of planets and their observed positions were infinitely more disturbing. For example, the calendar was in need of reform ; this was a religious problem in view of the dating of Easter, yet it could only be solved with astronomical assistance ; the Lateran Council of 1512 was forced to postpone consideration of possible correction of the calendar " for the sole cause that the lengths of the years and months and the motions of the Sun and Moon were not held to have been determined with sufficient exactness," as Copernicus reminded Pope Paul III.[5] Even more serious, the current tables of planetary positions, drawn up at the command of Alphonso the Wise in Spain at the end of the thirteenth century, were so grossly inaccurate as to inconvenience astrologers. For these and other reasons, astronomers were uneasy ; it is almost fair to say that the Copernican revolution was predicted a century before Copernicus published his great work. Even laymen knew that astronomy needed reform : thus the humanist Pico della Mirandola (1463–94), arguing against astrology on religious, philosophical and scientific grounds (it denied the omnipotence of God, it denied man's free will and it was strikingly inaccurate) pointed out that the astronomical basis of astrology would be shattered when astronomers altered their system, as he believed they would.

Because they were thoroughly imbued with the humanist point of view the astronomers of the fifteenth century naturally turned to the ancients for a clue to the way out of the astronomical labyrinth in which they found themselves, just as Copernicus was to do in the next century. This was the more reasonable because one of their main sources of disquiet was that astronomical representation no longer conformed to criteria established by Plato and continued by a long line of Greek astronomers. Plato, who had first suggested the search for a mathematical device that would interpret the observed planetary motions in terms of precise mathematical law, had also insisted that the law when found must express such motions in terms of uniform circular motion about a unique centre. Overtly in rebellion against the dead hand of the past in the person of Aristotle, humanists everywhere turned to Platonist (and neo-Platonist) doctrines, and stressed the importance of order, harmony and uniformity of circular motion throughout the astronomical universe. No one could claim that Ptolemaic astronomy was truly in conformity with this Platonist philosophy.

Eccentrics, as originally devised to explain the varying brightness of the planets (rightly interpreted as caused by varying distances from the Earth) began as circles not quite centred on the Earth, about whose circumference the planet was assumed to travel ; in the fifteenth century eccentrics had become the inner or outer surface of a crystalline sphere, whose shell was therefore of varying thickness. (Since the heavenly spheres nested one inside another to prevent empty space, the corresponding surface of the next sphere was also eccentric to its centre.) The eccentric sphere was at the same time combined with what had been for

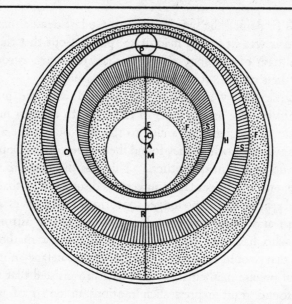

**FIG. 3. THE MATHEMATICAL SYSTEM OF THE WORLD
ACCORDING TO PEURBACH**

M is the centre of the world, A of the equant, C of the small circle,
E of the eccentric which is movable and describes P, the epicycle.
The two dotted spheres, F,F, carry the apogee of the equant; the
two spheres S,S, carry the apogee of the eccentric. The deferent is
the white sphere lying between S and S; its eccentric circle is OPHR

Greek astronomy a purely geometrical device, namely the epicycle and deferent, the epicycle a small circle carrying the planet,
itself travelling around the larger circle, the deferent, so that the
planet partook of the motion of both epicycle and deferent.
(The epicycle accounted for the " retrograde " motion of the
planets, when, as a result of the combined motions of Earth and
planet about the Sun, the planet appears to travel backward in a
great loop.) The deferent could be concentric or eccentric with
respect to the Earth ; its velocity might be uniform with respect

to its centre, or with respect to its equant. (The equant accounted for the fact that the velocity of a planet is not, in fact, uniform, being more rapid when the planet is near the Sun, as Kepler's Second Law explained.) The equant preserved uniformity of motion, but at the price of introducing a purely mathematical point into a physical system. How epicycle, deferent and equant could be worked into the system of planetary spheres is shown in figure 3, taken from Peurbach's *New Theory of the Planets*. The complex system of crystalline spheres was far from the mathematical system that Plato had suggested ; but it had the advantage of providing a kind of physical reality, and it certainly explained why the planets remained in their places in the sky, proceeding with a regular motion through their appropriate revolutions. God had created the spheres at the beginning of the world, and set them moving in the way that they had continued ever since.

Nevertheless, such divergence from perfect circularity and uniformity bothered fifteenth-century astronomers as acutely as did the divergence between theory and observation. Drastic action was needed ; yet few cared to go as far as Nicholas of Cusa (1401–64) in rejecting conventional ideas altogether. Cusa was a prominent ecclesiastic, eventually a cardinal in spite of a stormy career advocating violent reform in all aspects of life and thought, from reform of church government to reform of the calendar and reform of philosophy. His chief philosophical work is *On Learned Ignorance* (1440) : learned ignorance is the recognition of the inability of the human mind to conceive the absolute or the infinite. Its astronomical significance lies in the fact that the human mind is, strictly speaking, incapable of framing an ordered cosmology. There is, for Cusa, truly no simple harmony to be apprehended behind the apparent irregularity of the skies ; only a complexity whose order we cannot conceive. The universe is without boundary or limit ; not infinite, for then it would be coexistent with God, but indeterminate, a partial expression of God, representing in multiplicity what God is in unity. Nothing is fixed, all is relative ; the centre is everywhere and nowhere, for in an infinitely large circle the circumference coincides with the tangent and in an infinitely small circle, the circumference coincides with the diameter. All things are in motion, even the centre of the universe, which is also the circumference since it lies in God who is both at the centre and the circumference. Hence the Earth, properly speaking, is not at the centre ; indeed, no heavenly body is ever exactly at any fixed point for there are no constant spheres. Nor is there constant uniform motion, though there is relative motion everywhere and always, in the Earth as well as in the rest of the universe. Yet, at the same time, the Earth can be considered more or less at the centre for purely astronomical purposes ; and in an astronomical note scribbled on the flyleaf of a book dated 1444, Cusa explained how the motion of the Earth makes no difference to the total system. The motion of the Earth is a revolution about the poles of the world, every 24 hours from East to West ; meanwhile the sphere of the fixed stars turns from East to West in twelve hours, so that exactly the same effect is obtained as if the Earth stood still and the fixed stars rotated every 24 hours. Though Cusa worked out his system in some detail, he intended only to show the philosophical necessity for breaking with the concept of an ordered universe, a demonstration destined to appeal more to philosophers than to astronomers.

To fifteenth-century astronomers, the most obvious first step was to return to the fountainhead of the current cosmological system, and see whether the difficulties of their astronomy were not the result of the accumulation of errors over the centuries. It was to them eminently reasonable to suppose that Ptolemy must have been more accurate and more correct than were Moslem and mediaeval astronomers ; it was quite likely that many errors had been caused by miscopying and mistranslation, a natural belief in an age when humanists had just discovered the joys of a critical scholarship which could clear up ambiguities and clarify garbled texts. This was the view of George Peurbach (1423–69), as much humanist as astronomer, who lectured simultaneously at Vienna on astronomy and on Latin literature. Peurbach had begun with a careful study of the Arabic commentaries on Ptolemy, extending their work in spherical trigonometry. It was perhaps as the result of writing a textbook, the *New Theories of the Planets* (*Theoricae Novae Planetarum*), that he began to long for an accurate text of Ptolemy's *Almagest*. His own book was far superior as an introductory text to Sacrobosco's thirteenth-century treatise *On the Sphere* which was widely used in the universities; Sacrobosco had described the framework of the heavens, but he had merely named the devices of epicycle and eccentric and had not at all discussed their use or the motion of the planets. Peurbach gave a careful detailed description of the interlocking spheres of each planet, but he knew that the system was not as good as he would like. He became obsessed with the belief that the only way to improve astronomy was to study an accurate Greek text of the *Almagest* in place of the Latin translation derived from the Arabic which was all he had to work with. He had studied in Italy, and knew the wealth of manuscript material to be found there; he made preparations for the journey, planning to take with him a disciple, Johann Müller of Koenigsberg, known after the fashion of the times by a Latin form of his birthplace, Regiomontanus. Peurbach died before the journey could be completed, but Regiomontanus took his place.

Regiomontanus (1436–76) was, like his master, both humanist and astronomer, having already lectured on Cicero; when he arrived in Italy he copied the tragedies of Seneca while learning Greek in order to translate the *Almagest*, and later, the *Conic Sections* of Apollonios. He finally settled in Nuremberg and established a printing-press; one of the first books he published was the astronomical poem of Manilius (first century A.D.), in the twentieth century the special interest of A. E. Housman. Regiomontanus had ambitious plans for editions of most of the treatises of the great age of Greek science, plans which he never executed, though the proposed trade list has survived.[6] He prepared an astrological almanac, the *Ephemerides*, widely used at the end of the century. He continued the work of Peurbach in two directions : first, in trigonometry, writing *On Triangles* (a systematic account of principles) and secondly, in Ptolemaic astronomy, writing an *Epitome of Ptolemy's Almagest*, which is a summary of the mathematical as well as the descriptive parts. Regiomontanus partially deserted secular pursuits to become archbishop of Ratisbon, but it was his astronomical knowledge which caused his summons to Rome in 1476 for consultation on the reform of the calendar, where he died.

Between them, Peurbach and Regiomontanus had improved the teaching of elementary astronomy, had advanced the study of spherical trigonometry, and had rendered the detailed account of the *Almagest* more readily available to all who read Latin. The new interest they stimulated continued : their textbooks were printed and reprinted throughout the sixteenth century and served as models for other books. The first printed edition of the *Almagest* (1515) was the mediaeval Latin version which Peurbach and Regiomontanus had found so unsatisfactory ; but a new Latin version appeared in 1528, followed ten years later by the Greek text. Ironically, astronomers now realised that the immediate goal of Peurbach and Regiomontanus was futile ; no matter how purified and carefully edited, Ptolemy had nothing for reformers. When sixteenth-century astronomers followed the inspiration of Peurbach and looked to the ancients for aid, they sought it in the works of Ptolemy's predecessors, retaining only the mathematical sophistication of Ptolemaic astronomy. This was the method of Copernicus, so one can see that Peurbach's design was not altogether valueless. His greatest contribution was to raise the standard of astronomical consciousness by rendering Ptolemaic astronomy more accessible. It was necessary to ensure greater understanding of the existing system before further advances could be made. Fifteenth-century astronomers to their credit saw that astronomy would change, though they could not see in what direction.

Astronomy in the fifteenth century professed seemingly contradictory origins, deriving equally from humanist scholarship and practical demands. The same paradoxical blend was to run through all Renaissance science. Striving to master Greek scientific texts, while keenly aware of later technical progress, mathematicians, botanists and physicians, like astronomers, strangely combined reverence for the literal word of the remote past with a desire for novelty. Endeavouring to see in nature what Greek writers had declared to be there, European scientists slowly came to see what really was there.

CHAPTER II

THE PLEASURE AND DELIGHT
OF NATURE

The science of biology is an invention of the nineteenth century. Earlier ages generally knew only natural history—which satisfies man's inherent curiosity about the living world around him—and medicine, which in the fifteenth and sixteenth centuries comprehended both the physical nature of man and the search for naturally occurring remedies for man's seemingly unavoidable ills. Natural history in the fifteenth century still tended to cater for the same love of the marvellous that infuses mediaeval bestiaries, following a tradition derived through Christian moral tales from Pliny's *Natural History*. A new current was now to appear, fed on the one hand by a renewed interest in Aristotle's biological writings, and on the other by the new world of nature discovered in the American continents. At the same time, natural history continued to act as a subsidiary to medicine in the production of herbals, as the later Middle Ages had christened the works of descriptions of herbs and other plants useful in medicine ; these derived in an unbroken tradition from the works on materia medica written by Greek army doctors. The most famous Greek herbal, that of Dioscorides (first century A.D.) was copied and recopied incessantly until printed in 1478, still carrying the illustrations which had been an integral part of the book since the time of Dioscorides himself.

Much of the interest in natural history was a popular one, reflected equally in the gardens and menageries of the wealthy, and the picture-books, often written in the vernacular, provided for the less well-to-do. One of the most striking aspects of natural history in the fifteenth and sixteenth centuries, related to this popular interest, is the appeal of nature to the artist, who increasingly turned to plants and animals for models and studied them carefully in order to be able to paint them as accurately and as sympathetically as possible. Besides this, humanism provided a stimulus to the existing interest by furnishing new and better editions and translations of the works of the ancients—Aristotle and Pliny on animals, Theophrastos and Dioscorides on plants— and in each case the older and better work excited the greater interest. Humanists also carefully cultivated delight in nature, insisting that it should be understood and enjoyed for its own sake rather than, as St. Augustine had desired, for its value in interpreting the Bible or as an allegorical representation of the wonder of God and the truth of religion. This love of nature is well exemplified by the Swiss Conrad Gesner (1516–65) writing in praise of mountains and mountain climbing ; and even better by the German botanist Leonhard Fuchs (1501–66) who in the preface to his *History of Plants* (*De Historia Stirpium*, 1542) wrote lyrically :

> But there is no reason why I should dilate at greater length upon the pleasantness and delight of acquiring knowledge of plants, since there is no one who does not know that there is nothing in this life pleasanter and more delightful than to wander over woods, mountains, plains, garlanded and adorned with little flowers and plants of various and elegant sorts, gazing intently upon them. But it increases that pleasure and delight not a little, if there be added an acquaintance with the virtues and powers of these same plants.[1]

The charm of nature, the beauty of flowers and the collecting instinct all played their part, together with the pleasures of recognition : for one of the tasks of the natural historian was to identify in nature the animals and plants described by Aristotle or Theophrastos or Pliny or Dioscorides.

As so often happened, the humanists produced a paradoxical situation : for their emphasis on the ancients simultaneously advanced and retarded botanical and zoological knowledge. This is most readily apparent in botany. On the one hand, various works were shown by means of careful humanist scholarship to be falsely attributed to Aristotle, and really undeserving of the respect in which they were held, and the excellent *Enquiry into Plants* of Theophrastos (fourth century B.C.), little noted in the Middle Ages, was substituted ; it was translated and printed in Latin (1483) and Greek (1497). On the other hand, Dioscorides was also printed frequently in Latin, Italian, German, Spanish, French and Greek. This was not in itself a retrograde step—though it emphasised the medical aspect of botany to the detriment of broader interests—except that these editions were naturally illustrated with the traditional drawings, now direfully debased copies of successive copies of the delightful and accurate originals. (The strength of the original drawings is indicated by the fact that in a copy made in A.D. 512 the illustrations preserve a considerable degree of freshness and verisimilitude.) Historians have often been puzzled to account for the shocking difference between the crude and conventional woodcuts illustrating fifteenth-century herbals, and the accuracy and artistic merit of the work of painters and miniaturists of the same period. It is reasonable to suppose that the fifteenth century saw no conflict : the woodcuts were copied from the illustrations of the manuscript whose text was also faithfully copied ; the illustrations illustrated the text, not nature — a peculiar view, no doubt, but there was as yet no really independent botanical (or zoological) study.

That was to be the contribution of the sixteenth century, when herbalists stopped depending *only* upon Dioscorides and Theophrastos, zoologists on Aristotle and Pliny ; and natural historians began to believe that they could work on their own (a situation to be repeated in the study of human anatomy). First came uncritical acceptance of new or at least unhackneyed texts ; then critical appraisal ; finally emancipation and originality. For well-known texts the process was speeded up : thus Ermalao Barbaro (1453–93), a famous humanist who edited Dioscorides and Pliny, claimed to have found no less than five thousand errors in the standard Latin text, errors both of transcription and of fact, and he published his findings in a book aptly entitled *Castigations of Pliny* (1492–3). Others followed, and though Pliny found some defenders, he emerged with his reputation for reliability much impaired. The same thing was in danger of happening to Aristotle and even Theophrastos, until it was at last realised that the Greek scientists described the Mediterranean flora and fauna which they saw around them, and their descriptions consequently could not be expected always to fit the species of Northern Europe. This discovery relaxed the efforts of the conservative to see with Aristotle's eyes, and avoided a great deal of wasted effort ; it also suggested that it might be profitable to study the variations in flora and fauna in different parts of the globe.

One of the most interesting aspects of Renaissance books on botany and zoology is the wealth of pictures with which they are almost invariably filled, and which makes them as delightful to look at now as they were to their contemporaries. The fifteenth-century natural historian had seen in picture-books a method of appealing to an illiterate or barely literate audience ; the sixteenth century refined techniques to produce handsome volumes accurately illustrated by admirable artists. Botany, zoology, anatomy, engineering and invention all lent themselves readily to this process. The illustrations delighted the eye and supplemented the text ; but in botany and anatomy they did more, for they could convey what words, as yet insufficiently subordinated to technical needs, could not. There was as yet no technical language accurate in meaning and universally known, fit to explain in detail the necessary description of form ; in fact, botany dispensed with pictures when, in the eighteenth century, such a technical language was developed. In herbals especially it was true that pictures were often the only possible means of identifying a plant loosely or inaccurately described. Here a revolution took place as authors, in despair at the inadequacies of purely verbal description, sought the aid of skilled draughtsman and artists, trained to observe carefully and well.

How much credit should be given in such cases to the writer of the text and how much to the artist is difficult to determine ; for when a herbal is praised, it is more often than not the illustrations that are really being judged. There is remarkably little evidence (as in the parallel case of anatomy) of how closely artist and writer worked together : few were as meticulous as Fuchs who saw to it that the men who drew the plants from nature, copied the drawings on to woodblocks and cut the blocks were all given credit ; he is unique in praising them in his preface and including their portraits at the end of the book. Even Fuchs does not indicate how far, if at all, he directed the artist ; nor did his predecessor Brunfels (1488–1534), though his great work was called *Living Portraits of Plants* (*Herbarum Vivae Eicones*, 1530), and the pictures by Hans Weiditz are far superior to the text. In one respect the use of real artists was a disadvantage, for though the artist always drew directly and observantly from nature, he tended to draw exactly what he saw, " warts and all " ; hence Weiditz drew the specimen before him exact in every detail, broken leaves, wilted flowers and the ravages of insects. It was only slowly, as botanists learned the necessity of guidance, and chose illustrators rather than independent artists, that the drawings began to represent types, rather than individuals. Such problems hardly arose in zoology, for drawings of common animals had to have verisimilitude in their rough details at least, and no one could tell whether the portrayal of the exotic creatures said to exist in tropic and arctic regions were accurate or not. In any case, exact representation was less important, and verbal description easier than for plants. Consequently the artistic calibre of most zoological illustration is low, and all through the sixteenth century it is the text, not the pictures, that claims the reader's attention.

In botany, the existence of the herbal amply demonstrated the domination of Dioscorides, which only gradually relaxed as naturalists learned to add and substitute their own observations. Soon after the first printed edition of Dioscorides (1483) there appeared the first of a series of variations, themselves perhaps based upon a fifth- or sixth-century version of Dioscorides. These works, in Latin and various vernaculars, were called either *Herbals* or *Gardens of Health* (*Ortus* or *Hortus Sanitatis*), and were immensely popular. This popularity derived mostly from their usefulness in gathering appropriate "simples" for herbal remedies, though presumably they were also used by collectors. Sixteenth-century herbals all follow much the same pattern, and they were numerous in every country. Among the more famous herbals were those of Fuchs, Valerius Cordus (1515–44) and Camerarius (1534–98) in Germany, Plantin (1514–88), Dodoens (1517–85), Clusius (de l'Ecluse, 1526–1609) and Lobelius (de l'Obel, 1538–1616) in the Low Countries, Mattioli (1501–77) and Alpino (1553–1617) in Italy, the two Bauhins (1541–1612 and 1560–1624) in Switzerland, Ruel (1474–1537) in France, Turner (*c.* 1510–68) and Gerard (1545–1607) in England—the list indicates how numerous and how persistent this type of presentation proved to be. The interest of the herbalist was descriptive and utilitarian ; in fact the herbal is a handbook of the botanic garden which was soon to be an indispensable adjunct to any good medical school. Though there was much copying from one to another, each herbal has some peculiar merit of its own ; each described some new plants, and each improved the descriptions of some well known plants. Notable among them is the work of Fuchs, for his *History of Plants* (1542) showed an interest beyond the purely medical, and he took pains to be as comprehensive as possible. Though Fuchs listed his plants alphabetically (in part for ease of locating the descriptions) he did try to name and compare each part of the plant as clearly as he could, and to give some idea of such characteristics as the habit of growth, the form of the root, the colour and shape of the flower, and the habitat. But indeed the encyclopedic approach was the aim of almost all herbalists, though not all succeeded as well as Fuchs.

Though the major advances in botany are displayed in books, yet there were several new developments which in the long run were to have perhaps a greater influence. The first of these was the invention of the herbarium (or *hortus siccus*), a collection of dried plants preserved by pressing the specimens between sheets of paper. Dried flowers were known earlier (commonly for household use) but the first recorded herbarium, of some three hundred specimens, was made by an Italian botanist, Luca Ghini (d. 1556), professor at Bologna, a university long famous for medical studies. His pupils followed his example, and the oldest herbarium now in existence was made by one of them under his influence. By the time of Ghini's death, herbaria were well known in England and on the Continent. Other developments derive from the influence on medicine of the immense increase in botanical activity. Medical schools (beginning with Padua in 1533) introduced chairs of botany, whose holders were expected to lecture on the medicinal properties of plants ; and soon after this, schools of medicine also began to institute botanic gardens, to be looked after by the newly appointed professor of botany. The Paduan garden was begun in 1542, and has continued in flourishing existence ever since.

Perhaps because of the greater knowledge of the animal world, and the fact that animals did not have the same usefulness as plants, there are relatively fewer encyclopedic works comparable to the herbals and more works dealing with small groups of animals, like fishes or birds. Zoology as a whole was far less popular in the sixteenth and seventeenth century than botany, and only a few men wrote surveys as extensive as those of the herbalists. The most complete summary of the animal world was the *History of Animals* (1551–8) of Conrad Gesner, a great encyclopedia intended to replace Aristotle's work of the same title. Gesner was an almost universal scholar. Humanist, encyclopedist, philologist, bibliographer, zoologist, botanist, alpinist, linguist, an M.D., it is no wonder that he was later called a " monster of erudition." His *History of Animals* is comparable with Fuchs' *History of Plants*, but even more encyclopedic in character, for it was designed as a work of reference and included descriptions drawn up by others, to whom Gesner, editor as well as author, gave full credit. Gesner listed his animals alphabetically, though he did use such divisions as birds, fishes, insects, and so on, the same divisions used by Aristotle. Under each animal's name is a wealth of diverse information : names in all languages known to Gesner, habitat, description, physiology, diseases, habits, utility, diet, curiosities, all with careful references to authorities, ancient and modern. Gesner very sensibly drew heavily on the best zoological descriptions he could find, and where these were lacking he persuaded men like William Turner and Thomas Penny to write special accounts of their own investigations for his use. The catholic nature of Gesner's interests led him into dealing with certain ambiguously genuine creatures. This is especially true of marine animals, always a tricky subject, though Gesner appreciated the essential difference between fish and river and sea animals. Thus in volume IV (*On the Nature of Fish and Marine Animals*) there is a good description of the sea-horse, with appropriate illustrations (the *History of Animals* was copiously furnished with pictorial as well as verbal description). This is naturally followed by the hippopotamus, after which come the sea-man and (from Rondelet's *Marine Fish* of 1554) the bishop-fish, of which Gesner ambiguously comments " I here show the pictures of certain monsters : whether they truly exist or not, I neither affirm nor deny ".[2] Yet he was inclined to doubt the freaks of nature which some, he says, take for miracles ; and he dismisses Tritons and Sirens as inventions of the ancients ; while he includes them, one suspects it is because they make such admirable pictures. Besides, an encyclopedist must include all relevant information, whether he accepts it as true or not.

The same encyclopedic tradition which stimulated Gesner to the production of his *History of Animals* continued throughout the century. A famous example is the work of Ulysse Aldrovandi (1522–1605), M.D., professor of pharmacology at Bologna, first director of a museum of natural history, and later of the botanic garden which was founded in 1567. He was an indefatigable worker ; he professed to have investigated himself the subjects covered in what became fourteen published volumes. Even his long life did not permit him to complete the work ; only the volumes on birds and insects were published in his lifetime ; the remainder represents the work of his pupils, based on the volum-inous manuscript material he left on his death. (There was such a mass that it has never all been edited.) This posthumous pub-lication makes Aldrovandi liable to a false comparison. For though Aldrovandi's classification is better than Gesner's, and his books are more handsomely printed, he was far less critical even than Gesner.

The most interesting work in descriptive zoology in the sixteenth century is that of men who were content to explore limited groups of animals ; this enabled them to make a more thorough and original survey, and generally permitted the inclusion of anatomical as well as external features of the creatures described. This was new in the sixteenth century. Fish and birds were the most popular subjects, perhaps in part because Aristotle's treatment of marine life had been so full and (because he often dealt with species that do not exist outside the Mediterranean) so baffling. But there was a popular side to the interest in fish and birds as well, for the sportsman liked to know about his bag or catch. Izaak Walton's *Compleat Angler* (1653) is the ultimate successor of the more learned sixteenth-century treatises of Guillaume Rondelet (1507–66) and Pierre Belon (1517–64). Rondelet, like most naturalists of the time a physician, was professor at Montpellier which at that time rivalled Paris as a centre for medical studies. His great work, partly the result of travels in Italy, partly of his desire to vindicate and re-confirm the descriptions of Aristotle, is *On Marine Fish* (*De Piscibus Marinis*, 1554), a work aided by copious and accurate illustrations. Rondelet was able to detect certain peculiarities of marine life described by Aristotle, but regarded as improbable : for example, he illustrated Aristotle's account of the placental dogfish by a picture delineating the young just after birth. He also dissected the sea-urchin, and his illustration of this is the first of a dissected invertebrate. (Indeed invertebrates had been much neglected.) Though his standards were thus high, his credulity was strong, and his book provided the picture of the bishop-fish used by Gesner.

His younger contemporary Belon, also educated as a physician, managed to earn his living as a naturalist by acquiring powerful patrons : first Cardinal Tournon, and later the King of France. King Francis sent expeditions to the Near East, as well as diplomatic missions ; these were to bring back curiosities for his palaces, animals for his menageries, books for his libraries, plants for his gardens, and cartographic information for his merchant fleet. Belon accompanied one of these expeditions and made a thorough survey of the coasts of the Levant. He was an infatigable note taker and on his return had enough information for *L'Histoire Naturelle des Étranges Poissons Marins* (1551), *La Nature et Diversités des Poissons* (1555), *L'Histoire Naturelle des Oyseaux* (1555), besides a monograph on conifers and an account of his travels. (He might have written more had he not been murdered one night in the Bois de Boulogne on his way back from Paris to the king's palace, where he was a pensioner.) Though Belon does not give any account of anatomical dissections, he has excellent external descriptions, and he drew many of his own illustrations, which are generally more accurate than those of Rondelet. Like Rondelet, he confirmed many of Aristotle's descriptions of the generation of marine animals, like the viviparous sharks, and was discerning enough to realise that Aristotle was describing the marine population of the Eastern Mediterranean (which Belon had now studied), and to notice how much this differed from that of the coastal waters of Northern Europe. The work on birds is notable for the portrayal of their skeletal structure and includes a famous discussion of the homologies between the skeleton of a man and a bird.

Other specialised works in the sixteenth century include the book on dogs by the English medical humanist John Caius written for Gesner but first printed in London in 1570 ; and the *Theatre of Insects* compiled (and largely illustrated) by Thomas Mouffet from the work of Edward Wotton (1492–1555), the notebooks of the botanist Thomas Penny (1530–88), the compilations of others and even his own observations ; after many vicissitudes, the book was finally published (posthumously) in 1634, just before the introduction of the microscope was to transform entomology. Far superior to any of these, which are entirely works of natural history, is the monograph on the horse by Carlo Ruini, a senator of Bologna. *On the Anatomy and Diseases of the Horse* (1598) is a beautifully illustrated and strikingly accurate and comprehensive work, which entirely avoids the usual sixteenth-century practice of treating animal anatomy as a branch of human anatomy. (In fact, though there are some overt comparisons in the anatomical literature, very often the fact that animal and not human material had been used in dissection was not even mentioned, for the differences were by no means thoroughly appreciated.) Ruini's work was not bettered for several generations ; it is one of the first examples of what could be done when animals were studied for their own sake, not for the gratification of the love of the curious, or for the sake of their possible utility.

One other branch of zoology was embryology, fairly widely pursued in the sixteenth century, though without significant advance. The difficulties in the way of achievement were in fact great, though the embryologists of the period could not know that they were foredoomed to stagnation. They were stimulated to explore embryology (as the discussions of reproduction by Rondelet and Belon clearly demonstrate) by the desire to emulate Aristotle ; they very nearly succeeded in proving the validity of the extreme humanist position, that modern man could not hope to do more than know as much as the ancients had known. Lacking microscopes, sixteenth-century scientists could see little more than Aristotle had seen. They could, like him, merely open an egg day after day to observe the stages of development as far as they are visible to the naked eye. A number of them—Aldrovandi, his pupil Volcher Coiter (1543–76), and Fabricius of Aquapendente (1537–1619)—did it with great zeal, taking special joy in noting errors committed by Galen, though generally confirming the observations of Aristotle. The most thorough embryological treatise is *On the Formation of the Egg and Chick* by Fabricius of Aquapendente (1612), a massive and profusely illustrated volume, which is so Aristotelian that the account follows Aristotle's point by point, elaborating, discussing and occasionally refuting. Fabricius treated the egg—both hen's egg and insect egg—exhaustively. He was particularly careful to explain each of Aristotle's four causes in great detail, paying special attention to the final cause, the purpose for which each part of the egg exists. Perhaps the greatest embryological contribution of Fabricius, as of his predecessors, was that the knowledge left by Aristotle was thoroughly canvassed, and all discrepancies, together with a few errors, carefully noted. At least the way was clear for further advance in the fascinating problem of the generation of animals, though it was to be half a century before the use of the microscope made this advance possible.

Zoology and botany met on a descriptive level in the numerous accounts of the flora and fauna of the Americas. Primarily travel books, these works nevertheless opened up a veritable new world of plants and animals, providing much to stimulate curiosity and adding a vast list of herbs to the standard pharmacopoeias. The earliest examples in this genre were Spanish accounts of South America, especially of Peru, often intended as propaganda pieces. They are therefore not totally reliable, but they contain the first descriptions of such plants as tobacco, maize, potatoes, pineapples and so on, soon to be essential to European pharmacology, and later to the European diet. Thus there is the *History of the Indies* by Oviedo y Valdes (1478–1557), prepared soon after his return from forty-five years as an administrator ; this is an observant though credulous account of the new world of the Spanish colonies, describing the rubber tree, the potato of Peru, and tobacco, already being imported into Spain as a medicine. The first illustrated and reasonably detailed description of tobacco is to be found in the work of Nicholas Monardes (1493–1588), first published in 1569, but best known in its English translation of 1577 under the delightful title *Joyfull Newes out of the Newefound World*. This contains more natural history than most accounts, and described many useful medicinal plants, herbs and barks, as well as exotic animals like the armadillo. Monardes was a naturalist, rather than a traveller.

More complete than either of these, though more purely descriptive than Monardes, is the interesting and sympathetic account of Peru and Mexico by the Jesuit Jose d'Acosta. His *Natural and Moral History of the Indies* (1590) deals, as the title implies, with the country and its inhabitants ; botanical, zoological and human. Most of the exotic plants and animals of the region are described, including, for example, the cochineal insects parasitic on the prickly pear, which were later to provide red dye for European cloth. Acosta was also interested in peculiarities of climate, noting that in the Western hemisphere the Sun does not seem so hot on the equator as it does in Africa, and commenting upon the reversal of seasons found below the equator. Even more of a curious problem was the origin of the American Indians and indeed of the very existence of the New World. Acosta faced the problem squarely : there is, he says, no denying that the evidence of the senses flouts accepted authority ; for many of the Church fathers had undoubtedly denied the existence of the Antipodes, and the Ancients (with the notable exception of Plato) were ignorant of the existence of any continents other than Europe, Asia and Africa. One must accept the fact of their fallibility. Even worse was the case of the inhabitants of the New World, men and animals. To regard the Indians as pre-Adamite was to flout the authority of the Bible, too much for Acosta to contemplate ; and besides, Noah's Flood had undoubtedly covered the whole surface of the Earth, and must have wiped out all living things in America as well as elsewhere. Some had suggested that the Indians were the Ten Lost Tribes of Israel ; this Acosta denied on the basis of their cultural habits. For, he noted, they did not know how to write, as the ancient Hebrews did ; they did not practise circumcision ; and (for him the clinching argument) they had no great love of silver. Surely, he argued, even in thousands of years they would not have lost such characteristic and ingrained customs. So he concluded, sensibly enough, that the Indians had arrived from Asia over a land link which must be north or south of the coastal regions so far explored.

There is nothing for North America so detailed or so lively as the books of the Spanish in this period. There is, nevertheless, the account of Virginia written by the mathematician Thomas Hariot for Walter Raleigh, published in 1588 as *A Briefe and True Report of the New Found Land of Virginia* ; it includes something on the fauna and flora, but it was very brief. One of the first members of the Virginia colony was John White, whose water colours of the natives, the plants, the fish, birds, insects and reptiles, were not only delightful, but accurate ; engravings made from these drawings were used to illustrate the second edition of Hariot's book in 1590.

It should be obvious that most botany and zoology was still purely descriptive in the sixteenth century ; which is not surprising when so much still remained to do in this regard. It is true that some zoologists did good work in anatomy, though mainly in order to illuminate human anatomy, just as animal physiology was entirely subordinate to human physiology, which it was used to illuminate without any appreciation of possible differences either way. In botany, even taxonomy was not widely considered : most herbalists either listed plants alphabetically (in Latin or a vernacular) or divided them into a few rough groups. This was chiefly because the botanist's concern was to identify plants, to enable others to identify them as well, and to describe their uses. This strictly utilitarian approach left little room for pure science. Few were aware of any alternative approach ; but that there was some appreciation of the need for pure science was expressed by the Bohemian botanist Adam Zaluziansky, who in 1592 wrote

It is customary to connect Medicine with Botany, yet scientific treatment demands that in every art, theory must be disconnected and separated from practice, and the two must be dealt with singly and individually in their proper order before they are united. And for that reason, in order that Botany, which is (as it were) a special branch of Natural Philosophy, may form a unit by itself before it can be brought into connection with other sciences, it must be divided and unyoked from Medicine.[3]

Zaluziansky tried to follow his own precepts : he analysed plants into lower and higher forms (less and more ordered), and tried to categorise the degree of order in terms of the degree of elaboration of the leaf, but with only moderate success.

Others who felt that botany needed to be more than a purely descriptive science were a curious group of humanist reactionaries, more concerned to rescue botany from the presentation of Dioscorides in order to restore it to a (presumed) Aristotelian purity than they were to consider the problem *de novo*. The results were not without merit, but rather too remote from conventional botany. Most striking of these neo-Aristotelians was Andreas Cesalpino (1519–1603). Cesalpino received a medical education at Pisa, and after taking his M.D. in 1549 became professor of pharmacology there. His work is, however, only incidentally medical. His two most important books, *Sixteen Books on Plants* (1583) and *Peripatetic Problems* (*Questiones Peripateticae Libri* v, 1588) both treat botany in as Aristotelian fashion as is possible. In the *Peripatetic Problems*, Cesalpino developed a general theory of nature based upon an attempt to reform all branches of science —not just biological science—by rejecting later views in favour of Aristotelian doctrines : even Galen and Ptolemy were unsatisfactory. Convinced that the Aristotelian doctrines of form and matter could supply the all-embracing principle needed to organise nature, Cesalpino applied these doctrines to biology, endeavouring to establish even more firmly than Aristotle had done the absolute continuity of the scale of nature. Cesalpino's argument was that as living matter (whether highly organised or not) contains a single living principle, each living entity must have a single, undivided living principle, located in some definite spot in the organism. Higher animals clearly have their living principle in the heart ; which is their centre ; for without the heart they die. Lower animals and plants, which live when divided, must have a less centralised structure : Cesalpino had already decided for somewhat different reasons that the centre of a plant is the so-called collar of the root, the point at which stem and root join ; but this is the centre *in actu* only, that is, at any given moment, for *in potentia* the centre is anywhere, since cuttings will take root to form a new individual, whose centre of life will once again be where stem and root join. Again, since there is only one living principle for animals and plants, all living matter must be organised according to the same basic pattern, so that the organs of plants must correspond to those of animals : on this basis, the root corresponds to the digestive system, the pith to the intestines, stalk and stem to the reproductive system, and fruit to the embryo. (In spite of this, Cesalpino denied the sexuality of plants.) Each organ thus has its specific function and use, on which account Cesalpino concluded that the leaves exist to protect the fruit.

Though, obviously, Cesalpino has much to say that is interesting, it is equally obvious that the pattern of his thought is antiquated. Humanism not infrequently led to such excesses as those of Cesalpino ; in his case, the attempt to rescue Aristotle from the attacks of his detractors (growing increasingly vocal in the last third of the sixteenth century, in Cesalpino's old age) produced a work that was bound to be of minor importance, because it had so little to say that seemed relevant to his contemporaries. Very probably they were right ; Cesalpino seems original after four centuries, but at the time he seemed reactionary and was obscure if not actually obscurantist. It is difficult to see what use could have been made of his conclusions in botanical study. He was original and foresighted in trying to evolve a common pattern for botany and zoology, to create biology over two hundred years before Treviranus and Lamarck invented the word ; but his comparisons are naïve and must at the time have seemed excessively vitalistic and teleological, in an age that preferred its vitalism to be either more mystic or more muted. Botanists were probably well advised to continue producing ever better herbals and to begin the search for a successful method of classifying plants according to structure. In neither attempt could such physiology as Cesalpino's be of any great assistance. Anatomy —animal and plant—was the first and most fundamental requirement for the establishment of scientific zoology and botany ; there the seventeenth century was to benefit less from the natural historians than from physicians studying human anatomy, and using animal anatomy for comparison and assistance. More was learned about cold-blooded animals from Harvey's study of them in connection with the circulation of the blood (primarily a problem in human physiology) than from a century's inclusion of descriptions of serpents in natural history books. Medicine did, in the end, repay animals for their long scientific subservience to the needs of man.

CHAPTER III

THE COPERNICAN REVOLUTION

First of all I wish you to be convinced . . . that this man whose work I am now treating is in every field of knowledge and in mastery of astronomy not inferior to Regiomontanus. I rather compare him with Ptolemy, not because I consider Regiomontanus inferior to Ptolemy, but because my teacher shares with Ptolemy the good fortune of completing, with the aid of divine kindness, the reconstruction of astronomy which he began, while Regiomontanus—alas, cruel fate—departed this life before he had time to erect his columns.[1]

When Nicholas Copernicus was born in 1473, Ptolemy's *Almagest* had not yet been printed. When in 1496 Copernicus, his studies at the University of Cracow completed, set off for further study in Italy, the *Almagest* could still only be read in manuscript (most of the manuscripts were in Italy) and Regiomontanus' *Epitome of Ptolemaic Astronomy* was just being published. Copernicus was born into a world in which astronomers were groping for reform, and educated in a world in which only the first step—mastery of Ptolemy—had been taken. By the time that Copernicus had finished his preliminary training in astronomy, his teachers had begun to recognise that, although an intensive study of the *Almagest* was a necessary pre-requisite to further advance, to know only what Ptolemy knew could not suffice to rejuvenate astronomy.

Nevertheless, the humanist principle that all knowledge must lie with the ancients still appeared viable ; hence, when Ptolemy failed to give the required assistance, it seemed reasonable that the next step should be the examination of those notions of earlier Greek astronomy which the Ptolemaic system had in its day rendered obsolete. Nothing in his background or education inclined Copernicus towards drastically novel ideas. Everything combined to suggest that the badly needed reform in astronomy could be achieved by adherence to humanist principles. This was the legacy of Peurbach. It was fairly and clearly stated by Copernicus in the dedicatory preface of his great work, *On the Revolutions of the Celestial Spheres* (1543 ; generally known as *De Revolutionibus*) ; he said

> I was led to think of a method of computing the motions of the spheres by nothing but the knowledge that Mathematicians [i.e. astronomers] are inconsistent in these investigations ... I pondered long upon this uncertainty of the mathematical tradition in establishing the motions of the system of the spheres ... I therefore took pains to re-read the works of all the philosophers on whom I could lay my hands to find out whether any of them had ever supposed that the motions of the spheres were other than those demanded by the mathematical schools.[2]

This is not a revolutionary attitude ; indeed Copernicus never intended it to be. He was not a pioneer, and attempted nothing that others had not tried before, for many astronomers used ancient opinion to refute Ptolemy in the strange blend of iconoclasm and respect for authority that so often characterised the sixteenth century. Copernicus alone chose a system (the Pythagorean, he understood it to be) which had profound revolutionary implications, though it was not for another generation that these implications were to become apparent. Never, perhaps, has such a conservative and quiet thinker had such an upsetting effect upon men's minds and souls ; but seldom has such a conservative thinker been, even inadvertently, so bold in accepting the improbable.

Certainly there was nothing in his education to prepare him for a revolutionary role in that reformation of astronomy which, he learned, was needed. Son of a merchant of Cracow, Copernicus was while a boy adopted by his maternal uncle, Lucas Watzelrode, an ecclesiastic who later became a bishop. Naturally intended for an ecclesiastical career, Copernicus began his preparation at the University of Cracow. The professor of astronomy had published a commentary on Peurbach's *New Theory of the Planets*; the astronomy lectures were presumably sufficiently up to date to convey the problems confronting astronomers, though there can have been nothing very unorthodox. (Then or later Copernicus acquired some knowledge of the theories of Nicholas of Cusa, but he had no interest whatsoever in the daring flights of intellectual fancy demanded by the doctrine of learned ignorance.) After five years, Copernicus went to Italy to continue his studies at the University of Bologna : Greek, medicine, philosophy and astronomy. He was already considered a trained astronomer : the professor of astronomy at Bologna spoke of him as an assistant rather than as a pupil ; and in 1500 he went to Rome for an astronomical conference, probably to discuss the reform of the calendar. He was recalled to Poland to be installed as canon of the cathedral of Frauenburg, but was allowed to return to Italy to study canon law and medicine at Padua, and to take his doctorate in law at Ferrara. This accomplished, he returned to Poland, to act as his uncle's secretary and physician until 1512 when, at his uncle's death, he settled in Frauenburg. There he led an extremely busy life, as an active administrator, a practising physician, and a writer on economics and astronomy.

The history of the development of the Copernican system is obscure. Copernicus later said that, outdoing even Pythagorean caution, he had kept his work in abeyance for over thirty years ;

he presumably had constructed at least the outlines some time before he wrote his first sketch, the *Commentariolus* (*Little Commentary*), in 1512. This brief synopsis circulated among his friends and gradually acquired a wider audience ; it had reached Rome by 1533 and was discussed enough there to cause ecclesiastical pressure to be exerted on Copernicus to publish further. The Church, with an eye on calendar reform, was eager at this period to encourage mathematical astronomy. Yet probably nothing more would have happened, and Copernicus would have died leaving a mass of unpublished papers, if a young professor from the Protestant University of Wittenberg, Georg Joachim Rheticus (1514–76), had not determined to play much the same role for Copernicus that Halley was later to play for Newton. Rheticus had heard rumours of the new and exciting astronomical theories, and in 1539 arrived on the doorstep of the Frauenburg canonry, begging for astronomical enlightenment. Copernicus did not scruple to let the young man, Protestant though he was, have full access to his astronomical papers ; and barely two months later he permitted Rheticus to prepare for publication a brief account of the system, the *Narratio Prima*, published in 1540.

This *First Narration* (something of a misnomer, since the *Commentariolus* was in existence) is very brief, but it reached a wider audience than the *Commentariolus* with two editions in two years. Rheticus never named Copernicus, though he dated his work from Frauenburg ; he merely refers to " my teacher." No doubt Copernicus had stipulated anonymity ; very probably he wished to see what reception would be accorded to the theory by a general audience before he acknowledged it. Rheticus urged the publication of the years of work accumulated by Copernicus ; he promised to see the book through the press, and perhaps pointed out to the old man that the favourable reception accorded to the *First Narration* showed that the time was now ripe for publication. Copernicus yielded ; and though Rheticus

71

defected, handing his job over to a Lutheran pastor named Andreas Osiander, Copernicus let the printers continue. The book appeared in 1543 under the title *De Revolutionibus Orbium Coelestium Libri Sex* (*Six Books on the Revolution of the Celestial Orbs*), with somewhat strange prefatory material. Copernicus dedicated the work, naturally enough, to Pope Paul III, while Osiander, the Protestant, inserted an unsigned and wholly unauthorised note to the reader of his own composition, expressing his, not Copernicus' views on the physical reality of the system. The *De Revolutionibus* thus appeared in a curiously if anonymously Catholic-Protestant guise.* Tradition has it that Copernicus saw his great work only on his death-bed; he was certainly ill in the months immediately preceding its publication, and died soon afterwards.

Much controversy has raged over the question of Copernicus' avowed disinclination to publish in the years between 1512 and 1539, though there has been curiously little discussion of his apparent readiness to yield to the request of Rheticus. Why the hesitation, when he was prepared to give way in the end? Was he afraid of official censure? Was he jealous of his superior knowledge? Was he naturally over-secretive? Did he not truly believe in his system? Did he want to keep his discoveries to himself? Did he believe that learning was for initiates only?

Careful reading of the prefatory letter of dedication to Pope Paul III (hardly a sign in itself of fear of official disapproval) suggests that the reasons were derived from what he believed to be Pythagorean doctrine: that advanced scientific ideas should be discussed only among scientists, because non-scientists, misunderstanding, always travestied them. Copernicus was not thinking specifically of the popular misapplication of scientific

---

* The dedication to the Pope most probably explains why Copernicus did not mention the help given by Rheticus; not ingratitude, but policy, suggested the unwisdom of naming Protestants in 1543 in ecclesiastical circles. But the fact that Rheticus relinquished the editorship suggests that his help was not so great as one would expect.

theory which is of such common occurrence in modern science (though it was to happen with his theory when used by Bruno to support a pantheistic doctrine) but of the incredulity and scorn with which, he believed, new ideas must always be treated. He knew his theory was both novel and strange ; he feared lest it should be regarded as absurd as well, and he himself, as he put it, " hissed off the stage . . . Reflecting thus, the thought of the scorn I had to fear on account of the novelty and incongruity of my theory, nearly induced me to abandon my project."[3]

Fear of ridicule is not a very noble motive for withholding publication, perhaps, but it can be a very real one : Galileo felt something of this half a century later, and there seems no reason not to believe Copernicus when he plainly says it influenced him strongly. One should reflect that then, as now, publication of a book meant committing one's ideas to the mercy of a wide general audience. Copernicus had long before been willing to circulate his theory in manuscript (by means of the *Commentariolus*), a form equivalent to the modern method of reading a paper in technical form to a learned society where it will be judged by specialists only before venturing on formal publication. Actual publication of the *De Revolutionibus* exposed the Copernican system, as its author knew would happen, to comment and criticism by all and sundry—humanists, scholastics, astrologers, mathematicians, crackpots, ecclesiastics—for any educated man in the sixteenth century when astronomy was the most widely studied of the natural sciences fancied himself competent to pass judgement on astronomical theories. Remembering Luther's scornful comment about the fool who sought to turn science upside down, one can see that Copernicus had reason to fear ridicule. Attack by a Protestant rebel could hardly have affected his position in the world, but he obviously disliked personal ridicule of any kind ; and as a public figure of some note he was vulnerable to disdain from university professors and his superiors in the Church, quite capable of judging astronomical theory. But, as he remarked disarmingly in his dedication, the Pope, by his "influence and judgement can readily hold the slanderers from biting." [4] All he wanted was a fair hearing; though he did not live to see it, his book was indeed well received in the Church and used, as he had hoped, to further calendar reform. Ironically, little serious attention was given at first to the heart of the book, the new theory.

Looking through the *De Revolutionibus* one is immediately made aware of the fact that, true to his training, Copernicus had studied the *Almagest* very carefully indeed. For the *De Revolutionibus* is the *Almagest*, book by book and section by section, rewritten to incorporate the new Copernican theory, but otherwise altered as little as might be. Kepler was to remark later that Copernicus interpreted Ptolemy, not nature, and there is some truth in the comment; to Copernicus the way to nature lay in a re-interpretation of Ptolemaic astronomy, wrong in details, but right in conception. And it was indeed essential, if one wished to replace Ptolemy, to do everything he had done, only better. Copernicus did not wish to claim novelty, which had no appeal to him; he claimed to be doing no more than revive Pythagorean doctrines, especially those of Philolaus (fifth century B.C.), as described by Plutarch.* But he was well aware of the fact that, before his own, there was no system of astronomy comparable to

* It may seem odd that Copernicus did not claim to be reviving the doctrines of Aristarchos of Samos (fl. c. 270), "the Copernicus of antiquity" who gave the Earth both diurnal and annual motion. But our only real knowledge of this theory is derived from a couple of sentences in the *Sand-Reckoner* of Archimedes, combined with two brief references in Plutarch. Copernicus had probably never seen the work of Archimedes as it was first printed the year after his death. He originally concluded the first book of *De Revolutionibus* with a brief recapitulation of the difficulties of discussing complex scientific ideas for a general audience, and then remarked that according to some authorities, Aristarchos held the same opinion as Pythagoras. He later rejected these paragraphs.[5]

that of Ptolemy, simply because none offered a computational scheme and method that could replace that of Ptolemy. If one genuinely wanted to supersede Ptolemy (as Copernicus did) one needed to offer more than a qualitative cosmology ; one must, in addition, present a thoroughly worked out mathematical system, capable of giving results at least as good as those derived from the Ptolemaic system when used for computing planetary tables. In this Copernicus was eminently successful ; his mathematical theory was even to be used for computation of tables by astronomers who totally rejected his cosmological system.

The problem appears in sharper focus if one examines the most ingenious and detailed of anti-Ptolemaic systems proposed before Copernicus. In 1538 there appeared *Homocentrics*, dedicated, like *De Revolutionibus*, to Pope Paul III. The author was Girolamo Fracastoro, an Italian humanist, poet, physician and astronomer, who had been professor of logic at Padua when Copernicus was a student, but who spent most of his life as an author and physician. Fracastoro did not profess to have originated the central idea of *Homocentrics*, which was to replace Ptolemaic epicycles and eccentrics (see p. 434) with the concentric (or homocentric) spheres originated by Plato's pupil Eudoxos (*fl. c.* 370 B.C.) and elaborated by Aristotle. Fracastoro did abolish epicycles and eccentrics but at the price of a somewhat improbable system, certainly one even further removed from physical verisimilitude than the Ptolemaic system it sought to replace. Fracastoro assumed that every motion in space is resolvable into three components at right angles to one another, so that the motions of the planets could be represented by the motion of crystalline spheres with their axes at right angles, three for each motion. He further assumed, quite gratuitously, that while the outer spheres move the inner ones, the motion of the inner spheres does not affect the outer ones ; this permitted him to eliminate many of Aristotle's spheres—those that served to counteract the frictional motion caused by two spheres rubbing against one another—but at the same time to allow the diurnal rotation of the *primum mobile* to account for the rising and setting of the planets as well as the fixed stars. In this way, he required seventy-seven spheres in all! Fracastoro ingeniously eliminated the great disadvantage of the Aristotelian system, which was that if the planets are located on the equators of spheres concentric with the Earth there ought not to be any variation in their brightness. He explained the observed variation by assuming that the spheres (being material) were of variable transparency, because of variable density. This system (with which others experimented as well) shows how much Copernicus was following the fashion of the times in reviving ancient systems to replace that of Ptolemy ; it also shows the immense superiority of the conception of Copernicus. For, in spite of fairly detailed discussion, Fracastoro did not offer a replacement to the computational methods of Ptolemy. He knew and understood the *Almagest*, but he had neither the patience nor the mathematical skill to write it anew ; he was content to explain how one could eliminate epicycles and eccentrics, without pausing to explore completely the validity of his assumptions about the mathematical representation of motions by means of spheres.

Copernicus, on the other hand, wrote *De Revolutionibus* as a careful parallel to the *Almagest*, with the mathematical and computational methods revised for a different concept of planetary motions. Book I deals, as does Ptolemy's Book I, with a general discussion of the universe : the sphericity of the universe and of the Earth ; the circular nature of heavenly motion ; the size of the universe ; the order of the planets ; the question of the Earth's motion ; the basic theorems of trigonometry, are all discussed in both works. But where Ptolemy argued for a geocentric and geostatic universe, Copernicus argued that the Earth and all the planets circled the Sun, carefully refuting Ptolemy's counter-arguments one by one ; and he showed himself able to add to Ptolemy's work in trigonometry. Book II deals with spherical trigonometry, and the rising and setting of the Sun and the planets (now ascribed to the Earth's motion) ; Book III contains the mathematical treatment of the Earth's motions, Book IV of the Moon's motions, Book V of the motions of the planets in longitude, and Book VI, the motions of the planets in latitude. Or, as Copernicus explained it :

In the first book I describe all the positions of the spheres together with such movements as I ascribe to the Earth ; so that this book contains, as it were, the general system of the universe. Afterwards, in the remaining books, I relate the motions of the other planets and all the spheres to the mobility of the Earth, so that we may thus comprehend how far the motions and appearances of the remainder of the planets and spheres may be preserved, if they are related to the motions of the Earth.[6]

No one was going to be able to brush aside the work of Copernicus because he only sketched a system, as others (and he himself in the *Commentariolus*) had done before this time. By his plan and development he insisted on being judged on the same basis as Ptolemy ; there was nothing one could find treated in the *Almagest* that was not also treated in *De Revolutionibus*. Copernicus certainly desired nothing better than to be regarded as the Ptolemy of the sixteenth century ; he could see no higher aim than to explain by his own system the appearances of the heavens as known to Ptolemy. The Copernican would replace the Ptolemaic system, he believed, because it was simpler, more harmonious, more ingenious and more in keeping with the underlying philosophical basis, which demanded that the motions of the heavenly bodies, being perfectly circular, be represented by mathematical curves that were as nearly perfect circles as might be. It was on this that he wished to be judged.

At the heart of the Copernican system lies the point which required the most carefully reasoned argument : the attribution of motion to the Earth. It was this attribution that caused Copernicus to fear that astronomers would laugh and refuse to take him seriously. For to assume that the Earth moved, in the sixteenth century, required such a straining of well-assured fact as to amount to an absurdity of the degree that would be provoked by the contrary argument to-day. It is difficult in the twentieth century to understand this ; we are convinced that the Earth moves because we have been told so since babyhood, though relatively few people can readily offer proof of this motion. In the sixteenth century everyone knew, for similar reasons, that the Earth stood still ; and no one needed arguments to prove what the evidence of the senses confirmed. To be sure, scientists and philosophers conventionally offered various kinds of proof, logical and scientific, nearly all derived from Aristotle and Ptolemy. Thus, for example, it was habitually pointed out that the Earth belonged at the centre of the universe because, according to the tenets of Aristotelian physics, that was the natural place for the heavy element earth of which it was chiefly composed ; that it was inherently improbable that any such naturally heavy and sluggish object should move ; that the natural motion of the terrestrial elements was rectilinear, whereas the natural motion of the celestial element was circular ; that if the Earth did rotate on its axis, either the atmosphere, or else missiles and birds moving through it, would be left behind, and a stone dropped from a tower would not hit the ground at the foot of the tower.* These were points that it was essential to refute.

Copernicus did so, firmly and ingeniously, using Aristotelian arguments wherever possible. Thus, to the argument that the Earth should not be assumed to move, because to move was contrary to its nature, Copernicus retorted that it was easier to

* This last was a new argument in the sixteenth century.

imagine that the relatively small Earth moved, than that the great heavens hurled themselves around every twenty-four hours, a feat that must require truly enormous speed. (To call the Earth small, even relatively, required a leaping imagination that others could not encompass as readily as Copernicus did). Surely, so he argued, it was easier to imagine that the apparent motion of the heavens was really the result of the motion of the Earth, turning on its axis once every twenty-four hours. As for Ptolemy's fears that in that case the atmosphere would be left behind, surely these were groundless : for the atmosphere was a part of the whole terrestrial region, and as such would share, like things suspended in it, the motion of its central body, the Earth. Copernicus could not deny that circular motion was natural to the heavens, and rectilinear motion to the Earth and its regions ; so he was forced to modify the rigid distinctions between celestial and terrestrial physics which had for so long been essential tenets of the Aristotelian cosmos. He had first to modify Aristotelian physics to argue that rectilinear motion and circular motion might coexist in the same body, so that the whole might rotate, while the parts moved in straight lines. And he had secondly to argue that the spherical nature of the Earth fitted it to move in circles as much as did the spherical shape of the heavenly bodies.

All unconsciously, by denying one essential difference between the heavenly and terrestrial spheres, Copernicus began that encroachment on cosmical dualism that was destined to end fatally. Once astronomers began to consider heavens and Earth as one, there was logically a need to treat their dynamical problems as one. Copernicus was the first modern cosmologer to begin to break down the old-established barriers between the Earth and the celestial regions ; one by one these barriers were demolished until in the Newtonian universe modern physics allowed a return to the unified and uniform cosmos of the original pre-Socratic conception. Not that Copernicus argued like this; his method was thoroughly Aristotelian in spirit if not in content. He insisted, "We conceive immobility to be nobler and more divine than change and constancy."[7] So, if the heavens were nobler, they should be at rest, while the baser Earth moved. Since it was possible that the heavens were at rest and the Earth in motion, it was, to Copernicus, also probable, reasonable and fitting that this was in fact the case. And Copernicus felt that a probable argument was all that could be expected of him.

The rearrangement of the Ptolemaic system to form the Copernican required more than the assignation of motion to the Earth. In the outline sketched in the *Commentariolus*, Copernicus listed seven assumptions (his own word) required before serious consideration of the system could begin.[8] First, he had to assume that there was no one centre of motion for all the heavenly bodies. For although he was to postulate that the planets all revolved about the Sun, the Moon still clearly revolved about the Earth. This dichotomy was, at the time, considered a disadvantage, for one of the niceties of the Ptolemaic system was that all the heavenly bodies revolved around the same point. Second, Copernicus removed the Earth from the centre of the universe. He still necessarily retained it, as he said, as the centre of the lunar sphere; for whatever one might postulate, it was certainly a fact that heavy bodies were observed to fall to the Earth, just as it was a fact that the Moon circled the Earth. Here the Copernican system was again at a disadvantage, because physics and cosmology no longer supported one another. According to Aristotelian physics, heavy bodies fell to the Earth because it was the centre of the universe; when Copernicus made this explanation impossible, he left gravity as a mysterious or occult force, needing explanation in a way that it had not done before. He could only postulate that gravity was common to all the planets, without particularising further. The result was to raise a new and fertile problem to be tackled by later cosmologists.

The third assumption was that the centre of motion of the planetary system was in fact the Sun,* which was therefore the true centre of the universe, a conclusion which Copernicus declared to be "suggested by the systematic procession of events and the harmony of the whole universe." [9] This special position of the Sun seemed to Copernicus to explain much that had hitherto been mysterious : for it had always been a peculiarity that the Sun, a planet like Venus or Mars, but not very near the Earth, should be so distinguished from the other planets. The Sun alone shed light and warmth that fostered life itself ; its importance was obviously so much greater than the planet's astrological influence that it had always been given a special consideration. Now at last the unique properties of the Sun were recognised as corresponding to a unique position ; as Copernicus emphatically explained :

In the middle of all sits the Sun enthroned. How could we place this luminary in any better position in this most beautiful temple from which to illuminate the whole at once ? He is rightly called the Lamp, the Mind, the Ruler of the Universe . . . So the Sun sits as upon a royal throne ruling his children the planets which circle around him. [10]

Besides, it explained why all planetary motion contained a 365-day period.

The fourth assumption concerned the size of the universe ; it must, so Copernicus declared, be very large, so large, in fact, that the distance from the Earth to the Sun must be negligibly small compared with the distance of the Sun from the sphere of fixed stars. This was an extremely necessary postulate ; for it alone could account for the fact that the motion of the Earth is not

---

* As it worked out, since the Earth's sphere is eccentric to the Sun, the real centre of motion was the centre of the Earth's orbit. It was nevertheless true that the planets "went around" the Sun ; and in the Ptolemaic system the same thing occurred with respect to the Earth.

reflected in an apparent motion of the fixed stars, as it would otherwise be. The fixed stars in the Copernican system ought to exhibit the phenomenon of parallax ; that is, any one star should appear to move slightly to and fro against its background during the year as the Earth travels from one side of its annual path to the other, just as the photographer's view of a group drawn up before him varies as he walks to and fro in front of it. But the fixed stars did not appear to exhibit any parallax ; a fact which is hardly surprising since it continued to evade telescopic detection until 1838–9. It was a weak point in the Copernican system, since Copernicus could only insist that the parallax was there, but was too small, owing to the immense distance of the stars from the Earth, to be detectable.

The last three assumptions were concerned with the motion of the Earth : Copernicus assumed that the Earth's diurnal rotation produced the apparent rising and setting of the Sun, planets and fixed stars, and the Earth's annual revolution about the Sun produced the apparent annual motion of the Sun, and the apparent retrogradations of certain of the planets.* The resultant system is the familiar picture of the Copernican universe : at the centre, the Sun ; then the spheres of Mercury, Venus, the Earth with its Moon, Mars, Jupiter and Saturn, with the (now stationary) sphere of the fixed stars forming a boundary and limit to the universe as a whole.

By these means and with considerable felicity, Copernicus enhanced the order and harmony of the planetary arrangement. The motions of the Earth, as Copernicus insisted, explained much that had hitherto been a source of uneasiness and distress to astronomers, because it seemed so untidy. The Earth's diurnal motion

---

* To these two motions, Copernicus later added a third (unnecessary) one, a motion of the poles to account for the constancy of the angle of inclination of the Earth's axis which, when the Earth is carried around by a solid moving sphere, is in danger of a progressive tilt.

alone explained the daily rising and setting of stars, planets and Sun, which did not now themselves exhibit diurnal motion. Better still, the annual revolution of the Earth around the Sun was not a mere replacement of the Sun's yearly motion : it served as well to effect a regularising of the motions of the planets.

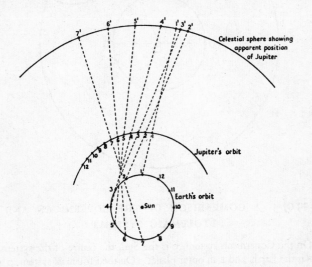

FIG. 4. THE APPARENT MOTION OF JUPITER RELATED TO ITS REAL MOTION

The size of the orbits is not to scale

In the Copernican system, the "retrograde" motions of the planets were shown to be merely *apparent* motions ; the *real* motion of each planet was always in the same direction about the Sun, though (because of the Earth's motion) it did not seem so. The Earth's motion (which is our own) causes us to see the planet against the background of fixed stars from different points of view as it (and we) travel in orbits about the Sun, in the same direction, but at different speeds.

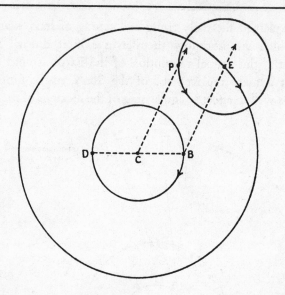

FIG. 5. A COMPARISON OF THE COPERNICAN AND PTOLEMAIC SYSTEMS

On the Copernican system, C is the Sun, the centre of the system, B is the Earth, and E an outer planet. On the Ptolemaic system, C is the Earth, D the Sun and E the centre of the planet's epicycle, the planet itself being at P. The line from Earth to planet in the second case will be parallel to the line from Earth to planet in the first case, and the angle between this line and the Earth-Sun line will be the same in each system. Consequently the apparent position of the planet is the same

Suppose we consider the apparent path of an outer planet like Jupiter over the period of a year. While the Earth is making one complete revolution about the Sun, Jupiter will move only about 30° of its orbit (here both orbits are assumed circular), since it takes nearly twelve years to circle the Sun. As shown in figure 4, the result will be that while the Earth makes its circuit in the twelve months, and Jupiter in fact travels on its orbit from the points marked 1 to 12, the path of Jupiter as seen from an apparently stationary Earth against the sphere of fixed stars will be as indicated by the figures 1' to 12', where at the beginning of the year the motion is backwards ; and if the apparent path of Jupiter is observed throughout its complete orbital period, the result will be a series of loops. The apparent path of Jupiter is thus the result of neglecting to differentiate between relative and absolute motion ; in fact its orbit is as smooth a curve as that of the Earth, and its motion always in the same direction. To Copernicus this discovery represented a profoundly satisfactory simplification, though it may be doubted whether to others it necessarily did so. Certainly it curiously separated appearance and reality, for all that Copernicus tried to argue calmly and persuasively that relative motion was a simple affair :

> It is but as the saying of Aeneas in Virgil—" We sail forth from the harbour, and lands and cities retire." As the ship floats along in the calm, all external things seem to have the motion that is really that of the ship, while those within the ship feel that they and all its contents are at rest.[11]

This was true enough, and it was true that many of the motions in the two systems which appeared to differ were in fact interchangeable, being identical though assigned to different bodies.* But in that case, why change ? Who could be sure that Copernicus really knew which was ship, and which shore ?

Copernicus was certain that he did know, and equally certain that he had answered Ptolemy's arguments against the stability of the Earth as convincingly as anyone could demand. It was certain that he had successfully turned Ptolemy inside out, removing the Earth from the centre of the universe (though not so far as to be uncomfortable ; for it was quite near the centre compared with its distance from the fixed stars), and setting the

---

* The case of an outer planet (like Jupiter) is shown in figure 5.

Sun there instead ; making the fixed stars truly fixed, and the Sun at rest ; using the motion of the Earth to explain several motions at once. By these means, as Copernicus argued, he had introduced a greater measure of simplicity, order, harmony and uniformity, which corresponded, far better than the Ptolemaic system could do, to Plato's original conception of a universe mathematically expressible in terms of circular motion. It is true that Copernicus still had to use eccentrics and epicycles and deferents which were hardly simple, though their use had long been interpreted as conforming to Plato's requirement of a combination of circular motions. But he had abolished the equant, which was a dubiously satisfactory contrivance, and physically meaningless ; and he had explained away the awkward retrograde motions as mere appearances. There was a mathematical ingenuity and elegance that must appeal to theoretical astronomers, Copernicus was sure ; it was truly a Pythagorean system designed for the appraisal of mathematicians. That it still used the familiar epicycles and eccentrics might be an advantage to the sixteenth century ; everyone knew how to manipulate these, and would have found a universe totally denuded of them naked indeed. Similarly, the fact that Copernicus retained the crystalline spheres not only explained what kept the planets in their positions in the universe, and why their motions were circular, but also preserved the familiar Aristotelian concept of the universe as a nest of concentric spheres. The universe had grown larger, but was still enclosed by the outer rim of the sphere of fixed stars, now truly fixed and immovable. In addition, the mathematics of the Copernican was a little easier to manage than the mathematics of the Ptolemaic system.

Were these advantages enough ? Could one readily accept a system so reasonable, but so unprovable ? Was it, in fact, really so reasonable ? There were advantages in the Copernican system admittedly, but were these advantages great enough to encourage men to destroy the work of centuries, and substitute Pythagoras and Copernicus for Aristotle and Ptolemy ? Even the advantages of uniformity were not attained without corresponding disadvantages. Though the motion of the Earth about the Sun created certain simplicities, it failed to explain why, when the Moon revolved about the Earth, it was not reasonable to suppose that the planets did also. It was comforting, to be sure, to find that the Earth was not totally reduced to the status of a planet, for the Earth was still unique in having the Moon as a companion ; but this was not sufficient compensation for its degradation.

Even granted the cogency of the arguments which Copernicus presented, there was no proof. It was true that no one could think of an experiment to test the rotation of the Earth on its axis ; but everyone knew that the annual revolution should produce stellar parallax, and equally that Copernicus did not claim that it did so, but only said that the distance was too great for this phenomenon to be observed. The argument from the well-known fact of the peculiarities of relative motion was plausible, but hardly conclusive ; however much it was true that the passenger on the moving ship *could* imagine that he stood still while the land moved, the passenger in fact recognised almost immediately that his senses were playing tricks with him. There was absolutely no sign in everyday experience that the inhabitants of the Earth needed to readjust their ideas ; for everything showed the Sun moving and the Earth standing still. If the Earth were not at the centre of the universe, what about gravity ? For one need not accept the argument of Copernicus. Worse than this : if the Earth were not at the centre, what happened to the dignity of man ? Had not God created the universe for man's enjoyment, and put the Earth at the centre to prove it ? Certainly the Earth was the only abode of man (one did not need to consider the wild remarks of Epicurean atheists who thought otherwise) ; and this proved that it was unique, and ought to occupy a unique position. Why did the motions of the planets influence the Earth and its inhabitants if the planets in fact circled the Sun ? To us, knowing that Copernicus was right makes the opposing arguments seem trivial. We falsify both the achievement of Copernicus and the difficulties in his way if we do not realise that it was not so simple ; he had reason to fear scorn because his position seemed at the time so untenable as to approach the ridiculous.

And yet, Copernicus was only following humanist precepts : he was trying to replace Aristotelian authority, which to the sixteenth century represented the outmoded intellectual pattern of the Middle Ages, with a system equally derived from Greek authority, which had the added advantage of being consonant with Platonic doctrines, so much more highly esteemed now than those of Plato's pupil Aristotle. And in doing so Copernicus achieved a good and interesting hypothesis. But to carry conviction an astronomical system required more than probable hypotheses, which were readily available ; it needed to present a true physical picture of the universe. Did Copernicus intend to supply this as well as a mathematically useful hypothesis of planetary motions ? There is every reason to suppose that he did. At the time his book appeared, this was by no means clear, for it was prefaced by an *Address to the Reader* " Concerning the Hypotheses of this Work " which tried to imply that astronomy was to be regarded as an intellectual exercise, in which the astronomer, incapable of attaining physical truth, must content himself with presenting any ingenious hypothesis which pleased him and fitted the facts. Careful reading of the whole discussion readily suggests (what was indeed the case) that the writer was not Copernicus. Later sixteenth-century astronomers knew that it was Osiander.[12] At first no one paid great attention to this prefatory disclaimer, because no one cared how deeply committed Copernicus was to the truth of his system. It was later Copernicans like Kepler who felt that it was imperative to establish that their master had meant what he said to be taken as physical fact, not mathematical hypothesis. In this they appear to have been correct. Certainly Copernicus thought that his assumptions about the motions of the universe were valid, and that his hypotheses were reasonable enough to be probably true. He did not expect observation to confirm this, because his own experiences suggested that a high degree of accuracy could not be expected. This was all the more reason for relying on mathematical argument when attempting to create a new cosmos. In his own eyes, Copernicus was not revolutionising astronomy ; he was using a different philosophical basis, the Pythagorean, to arrive at a truer picture of the universe than that which Ptolemy had ; but his universe was recognisably akin to Ptolemy's, and none the less valid for that. The framework was the same, though the structure differed. Copernicus had no wish to create a new heaven and a new Earth. For him it was better to explain the nature of the old ones more exactly.

CHAPTER IV

THE GREAT DEBATE

As I happened from time to time to meet anyone who held
the Copernican opinion, I asked him whether he had always
believed in it.   Among all the many whom I questioned, I
found not a single one who did not tell me that he had long
been of the contrary opinion, but had come over to this one,
moved and persuaded by the force of its arguments.   Examining them one by one then, to see how well they had mastered
the arguments on the other side, I found them all to have these
ready at hand, so that I could not truly say that they had for-
saken that position out of ignorance or vanity or, so to speak,
to show off their cleverness.   On the other hand, so far as I
questioned the Peripatetics and the Ptolemaics (for out of
curiosity I asked many of them) how much they had studied
Copernicus' book, I found very few who had so much as seen
it, and none who I believed understood it.[1]

It is peculiarly difficult to judge fairly of the effect of a new
scientific idea in the days before book reviews and formal scientific meetings.   One is entirely dependent on an appraisal of
comments, for and against : how does one balance, say, a lukewarm estimate by a scientist against one ardent defence and one
virulent attack, both by non-scientists ?   One can only try to
weigh and interpret the evidence imaginatively, remembering
that to receive any mention at all, even an unfavourable one, is in
itself a sign of achievement.

In the case of Copernicus there is the further complication

that his theory had been known in certain circles for many years before the publication of *De Revolutionibus* in 1543 : through the *Commentariolus*, through rumour and through the *First Narration* of Rheticus. He was held in high esteem in astronomical circles during his lifetime ; he was even spoken of as the potential saviour of astronomy. (Ironically, few who looked forward so eagerly were receptive to the new theory when at last it was made public.) Historians have sometimes tended to treat with surprise and sorrow the fact that not all astronomers were immediately converted, and to be shocked that some even wrote against the new system. The wonder should rather be that so many took the pains to try to assimilate a new and complex theory, whose proper appreciation required a high degree of mathematical skill.

In fact, *De Revolutionibus* was fairly widely read, enough, at least, to warrant a second edition's appearing in Basle in 1566, with the *First Narration* (now in its third edition) as an appendix. Of course, many must have learned more from Rheticus than from Copernicus, and presumably not all who talked so glibly about what Copernicus had or had not done ever looked into his great work. Yet there were many astronomers who were capable of utilising mathematical astronomy, and, however slow the pace of new ideas in the sixteenth century, the Copernican theory was used within half a dozen years of its publication. Popular discussions soon followed : by the end of the century even literary writers like Montaigne knew enough about the system to mention its implications. One might expect that its spread would be quickest in Germany, the centre of the astronomical instrument trade (and of astrology) and possessing new universities like Wittenberg, where Rheticus had taught. But by some perversity of intellectual development, countries like England and Spain, previously backward in cultural and especially scientific advances, were quick to notice the new astronomical ideas. Perhaps this was because they were not in firm possession of the old ones.

Rather oddly, much of the early praise for Copernicus is as an observational astronomer ; oddly, because he made very few observations as far as is known, and professed a low estimation of attainable observational accuracy. Even Tycho Brahe, the greatest observational astronomer between Hipparchus and Herschel, treated the observations of " the incomparable Copernicus " with respect, though he was puzzled to find them so crude.[2] It is true, however, that this emphasis on the observational aspects of the Copernican achievement was in part the result of the first use made of the new system, in the computation of planetary tables. Copernicus had given rough tables in *De Revolutionibus* ; now Erasmus Reinhold (1511–53), professor of astronomy at Wittenberg, drew up new, improved tables, complete enough to take the place of the now hopelessly out-of-date Alphonsine Tables. Reinhold called his tables " Prussian " in honour of his patron, the Duke of Prussia ; they are usually known under the semi-Latin designation of the Prutenic Tables (1551). Reinhold's relationship to the Copernican theory is peculiar. When he edited Peurbach's *New Theory of the Planets* in 1542, he declared (presumably on the basis of the *First Narration*) that Copernicus was to be the restorer of astronomy and a new Ptolemy.[3] When *De Revolutionibus* appeared, he immediately saw that the Copernican system could form the basis for calculating new tables. Yet he was never a Copernican ; for him it was enough that Copernicus had imagined a convenient mathematical device which simplified calculation.

The position of Reinhold was that of many other computational astronomers. Reinhold's Prutenic Tables were widely used; indeed, they appropriately helped to fulfil the reform of the calendar, as Copernicus had hoped a restored astronomy might do. They were frequently revised for other countries, and often expanded. The first such case was in 1556 when there appeared a work entitled *Ephemeris for the Year 1557 according to the Principles of Copernicus and Reinhold for the Meridian of London* ; its author, John Feild, did not have anything to say about the merits of the Copernican system (or, apparently, of anything else, for he is otherwise unknown). The preface was by the mathematician, astrologer, advocate of experimental science and spiritualist, John Dee (1527–1608) ; he there explained that he had persuaded his friend to compile these tables because he thought that the work of Copernicus, Reinhold and Rheticus had rendered the old tables obsolete ; but he did not think that a preface was a suitable place in which to enter into a critical discussion of the merits of the Copernican system. Nor did he ever commit himself : very possibly he had no desire to accept the physical reality of a computational and hypothetical system.

All astronomical computers had to reckon with Copernicus after Reinhold's work. So Pontus de Tyard (though actually a Copernican) in his *Ephemeris of the Eight Spheres* (1562) praised Copernicus as " the restorer of astronomy " purely because of his contributions to astronomical calculation. All these tables were an improvement on the older ones, and not only because they were up to date : how superior may be judged by the experience of Tycho. Wanting to observe a conjunction of Saturn and Jupiter, he found the prediction in the Alphonsine Tables to err by a whole month, while the Prutenic Tables erred by a few days only : too much, but still vastly better.[4]

Although there were a good many references to the Copernican system by non-professionals throughout the sixteenth century, there were few easy ways of getting a clear idea of its contents. Except for the work of Rheticus, there were almost no elementary presentations. Only one university curriculum even possibly included it : the statutes of the University of Salamanca were revised in 1561, and stipulated that mathematics (read in alternate years with astrology) was to consist of Euclid, Ptolemy or Copernicus at the choice of the students.[5] There seems to be no record of whether they did decide on Copernicus during the sixty years before they could no longer choose him. That the Copernican system was not otherwise taught in the universities is by no means surprising ; astronomy was an elementary subject, and the professors were expected to teach the basic elements as part of the general education of an arts student. For the future physicians who needed a competence in medical astrology a grounding in the Copernican system might well have proved an embarrassment, since astrological tables and instructions were Ptolemaic. So too were the everyday and literary references to astronomy. Besides, even to-day one does not begin science instruction by discussing the latest developments in nuclear physics ; nor were schoolboys fifty years ago started on Einstein before they understood Newton.

This was the point made by Robert Recorde in his *Castle of Knowledge* (1556) one of his series of treatises in the vernacular on mathematics, pure and applied. Recorde had been at both universities ; having graduated in medicine at Cambridge, he taught mathematics in London, a trade currently in much demand because of the lively interest in navigation. In the *Castle of Knowledge* he developed a dialogue between the Master and the Scholar which indicates not only the esteem in which he held Copernicus, but also his judgement that it took an advanced astronomer to weigh the arguments fairly and fully. The Master professes to believe that he need not discuss whether the Earth moves or not, because its stability is " so firmly fixed in most men's heads, that they account it mere madness to bring the question in doubt." This naturally provokes the Scholar into an incautious generalisation : " Yet sometimes it chanceth, that the opinion most generally received, is not most true," which in turn permits the Master to retort

And so do some men judge of this matter, for not only Heraclides Ponticus, a great Philosopher, and two great clerks of Pythagoras school, Philolaus and Ecphantus, were of a contrary opinion, but also Nicias Syracusius, and Aristarchus Samius, seem with strong arguments to approve it : but the reasons are too difficult for this first Introduction, & therefore I will omit them till another time. . . . howbeit, Copernicus, a man of great learning, of much experience, and of wonderful diligence in observation, hath renewed the opinion of Aristarchus Samius, and affirmeth that the earth not only moveth circularly about his own centre, but also may be, yea and is, continually out of the precise centre of the world 38 hundred thousand miles : but because the understanding of that controversy dependeth of profounder knowledge than in this Introduction may be uttered conveniently, I will let it pass till some other time.[6]

There is no doubt that Recorde believed that the young scholar was in no position to judge from the evidence, and might as easily turn against the new system as not; indeed, his Scholar thought it all vain conceits, and the Master was forced to rebuke him, telling him that he was far too young to have an opinion. This was fair enough ; but few ever did have the knowledge to have an opinion.

Many besides Recorde judged favourably of Copernicanism, but did not regard it as a sufficiently settled part of accepted astronomy to include in an elementary presentation. A typical example is the case of Michael Maestlin (1550–1631), professor of astronomy at Tübingen. A generation younger than Reinhold, he found it possible to accept the Copernican system without at first trying to advocate it publicly. His textbook *Epitome of Astronomy* (1588) very probably reflects his university lectures, and is strictly Ptolemaic ; but later editions contain Copernican appendices. The fact that Kepler (1571–1630) was his pupil, and treated Maestlin as his master, shows that with advanced students he did discuss the new doctrine ; for Kepler was a Copernican almost before he was a competent astronomer, and later remembered defending Copernicanism publicly. In 1596, Maestlin attended to the publication of Kepler's first book and, of his own accord, appended the *First Narration* of Rheticus, with a preface in praise of Copernicus. Whatever his beliefs may have been before this time, he was clearly a convert in the 1590's; and after the condemnation of the Copernican system by the Catholic Church, Maestlin, a Protestant, proposed a new edition of *De Revolutionibus*, though he got no farther than writing the preface. Maestlin's position is different from that of Christopher Rothmann, astronomer to the Landgrave of Hesse-Cassel, who carried on a long correspondence with Tycho Brahe in which he ardently defended Copernicus and earnestly refuted Tycho's counter-arguments, yet published nothing on the subject. Although there may be many reasons for the silence which some astronomers maintained, it was not usually want of conviction; it was perhaps in many cases merely that they saw no need to take a stand; so, long before the condemnation of Galileo, there was no need to stand up and be counted. In any event, one clearly cannot judge the influence of Copernicanism by the lack of elementary treatments in textbooks; even Galileo chose to lecture publicly only on Ptolemaic astronomy.

On the other hand, public commitment to the Copernican theory had a great appeal for the radical thinkers of the sixteenth century. Seeking escape from what they regarded as the trammels of scholastic Aristotelianism, they turned eagerly to any theory supporting their desire for innovation. Many discussions of Copernicanism are set within the framework of anti-Aristotelianism, and one sometimes gets the impression that the defence of Copernicus is partly a response to the intellectual delights of novelty and perversity. If one wanted to attack Aristotle in any case, what better way than to upset the cosmological basis of his natural philosophy? This anti-Aristotelianism

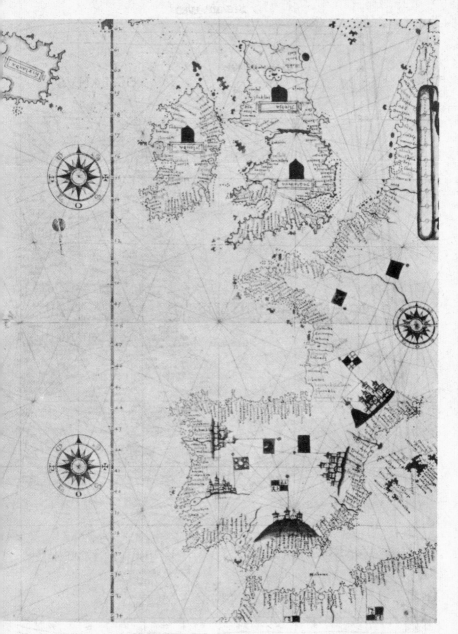

A portolan of the North Atlantic coast. From Portolan Atlas (1572) by J. Martinez. As the lettering shows, this was made to be folded, so that the map of the British Isles must be read with South at the top. The compass roses and compass lines are characteristic of all portolans

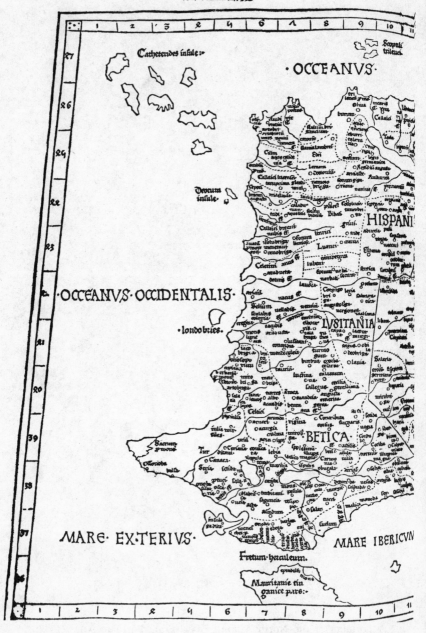

The Atlantic coast of the Iberian peninsula and the Straits of Gibraltar, according to Ptolemy. From *Cosmographia,* printed at Ulm in 1486

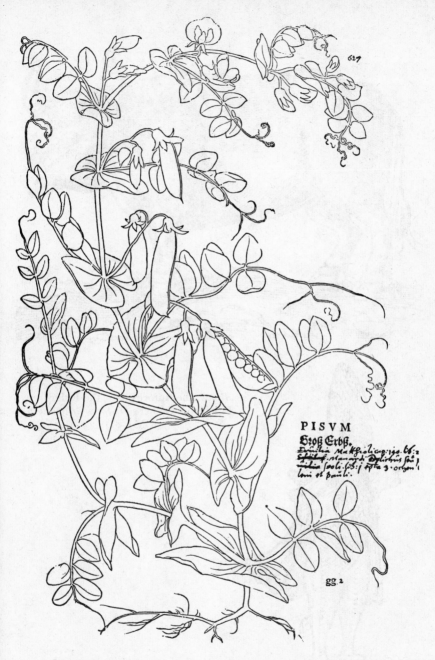

PISVM

The pea, from Fuch's *De Historia Stirpium* (Basle, 1542). The vegetables illustrated by Fuchs include the asparagus and several varieties of cabbage.

Above: the Aurochs; *left:* the Bishop
Fish. From Gesner's *Historia Animalium*
(1551–87)

Francis Bacon, Studio of P. Van Somer. *By courtesy of the National Portrait Gallery, London*

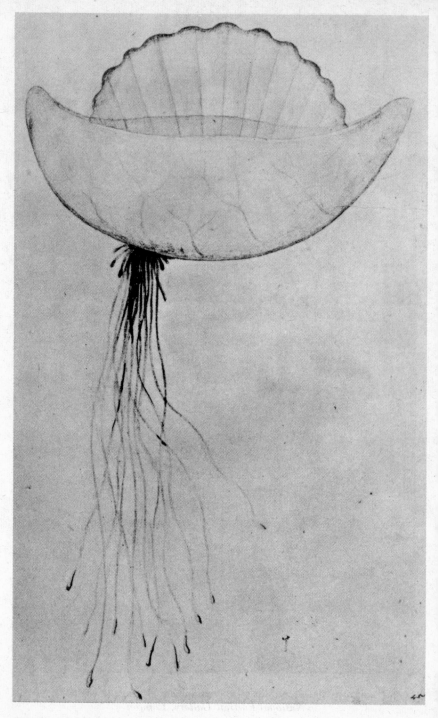

The Portuguese Man-of-War, from a watercolour by John White

Tycho Brahe and his great mural quadrant. Tycho is taking observations, which an assistant writes down. In the foreground and background, are scenes illustrating the normal work of Uraniborg: Tycho's assistants are working with various instruments and operating the printing press from which issued *Astronomiae Instauratae Mechanica* (1598), the source of this illustration

Mathematical instruments of the early 16th century (detail from Holbein's Ambassadors). On the table are a celestial globe, a shepherd's dial, a shadow scale, a quadrant, a block dial and a torquetum; below, a terrestrial globe, a rule, a lute, dividers, music book and map cases

An anatomy demonstration as conceived in the fifteenth century, from Mondino's *Anathomia* (Venice, 1493). The professor comments on the text, while the demonstrator displays the appropriate organs in the visceral cavity

Vesalius demonstrating the muscles of the arm, from his *De Humani Corporis Fabrica* (Basle, 1543)

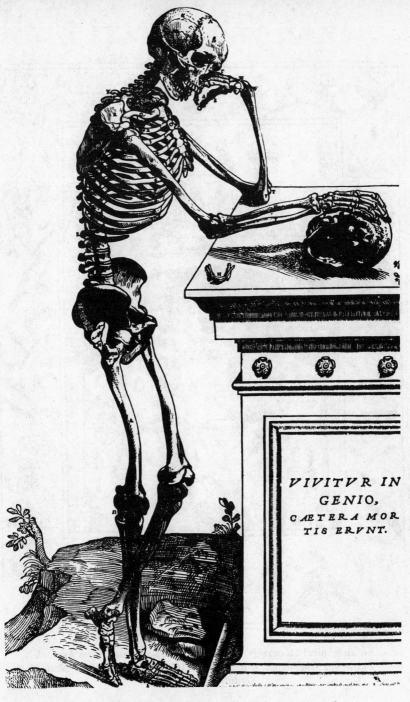

VIVITVR IN
GENIO,
CÆTERA MOR
TIS ERVNT.

One of the figures showing the whole human skeleton from
Vesalius's *De Humani Corporis Fabrica*

An alchemical laboratory, with seven furnaces. From Elias Ashmole's
Theatrum Chemicum Britannicum (1652), which reprints Norton's
Ordinall of Alchimy

A pump designed by Jacques Besson, from his *Theatres des Instrumens* (Lyon, 1579). The elaborate machinery seems unnecessarily complicated for the simple domestic task shown, and suggests the imaginative element in many Renaissance engineering books

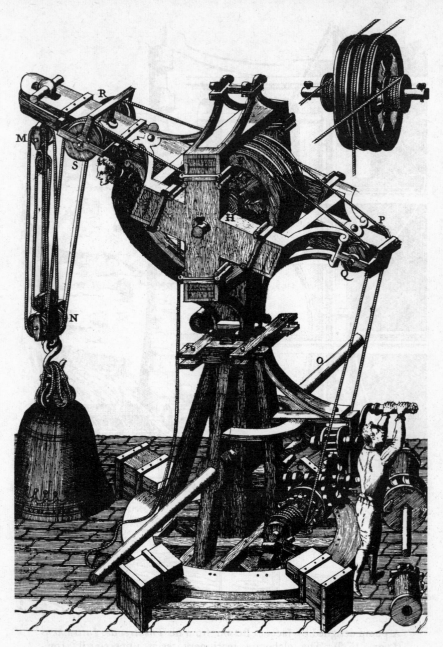

A crane, from Ramelli's *Le Diverse et Artificiose Machine* (Paris, 1588), illustrating the Renaissance engineer's love of complex gearing and pulleys

The House of Astronomy, from Kepler's *Rudolphine Tables* (1627).
Hipparchus holds his catalogue of fixed stars, and Copernicus his
De Revolutionibus (1543). Tycho points to a diagram of his system of
the world, while propped on the pillar is his *Astronomiae Instauratae
Progymnasmata* in which it was announced; Ptolemy is at work on a
mathematical problem with his *Almagest* beside him. On the pillars
hang an armillary sphere, a celestial globe, a rectangulum, two of
Tycho's quadrants, an astrolabe and a lunar eclipse dial. On the base,
Kepler is shown seated somewhat gloomily at table; the centre panel
contains a map of Tycho's island of Hveen; on the right is a printing
shop. Above the structure are figures symbolising the mathematical
sciences, while over all an imperial eagle (symbol of Tycho's and
Kepler's patron, Rudolph) scatters a meagre shower of gold

Galileo, from a portrait presented to the University of Oxford
by his last pupil, Viviani

perhaps explains why so many favourable references to Copernicus were made by men who were not astronomers, or even scientists at all, as well as why it is often associated with free thought or the wilder reaches of Lucretian Epicureanism. An interesting and not very well known example of this occurred in the " Academy " organised by and around various members of the French *Pléiade*. There were actually several academies, some informal, some formally associated with the Court, which existed more or less continuously from before 1550 throughout the century. (It is odd to think of Henri III in the dark days of the religious wars listening to the poets of the *Pléiade* discuss the modes of Greek music.) These groups, though organised by poets and originally literary in intent, expanded from poetry to music, and thence, in the Pythagorean spirit, to mathematics and natural philosophy. There were discussions of the state of astronomy, and the possible value of the new theories of Copernicus : their opponents cited these discussions as evidence of the wild speculative freedom of thought in which the *Pléiade* indulged.

In 1557 there was published a work with the title *Dialogue of Guy de Brues, against the New Academies* ; here de Brues, using as speakers actual members of the *Pléiade*, attacked the novelty of their opinions, including those on science. According to de Brues, Ronsard believed that astronomy must represent physical truth, and hence he could not accept the mobility of the Earth, for which there was no empirical evidence ; whereas Baïf regarded astronomy as merely a series of hypotheses, and could therefore argue

In Astronomy . . . there is no assurance of principles. . . . For example, as to whether the Earth is immobile : for notwithstanding that Aristotle, Ptolemy and several others have thought it to be so, Copernicus and his imitators *

* Evidently the reader of 1557 was supposed to know that there were those who accepted Copernican doctrines.

have said that it moves, because the heaven is infinite and
therefore immobile : for (says he) if the heaven is not
infinite, and if there is nothing beyond the heaven, it
would follow that it is contained by nothing, which is impossible, since everything which has being is in some place.
If, then, it is infinite, it must be immobile, and the Earth
mobile.[7]

One very interesting aspect of this attack is the attribution to
Copernicus of the belief (which in fact he did not hold) that the
universe is infinite ; clearly there was here a confusion of radical
ideas, for the Academicians were said to be Epicureans as well as
Copernicans, and it is easy for a non-scientist to confuse the
Copernican argument that the sphere of the fixed stars must be
enormously large with the Epicurean argument that the universe
must be infinite.

Whether in fact Ronsard and Baïf argued about the merits of
Copernicanism as well as about the relative merits of Latin and
vernacular poetry and of new and old poetic styles is uncertain ;
but astronomical questions did interest other "academicians." In
the same year as the Dialogues of de Brues there was published
The Universe (L'Univers) of Pontus de Tyard (c. 1521–1605), a
competent astronomer and a churchman, destined to become
Bishop of Châlons. The Universe consists of two dialogues, the
first of which deals with the state of philosophical opinion. Here
Tyard discusses the Copernican system in some detail : after
explaining the Greek sources of the theory, he gives a French
translation of Copernicus' description of the spheres, and uses
Copernicus' own arguments in favour of the Earth's mobility ;
in fact, the major arguments of the First Book of De revolutionibus were treated. Nevertheless, in spite of his full exposition,
Tyard declined to commit himself ; the most he would say was
that this was an interesting speculation, which was important
mainly for astronomers. For, as he said,

In truth his demonstrations are ingenious, and his observations exact, and worthy of being followed. Nevertheless, whether or not his disposition is true, the knowledge of the being of the Earth, so far as we are able to know it, is not in any way troubled thereby : and it does not prevent us from believing that it is a heavy, cold and dry Element, the which from received, vulgar, and, as it were, religious opinion we believe to be immobile.[8]

Though cautious, this was a fair rendering of the position ; Tyard was accustomed to free speculation, but this did not mean that he wished to flout received religious opinion, nor that he did not himself feel these opinions to have weight.

A physicist bent on attacking Aristotle's theory of motion could hardly fail to appreciate the advantages inherent in pressing the attack on Aristotle's cosmology as well. This was the case with G. B. Benedetti (1530–90) whose *Book of Diverse Speculations on Mathematics and Physics* was an anti-Aristotelian treatise. Benedetti was a mathematical physicist, not an astronomer ; but he was warm in praise of " the theory of Aristarchus, explained in a divine manner by Copernicus, against which the arguments of Aristotle are of no value,"[9] mainly, one suspects, because it was one more blow at Aristotle's authority. In a similar vein, Richard Bostocke, an obscure English writer, in *The Difference betwene the auncient Phisicke . . . and the latter Phisicke* (1585) found it natural to compare the physician Paracelsus and the astronomer Copernicus. Admittedly, Paracelsus was not the first to proclaim his ideas ; he was but the restorer of ancient and true doctrines. As Bostocke put it, Paracelsus was no more the " author and inventor " of medical chemistry *

then Nicholaus Copernicus, which lived at the time of this Paracelsus, and restored to us the place of the stars according to the truth, as experience and true observation doth teach, is

* Cf. below, p. 159f.

to be called the author and inventor of the motions of the stars, which long before were taught by Ptolemeus Rules Astronomicall, and Tables for Motions and Places of the stars.[10] Whether Bostocke was a Copernican is not of much importance, and he obviously had no exact notion of what it was Copernicus had done. But it is significant that in England, as in Italy, if one wished to attack Aristotle and defend scientific novelty in 1585, one appealed to Copernicus as an example and as a weapon. By 1585 any scientific audience, mathematical, physical or medical, could be expected to know something of the Copernican theory. And, clearly, there was no bar to a free discussion of the theory if one felt inclined.

Just as scientific radicals hailed the Copernican theory as an important one because it displaced Aristotelian authority, so, on the other hand, if one disliked scientific novelty, one attacked the Copernican theory. In the sixteenth century, as in the twentieth, non-scientists were apt to find scientific theories upsetting, and scientists restless fellows always trying to disturb the established order of things. The most violent attacks on Copernicus in the sixteenth century come from non-scientists, and they nearly always indicate that the basis of their attack is the fear of novelty. Educated in one system, such critics hated the idea of having to accept another or, even worse, having to balance the merits of one system against another. This was especially true when the new system involved a violation both of common sense, and of the apparent order and harmony of the universe. For once astronomers came to accept the heliostatic universe, the scientist had embarked upon that separation of the world of science from the world of common sense experience which is the basis of so much antagonism to science. There were now two worlds : the astronomer's, in which the moving Earth emulated the planets in circling the Sun ; and everyman's, patently geostatic and geocentric. The Copernican system was bound to provoke hostility from those uninterested in scientific analysis; for it raised the uncomfortable question of the reliability of familiar sense experience. It is this which is reflected in the malaise expressed in popular criticism of Copernicanism, especially among the poets, throughout the late sixteenth century, a malaise which only vanished when, in the late seventeenth century, science appeared to be restoring order and stability again.

By the last quarter of the sixteenth century, the Copernican system, though it had gained few adherents, was widely known; after thirty years of debate and discussion, non-scientists were familiar with the fundamental problem. And they were coming to resent the astronomers who seemed intent on disturbing their philosophical peace, even as the physical peace of the heavens was being disturbed by strange portents. Indeed, events in the heavens —a new star (nova) in Cassiopeia in 1572, and a long and apparently continuous series of comets between 1577 and the early seventeenth century—naturally called everyone's attention to astronomy, and to the heated discussions raging among astronomers who seemed to be taking a perverse delight in defending absurdities. This point of view was perfectly expressed by Guillaume du Bartas, whose influential work, The Week, or Creation of the World (La Sepmaine, ou Création du Monde, 1578), was one of the most widely read of all didactic poems in the late sixteenth century, and was partially translated into English many times. Although familiar with ancient sources, and not above borrowing from Lucretius, especially on literary points, du Bartas was fiercely opposed to whatever appeared to him to contradict his rather narrow view of orthodox cosmology: even Aristotle was attacked for his views on the infinite duration of the world. In his eyes, the age was wilfully determined to toy with novelties, and scientists in particular would adopt even absurdities if they were but new. After discussing God's Creation of the World, the elements and the geography of the Earth, he came to describe the glorious heavens, shining with lights, marred only by the peculiar views held by modern scientists :

. . . some brain-sicks live there now-a-days,
That lose themselves still in contrary ways ;
Prepostrous wits that cannot row at ease,
On the Smooth Channel of our common Seas.
And such are those (in my conceit at least)
Those Clerks that think (think how absurd a jest)
That neither Heav'ns nor Stars do turn at all,
Nor dance about this great round Earthy Ball ;
But th'Earth itself, this Massy Globe of ours,
Turns round about once every twice-twelve hours :
And we resemble Land-bred Novices
New brought aboard to venture on the Seas ;
Who, at first launching from the shore, suppose
The ship stands still, and that the ground it goes.
So, twinkling Tapers, that Heav'n's Arches fill,
Equally distant should continue still.
So, never should an arrow, shot upright,
In the same place upon the shooter light ;
But would do (rather) as (at Sea) a stone
Aboard a Ship upward uprightly thrown ;
Which not within-board falls, but in the Flood
A-stern the Ship, if so the Wind be good.
So should the Fowls that take their nimble flight
From Western Marches towards Morning's light ;
And Zephyrus, that in the summer time
Delights to visit Eurus in his clime ;
And bullets thundered from the cannon's throat
(Whose roaring drowns the Heav'nly thunder's note)
Should seem recoil : sithens the quick career,
That our round Earth should daily gallop here,

Must needs exceed a hundred fold (for swift)
Birds, Bullets, Winds ; their wings, their force, their drift.
    Arm'd with these Reasons, 'twere superfluous
T'assail the Reasons of Copernicus ;
Who, to salve better of the Stars th'appearance
Unto the Earth a three-fold motion warrants :
Making the Sun the Centre of this All,
Moon, Earth, and Water, in one only Ball.
But sithence here, nor time, nor place doth suit,
His Paradox at length to prosecute ;
I will proceed, grounding my next discourse
On the Heav'ns motions, and their constant course.[11]

Du Bartas knew well enough the simpler arguments against
the Copernican system, and was certainly not alone in regarding
them as absolutely destructive of the foolish novelties of the new
astronomy. Nor was he alone in thinking that the best way to
dispose of this absurd scientific idea was by ridicule. A similar
attack, in less lively vein, is found in the *Theatre of Universal
Nature* (1597) of Jean Bodin. In this work the French political
theorist and scourge of witches treated encyclopedically the whole
natural world. There he referred to Copernicus as one who had
"renewed" the opinions of "Philolaus, Timaeus, Ecphantus,
Seleucus, Aristarchus of Samos, Archimedes and Eudoxus," led
thereto merely because the human mind finds it so difficult to
comprehend the incredible speed of the heavenly spheres, and so
arrogantly denies it. Bodin clearly knew less of the Copernican
system than did du Bartas ; writing nearly twenty years later, it
was easier for him to speak from mere hearsay. He thought
that Copernicus had abolished epicycles, and did not know that
Copernicus had used the argument that rest is nobler than motion
(so that the nobler heavens should rest, while the baser Earth
moves), for he recommends it as a good argument for Copernicans to use ! Bodin thought the whole theory absurd ; and

anyway " if the Earth were to be moved, neither an arrow shot straight up, nor a stone dropped from the top of a tower would fall perpendicularly, but either ahead or behind." [12]

The attack is a poor one ; but it illustrates both the discomfort raised in non-scientific minds, and the fact that at the end of the sixteenth century even an elementary discussion of astronomy required a reference to Copernican ideas. Only a sceptic could shrug off the whole problem of choice between Ptolemy and Copernicus, and declare with Montaigne

> What shall we reape by it, but only that we need not care, which of the two it be ? And who knoweth whether a hundred years hence a third opinion will rise, which happily shall overthrow these two praecedent ? [13]

To most thoughtful men, it was little comfort to think that the unsettled state of astronomy might continue unabated. Most preferred to look back to a time (often before they were born) when all had been certain, the Earth had stood firm beneath men's feet, and the heavens were truly as they appeared. The position was immortalised by Donne ; though his lines were written in 1611, when the heavens had been further disturbed by the revelations of the telescope, they are perfectly in keeping with the complaints of a preceding generation :

> And New Philosophy calls all in doubt,
> The Element of fire is quite put out ;
> The Sun is lost, and th'Earth, and no man's wit
> Can well direct him where to look for it.
> And freely men confess that this world's spent,
> When in the Planets, and the Firmament
> They seek so many new ; then see that this
> Is crumbled out again to his Atomies.
> 'Tis all in pieces, all coherence gone ;
> All just supply, and all Relation.[14]

If this was the way in which the Copernican doctrine affected poets, no wonder they rejected it. Especially in an age when all was doubt, decay and dissension in the religious and political spheres in any case. Why should they welcome chaos among the stars as well ?

At the same time, many natural philosophers, especially mathematicians, found the Copernican system liberating to the spirit, and rather welcomed the freedom from the bondage of a tiny world which it offered, than feared the loss of a cosy certainty. Such bolder and more soaring spirits not only welcomed Copernicus ; they tried to improve upon him. And as they did so, the Copernican theory was stretched to breaking point. One of the first astronomers to enlarge the Copernican universe was Thomas Digges (d. 1595), an Englishman born about the year in which *De Revolutionibus* was published. His father, Leonard Digges was, though a gentleman, a practical surveyor and wrote much on applied mathematics, including astrology ; having taken part in Wyatt's Rebellion he had some difficulty in printing his works and left many of them unpublished when he died in 1558. He requested his friend, John Dee, to undertake the education of his son, and the younger Digges, in a characteristically Renaissance phrase, called Dee his second father in mathematics. Thomas Digges followed in the footsteps of both his fathers, and was active in the movement to teach practical mathematics to the unlearned. He was also an observational astronomer of some merit : along with other leading astronomers of the day (including Dee, but Digges's work was published earlier, and was better) he made a series of observations on the strange new star (nova) which appeared in the familiar constellation of Cassiopeia late in 1572. His observations were published the next year with the punning title *Mathematical Wings or Scales (Alae seu Scalae Mathematicae*, 1573) ; the " scales " were trigonometric theorems required for the determination of stellar parallax, for Digges accepted the nova as a new fixed star, and thought its appearance gave an unparalleled opportunity for testing the Copernican theory. (Digges mistakenly believed that the decrease in magnitude of the star after its first sudden appearance would be periodic, and hoped it might be parallactic in origin, the result of apparent motion.)

Though he was unable to use the star in this way, Digges had no doubt of the truth of the Copernican system. So convinced was he, that he found it necessary to forsake filial piety. In 1576, revising a twenty-year-old work of his father called *A Prognostication Everlasting* (a perpetual almanac, especially concerned with meteorological prediction) Digges could not bear to think that yet another work based upon " the doctrine of Ptolemy " should be given the public, now that

> in this our age one rare wit (seeing the continual errors that from time to time more and more have been discovered, besides the infinite absurdities in their Theorickes, which they have been forced to admit that would not confess any mobility in the ball of the Earth) hath by long study, painful practice, and rare invention delivered a new Theorick or model of the world.[15]

Since Copernicus had been led to his new model of the world by " reason and deep discourse of wit " so it was fitting that " such noble English minds (as delight to reach above the baser sort of men) might not be altogether defrauded of so noble a part of Philosophy." He wanted as well to show that Copernicus had intended not merely a mathematical hypothesis, but a true physical picture, for Osiander's Preface was being found out. So Digges appended to the *Prognostication Everlasting* a short work with a long Elizabethan title, *A Perfit Description of the Caelestiall Orbes according to the most auncient doctrine of the Pythagoreans, lately revised by Copernicus and by Geometricall Demonstrations Approved.*

This " perfect " description is mainly a translation of the

essential part of Book I of *De Revolutionibus* (in fact what every-one has chosen to translate ever since), but with a significant new concept of the translator's added. For to the Pythagorean doctrines expressed by Copernicus, Digges added a new dimension to the celestial sphere. Because of the lack of stellar parallax, Copernicus had postulated a very large celestial sphere, with huge stars. To Digges, this was an indication of the wonder and majesty of God; but why should not God have continued this sphere upwards until it met the firmament? Physically, this produced some interesting reflections. If, as Digges postulated, the sphere of the fixed stars were " garnished with lights innumerable and reaching up in spherical Altitude without end," then the stars must be at varying distances from the Sun and the Earth. They were all necessarily very large, but very probably their varying magnitude indicated merely differences in distance from the Earth, not different intrinsic size. And there must be an infinite number of stars, far more than we can see. For

of which lights Celestial it is to be thought that we only behold such as are in the inferior part of the same Orb [the sphere of fixed stars], and as they are higher, so seem they of less and lesser quantity, even till our sight being not able farther to reach or conceive, the greatest part rest by reason of their wonderful distance invisible unto us. And this may well be thought of us to be the glorious court of the great God, whose unsearchable works invisible we may partly by these his visible conjecture, to whose infinite power and majesty such an infinite place surmounting all others both in quantity and quality only is convenient.

The universe of Digges is no longer the closed world of Copernicus; the starry sphere is now unbounded on its upper regions. But more than that; for with a mystical daring char-acteristic of Dee's teaching, Digges has carried the astronomical heavens into contact with the theological Heavens. In breaking the bounds of the finite universe and wiping out the upper boundary of the celestial sphere, Digges has conceived the abolition of the boundary between the starry heavens and the firmament as well. If one could fly through the stars (which are only like our Sun) he would arrive straight in Paradise. This is made even plainer by the illustrative diagram which Digges included ; it shows an "orb" of fixed stars, but the stars are scattered outside the orb, right out to the edge of the picture, in fact. Within the sphere Digges has written

> This orb of fixed stars extendeth itself infinitely up in altitude spherically, and therefore immoveable : the palace of felicity garnished with perpetual shining glorious lights innumerable far excelling our sun both in quantity and quality ; the very court of celestial angels devoid of grief and replenished with perfect endless joy the habitacle for the elect.[16]

Mystical though this may be, Digges was indubitably stretching the real physical world ; the stars have burst their bonds and are no longer hung on the vault of heaven, but are scattered through immense space, and are themselves of an almost inconceivable size.

This is the first in a series of steps that fractured the tidy world of the ancients. At the time it may not have seemed particularly novel ; for many lumped all novelties under the heading "Epicurean," and confused immensity with infinity ; and Digges could have been "reviving" the opinions of Democritos, Epicuros and Lucretius. Certainly English readers now had the arguments of Copernicus readily available in the vernacular, though one may wonder how many, consulting *A Prognostication Everlasting* for a hint of next winter's weather, paused to study the Copernican appendix. Yet whether because of a confusion, or whether because of Digges, it did become common in the later sixteenth century to assume that the Copernican universe demanded an indefinitely large, if not an infinite universe, and many believed that infinity was demanded.

The next radical revision of the Copernican universe was quite different in origin from that of Digges. It was derived entirely from astronomical observation, not based upon mystical speculation ; and it was the work of a non-Copernican. Though Tycho Brahe never accepted the Copernican system, and though his own (Tychonic) system was designed as a rival, yet a number of radical concepts developed by him were generally adopted by Copernicans, and in the long run Tycho advanced the acceptance of the Copernican universe far more than many convinced Copernicans.

Tycho Brahe (1546–1601) began his interest in astronomy by observing the heavens, and it was in observational astronomy that, ultimately, his greatest contributions were to lie. It was a natural bent, for Tycho had no masters, and became an astronomer against the wishes of his relatives. His father, according to Tycho, did not even wish him to learn Latin (not a necessary accomplishment for a Danish noble) ; but he was brought up by his uncle, who saw that he had a proper Latin education and sent him at the age of fifteen to the University of Leipzig under the care of a tutor. Tycho seems not to have pursued the regular University course, for he insisted in his autobiography (grandly entitled *On that which We Have Hitherto Accomplished in Astronomy with God's Help, and on that which with His Gracious Aid has yet to be Completed in the Future*[17]) that he had taught himself astronomy and pursued it independently and secretly. His interest had begun with astrology and his first instruction came from astrological ephemerides ; this interest remained with him always, though it was gradually overshadowed by a preoccupation with the observations themselves. His first real observations were made in 1563, at the age of sixteen, with improvised instruments ; as he remembered bitterly thirty-five years later, his tutor refused him money to buy proper ones. These early observations were on the conjunction of Saturn and Jupiter, and the discrepancies between his observations and the predictions in the Alphonsine and even the " Copernican " Tables, convinced him thus early that the chief requisite in astronomy was more, and more accurate, observation. For this he needed and soon acquired, better, professionally made instruments, as he went from Leipzig to the astronomical centre of Augsburg. Here alchemy absorbed him as well as astrology— " terrestrial astronomy " he called it—and returning home, he very nearly became totally absorbed in alchemical experiment. But the sudden appearance of the new star in Cassiopeia in 1572 determined his career once and for all. This new phenomenon called forth all his resources as an observer, and the resultant account (*On the New Star*, 1573) attracted the attention of the King of Denmark who, anxious to keep such a promising scientist at home (national prestige demanded intellectual as well as military success), gave Tycho the feudal lordship of the island of Hveen. This magnificent generosity persuaded Tycho not to emigrate to Basle, as he had planned ; instead he spent twenty-one years on Hveen, which he made a centre for astronomical observation. Here he built the fantastic castle of Uraniborg, with its observatories and laboratories ; here he constructed new instruments of enormous size (the only way to attain accuracy before the invention of the telescope) ; and here he trained a whole series of younger men, who came begging to work with the greatest observational astronomer since Hipparchos.

Like Hipparchos, Tycho felt that the appearance of a new star demanded the drawing up of a new star catalogue, a project to which he devoted much energy in the next twenty years. But he was also profoundly interested in the nova for its own sake. Here was an amazing phenomenon : a new star in a well-known constellation, and when it was first observed as bright as Jupiter.

Tycho, Digges, Maestlin, Dee and many more studied it in wonder and perplexity. Tycho, Digges and Maestlin (still an amateur astronomer) all tried to measure the parallax of the new star ; not to test the Copernican theory, but because the new star was at first thought to lie in the sublunar sphere. It must naturally be a meteorological phenomenon, like rainbows, meteors and comets, for change belonged to the terrestrial regions, and the perfect heavens of Aristotelian cosmology were perfect because eternal and immutable. Anything that was located below the Moon must reveal its relative nearness by its apparent shifts of position in relation to the backdrop of the stars.

But the new star obstinately refused to yield any parallax to the most careful and attentive study. Tycho, Digges and Maestlin all concluded that it lay, consequently, in the sphere of the fixed stars. Thereby all were committed to admitting that the heavens did change, and were therefore not perfect. Not all astronomers could face this ; indeed not all astronomers agreed with the observations. Some insisted that the nova showed a parallax ; some, like Dee, ingeniously argued that it was moving in a straight line away from the Earth, which accounted for the fact that it grew progressively dimmer ; others, including Digges, related it to comets. Tycho was the boldest in accepting the inevitable conclusions, perhaps because he was the most firmly convinced of the reliability of his observations. He was at a loss to explain the new star's variation in brightness and colour (like all novae, it changed from white to reddish-yellow to red) ; but he was absolutely convinced that it lay in " the aetherial orbs." What its astrological meaning might be he discussed at great length, for so rare an event naturally had a strange and wonderful significance. Its astronomical significance was, however, equally great, and determined him to settle quietly where he could " lay the foundation of the revival of Astronomy "[18] by long and careful observation.

At Uraniborg Tycho observed, year after year, the location of the fixed stars, and the changing positions of the planets, the Sun and the Moon, developing new and better instruments and techniques until he had attained an accuracy far beyond that of any previous astronomer. Tycho's observations came to be pretty consistently accurate to about four minutes of arc, the limit of naked eye accuracy.* Tycho was well aware of the superiority of his methods, and kept himself to a high standard. As he wrote after he had left Uraniborg, he judged his observations as

> not of equal accuracy and importance. For those that I made in Leipzig in my youth and up to my 21st year, I usually call childish and of doubtful value. Those that I took later until my 28th year [i.e., until 1574] I call juvenile and fairly serviceable. The third group, however, which I made at Uraniborg during approximately the last 21 years with the greatest care and with very accurate instruments at a more mature age, until I was fifty years of age, those I call the observations of my manhood, completely valid and absolutely certain, and this is my opinion of them.[19]

Ironically, however, these very accurate observations served no purpose in Tycho's own theoretical work. Though he declared that " it is particularly these later observations that I build upon when I strive by energetic labours to lay the foundations of and develop a renewed Astronomy," his own use of them was negligible. He did indeed develop a new astronomy, based upon observations, but it was all based upon observations of 1572 and 1577 ; later observations on comets merely confirmed what he already knew, and his planetary tables were not needed in the only sketch he made of his system. Yet the great mass of accumulated and accurate data was not wasted : for Kepler was to use

* The naked eye cannot resolve points with an angular separation of less than approximately two minutes of arc.

this data in the laborious calculations on which he based an astronomical theory remote from Tycho's, though in many ways
derived from it.

The observations on the great comet of 1577 were the real
basis for the development of the Tychonic system; the only
description of its details that he ever made is inserted into an
account of cometary orbits. As before, in 1572, Tycho observed
the new phenomenon with the utmost care. Once again he tried
to measure its parallax, only to find that it was too small to be
consonant with a position in the atmosphere. Comets must,
then, like the new star, lie in the aetherial regions, now shown to
be capable of yet another change. This was confirmed when
other comets appeared; as Tycho put it, " all comets observed
by me moved in the aetherial regions of the world and never in
the air below the moon, as Aristotle and his followers tried
without reason to make us believe for so many centuries ".[20]
The observations on comets were to provoke Tycho to even
greater disturbances of the Aristotelian heavens. If the geocentric
universe were filled with crystalline spheres, where were the
comets to fit in ? Especially since Tycho believed that their
centre of motion was the Sun. Their special connection with the
Sun had already been noted : for example the applied mathematician Peter Apian * (1495–1552), observing a series of comets in
the 1530's, had been struck with the fact that the tails point away
from the Sun. Yet in the Ptolemaic system the area above and
below the Sun is completely filled by the spheres of the planets,
and even the introduction of a new sphere would not help.

Tycho, noting that however he arranged the spheres of the
planets, the paths of the comets would intersect them, decided

---

* His original name was Bienewitz; adoption of the name Apian (bee) is
a delightful example of the Renaissance tendency to use Latin surnames. Apian
was primarily a geographer; he showed no interest in astronomical theory,
and in any case his major work on cosmography was published in 1539.

that since comets were indubitably located above the Moon, there could be no crystalline spheres supporting and moving the planets. This revolutionary decision he made with complete equanimity. As he wrote in 1588 in a great survey of his study of the comet of 1577 (its title, *On the Most Recent Phenomena of the Aetherial World* is itself a challenge to orthodoxy and a manifesto of the new astronomy:

> There are not really any Orbs [spheres] in the Heaven . . . those which Authors have invented to save the appearances exist only in imagination, in order that the motions of the planets in their courses may be understood by the mind, and may be (after a geometrical interpretation) resolved by arithmetic into numbers. Thus it seems futile to undertake this labour of trying to discover a real orb, to which the Comet may be attached, so that they would revolve together. Those modern philosophers agree with the almost universal belief of antiquity who hold it as certain and irrefutable that the heavens are divided into various orbs of hard and impervious matter, to some of which stars are attached so that they revolve with them. But even if there were no other evidence, the comets themselves would most lucidly convince us that this opinion does not correspond with the truth. For comets have already many times been discerned, as the result of most certain observations and demonstrations, to complete their course in the highest Aether, and they cannot by any means be proved to be drawn around by any orb.[21]

So blandly to deny the reality of the crystalline spheres—to change the meaning of the word "orb" from "sphere" to "circular path" or "orbit"—was a most revolutionary measure, as revolutionary in its own way as the displacement of the Earth from the centre of the universe. Since the fourth century B.C., astronomers had unhesitatingly accepted the reality of solid spheres, in which the planets were firmly embedded. What else kept the planets fixed in the heavens, and how else could one give physical reality to a mathematical representation ? With the abandonment of the crystalline spheres came the imperative need to search for something else which kept the planets in their paths ; but not, apparently for Tycho, who never mentions the problem.

Now that the solid spheres were gone, all that was necessary was to rearrange the Ptolemaic spheres to make room for the comets to move around the Sun. As Tycho put it in the ornate style he affected,

> Because the region of the Celestial World is of so great and such incredible magnitude as aforesaid, and since in what has gone before it was at least generally demonstrated that this comet continued within the limits of the space of the Aether, it seems that the complete explanation of the whole matter is not given unless we are also informed within narrower limits in what part of the widest Aether, and next to which orbs of the Planets [the comet] traces its path, and by what course it accomplishes this.[22]

The Ptolemaic system as it stood was impossible : cumbersome, loaded with equants and superfluous epicycles, and too full to leave room for the comets. " That newly introduced innovation of the great Copernicus " was elegantly and beautifully mathematical, but presented even greater difficulties. For, as Tycho put it,

> the body of the Earth, large, sluggish and inapt for motion, is not to be disturbed by movement (especially three movements), any more than the Aetherial Lights [stars] are to be shifted, so that such ideas are opposed both to physical principles and to the authority of Holy Writ which many times confirms the stability of the Earth (as we shall discuss more fully elsewhere).

As other arguments against the motion of the Earth (apart from its unfitness for motion and the enormous space between the orb of Saturn and the fixed stars, evident from the absence of parallax) Tycho instances the great size of the stars, necessitated by their apparent diameters * and their presumed distance in the Copernican system ; and his belief that a stone dropped from a tower would never hit the ground at the foot of the tower, if the Earth were moving. Both these were convincing arguments, though both were based on erroneous physics. But the erroneousness of such physics was first clearly demonstrated only by Galileo.

Faced with these problems, said Tycho,

> I began to ponder more deeply within myself, whether by any reasoning it was possible to discover an Hypothesis, which in every respect would agree with both Mathematics and Physics, and avoid theological censure, and at the same time wholly accord with the celestial appearances. And at length almost against hope there occurred to me that arrangement of the celestial revolutions by which their order becomes most conveniently disposed, so that none of these incongruities can arise.

What Tycho wanted was a system with the advantages of the Copernican and without the disadvantages of a stationary Earth ; and the elimination of Ptolemaic complications. Like Copernicus, Tycho turned to the ancients for a suggestion ; of a different temper and generation from Copernicus, he never mentioned that his was essentially the system of Heraclides of Pontus (*fl.* fourth century B.C.). This system is really very simple. The Earth remains at rest, at the centre of the universe, and every twenty-four hours there turns around it " the most remote Eighth sphere, containing within itself all others " (the only solid sphere which Tycho retained) to account for the daily rising and setting of the stars. The Sun revolves annually about the Earth, while the

---

* Before the use of the telescope, it was thought that stars must have discs like planets, and exaggerated ideas existed about the size of their apparent diameters.

planets revolve about the Sun, and can only be said to revolve about the Earth because they accompany the Sun. As Tycho declared, " I shall assert that the other circles guide the five Planets about the Sun itself, as their Leader and King and that in their courses they always observe him as their centre of revolution." This system, as Tycho pointed out with pride, explained as well as the Copernican theory why Venus and Mercury were never far from the Sun ; why the Planets appeared to show retrograde motions ; why they appeared to vary in brightness ; and why the motion of the Sun was always mixed with that of the planets. At the same time it abolished any need for equants ; Tycho thought it could eliminate all, or almost all, of the epicycles and reduce the number of eccentrics, but in fact he never worked out the mathematical representation of the system.

One new complication only was introduced, readily apparent from the diagram : the orbit of Mars about the Sun is here seen to cross the Sun's orbit about the Earth. If the orbs are solid spheres, this is of course impossible. But Tycho had rejected solid spheres ; he knew

that the machine of Heaven is not a hard and impervious body full of various real spheres, as up to now has been believed by most people. It will be proved that it extends everywhere, most fluid and simple, and nowhere presents obstacles as was formerly held, the circuits of the Planets being wholly free and without the labour and whirling round of any real spheres at all, being divinely governed under a given law.

The fact that there were apparent intersections of the orbits (really the result of trying to represent three dimensions in two) was irrelevant. In fact this new arrangement had the advantage that it explained why Mars in opposition was at its brightest ; for it was then nearer to the Earth than to the Sun.

As all this was in the nature of a digression in his book on comets, Tycho did not explore the workings of his "machine of nature" further, but went on to deal with cometary motion. In the newly arranged universe, there was now room for a comet to circle the Sun in the space between the orbits of Venus and Mars ; * it could behave like "an adventitious and extra-ordinary planet" and display a path not totally dissimilar to that of the planets. True, it moved at a variable velocity, and its path was curious, but this was to be expected from the nature of comets :

> For it is probable that Comets, just as they do not have bodies as perfect and perfectly made for perpetual duration as do the other stars which are as old as the beginning of the World, so also they do not observe so absolute and constant a course of equality in their revolutions—it is as though they mimic to a certain extent the uniform regu-larity of the Planets, but do not follow it altogether. This will be clearly shown by Comets of subsequent years, which will no less certainly be located in the Aetherial region of the world. Therefore either the revolution of this our Comet about the Sun will not be at all points ex-quisitely circular, but somewhat oblong, in the manner of the figure commonly called ovoid ; or else it proceeds in a perfectly circular course, but with a motion slower at the beginning, and then gradually augmented.

This is the first serious suggestion that a heavenly body might follow a path that was neither circular, nor compounded of circles (though Tycho clearly did not think of comets as having a closed path). It is significant that Kepler, when he began to look for a non-circular path for Mars turned to the figure suggested by Tycho for comets, though he introduced it into the Copernican, not the Tychonic world.

* The path of the comet will clearly intersect the paths of various planets, but this (as in the case of Mars) was no longer a troublesome problem.

The advantages of the Tychonic system were enormous, for it had very nearly all those of the Copernican system (to which it is mathematically equivalent) without the awkwardness of a moving Earth. It did in fact become a popular and fairly long-lived rival to the Copernican system, and seventeenth-century astronomers who were not Copernicans more often accepted the Tychonic than the Ptolemaic universe (though some of them compromised and introduced a diurnal rotation of the Earth).* If one accepted Tycho's observational evidence for the non-existence of crystalline spheres, his system was highly acceptable. Many Copernicans followed him in rejecting spheres, and thereby began a fundamental change in the Copernican universe; especially as such men usually eliminated the sphere of the fixed stars as well, which was not needed if the stars were stationary. Such a universe, a combination of the ideas of two totally different systems, was destructive of Aristotelian cosmology in a way that might have alarmed Copernicus. It is not surprising that after Tycho's work it is often difficult to tell whether a man is a Copernican or not, and Copernicanism itself comes to include many different concepts.

If one followed Tycho, and dispensed with crystalline spheres, it did logically become necessary to consider what kept the planets in their orbits. No really satisfactory solution was to be found until much later; earlier attempts were crude, and often dismayingly mystical. The best-known consideration was

* This modification was first published by an obscure astronomer, Nicolas Reymers, in 1588 in *Fundamentum Astronomicum*. Tycho then began a long and virulent public controversy with Reymers, in which each accused the other of plagiarism. Tycho claimed to have invented his system in 1583, and to have described it to Reymers when Reymers visited Uraniborg. This Reymers stoutly denied. Reymers' works are very rare, and there are few contemporary references to him; whatever the merits of the case, it was certainly from Tycho that contemporaries learned of the system.

undertaken by Kepler,* who drew much inspiration from the English scientist, William Gilbert (1540–1603). Like Digges—possibly influenced by the atmosphere created by Dee—Gilbert combined rational science and mysticism in a peculiar blend, in which neither interfered with the other. Gilbert was a physician, not an astronomer, a university graduate who was highly regarded as a medical practitioner; he also associated with London's practical mathematicians, especially with the navigational instrument makers. Outwardly, his great work *On the Magnet* (*De Magnete*, 1600) was intended as an aid to navigation, an impression strengthened by the fact that it carried a preface by Edward Wright (1558–1615), the foremost English applied mathematician of the day. In fact, about a third of the work is devoted to navigational problems; this is the least valuable part, for its premises proved erroneous, and its methods impracticable. The earlier parts of the book remain the most valuable, because they contain the bulk of the experimental work. The last part is different again, for it is an astronomical section, devoted to offering evidence for the diurnal rotation of the Earth.

Gilbert believed that he had strong experimental evidence for the Earth's diurnal rotation, and this evidence was magnetic in character. He had already established that the Earth was a great magnet; and he had found that a spherical loadstone would rotate when its pole was displaced from the North, showing that a portion of the Earth naturally displayed circular motion. Therefore, he argued, it is reasonable to suppose that the whole Earth rotates as well. True, Aristotle had said that only the heavens were animate (that is, self-moved), while the terrestrial globe is inanimate and therefore stationary, but Aristotle was wrong. The Earth no less than the planets is animate, because it possesses a magnetic virtue, which is equivalent to a moving impulse. Having established that it is the nature of the Earth to

* See below, ch. x.

move, Gilbert argued that it is impossible that the heavens should do so, for "who . . . has ever made out that the stars which we call fixed are in one and the same sphere, or has established by reasoning that there are any real and, as it were, adamantine sphaeres?" [23] This combination of Digges and Tycho made Gilbert reject the idea of a Primum Mobile as well. This being so, it was more reasonable to suppose that the Earth (which as a sphere has the same aptitude for moving as the planets) rotates diurnally, than that the heavens do so.

This fact established to his own satisfaction, Gilbert did not go on to try to establish the annual revolution of the Earth, which indeed he appeared to reject, remarking that "it by no means follows that a double motion must be assigned to the Earth".[24] He did, however, go beyond Tycho in considering the question of what kept the planets in their orbits. In the posthumously published *New Philosophy of our Sublunary World* (1651), he extended the magnetic force of the Earth as far as the Moon, and argued that it was this magnetic force which kept the Moon circling the Earth, and which, as well, accounted for the Moon's influence upon the tides.

Gilbert thus occupies a peculiar place in astronomical thought : not an astronomer, he developed several new astronomical ideas ; not a true Copernican, he was warm in praise of what Copernicus had done. For Gilbert rated Copernicus as "the Restorer of Astronomy" not only for his bold ideas, but for his mathematical penetration.[25] Yet the Platonic harmonies that appealed to Copernicus have no interest for Gilbert ; he is rather concerned with an animate mysticism which endows the Earth with a living force, and accounts for physical rotation, and for eternal perfection :

The human soul uses reason, sees many things, inquires about many more ; but even the best instructed receives by his eternal senses (as through a lattice) light and the beginnings of knowledge. Hence come so many errors and follies, by which our judgments and the actions of our lives are perverted : so that few or none order their actions rightly and justly. But the magnetick force of the earth and the formate life or living form of the globes, without perception, without error, without injury from ills and diseases, so present with us, has an implanted activity, vigorous through the whole material mass, fixed, constant, directive, executive, governing, consentient ; by which the generation and death of all things are carried on upon the surface. For, without that motion, by which the daily revolution is performed, all earthly things around us would ever remain savage and neglected.[26]

This mystic strain among English astronomers—astrological with Dee, theological with Digges, magnetical with Gilbert—perhaps explains in part why the mystic philosopher Giordano Bruno found in London a stimulating atmosphere which encouraged him to produce his most important philosophical work. There is no evidence that he met any of the English scientists ; yet he may have heard of the Copernicanism of Digges and Dee, which his own even more mystical Copernicanism resembles. It was a long series of events which brought Bruno, born about 1548 in Nola, near Naples, to London for a few brief years in the 1580's : early education at the University of Naples ; entrance into a Dominican monastery ; a stormy and perverse eleven years as a monk who insisted on reading Erasmus ; finally flight from the monastery and a restless wandering about European capitals. He was always welcome wherever he went, for he had developed a system of mnemonics, probably based on such mediaeval systems as the so-called " art " of Raymond Lull, that was in much demand ; * but his contentious and restless personality was such that he always moved on in search of other and more congenial circles. It was during his English visit that

* Memory, like knowledge, was held to be power.

122

he first wrote on cosmological problems. The basis of his belief was the Epicurean theory, which he derived from Lucretius, of an infinite universe with a plurality of (inhabited) worlds. Bruno's was not merely an indefinitely large universe, like that of Nicholas of Cusa (whose ideas did influence him) but a truly infinite one ; indeed Bruno was probably the first philosopher who really comprehended the possibilities inherent in the idea of infinity. With Lucretius, Bruno blended the Platonic concept of the world-soul, and, from Nicholas of Cusa, a pantheistic concept of the relation of God and the universe.

Among astronomers, Bruno drew particularly on Copernicus and Tycho. The latter gave him arguments for the idea that all heavenly bodies are in motion, in confirmation of the doctrine of Nicholas of Cusa ; the former for an extension of the idea that there is no centre of the universe. The fact that the Copernican universe was very large helped him with physical arguments ; and the Copernican development of the concept of a solar system seemed to confirm the Epicurean notion of a plurality of worlds. (Bruno distinguished between " world " and " universe " ; the former means the solar system and the fixed stars, which is one system among many like it ; the universe is the totality of these worlds.) These worlds were like our own, each with its Sun, planets, inhabited Earth, and so on ; our own Earth could be anywhere in the universe, but certainly not at the centre. This was not a scientific system ; as Bruno said in the dialogue *On the Infinite Universe and Worlds*, " No corporeal sense can perceive the infinite ".[27] He had no interest in a scientific system ; he was a mystic prepared to push mysticism to its utmost power. He had nothing but scorn for those who could not accept his daring flights of intellectual fancy. For him, even more than for Nicholas of Cusa, God was everywhere, the infinity of the universe blended with the infinity of God, and there was one mystic whole. As he wrote defiantly, " It is

Unity which doth enchant me. By her power I am free though thrall, happy in sorrow, rich in poverty, and quick even in death." [28] In the mystic contemplation of the One lay the true liberation of the mind and soul.

This has little enough to do with astronomy. Yet the mystic vision of the potentialities of infinity attracted such minds as Gilbert and Kepler. The use of natural science in philosophy was familiar to all, for it was a large part of the force of Aristotle's philosophy that he covered all aspects, from natural philosophy to metaphysics. No wonder that, faced with the manifest heterodoxy of Bruno's philosophy, there was a tendency to feel that the associated astronomy was equally heterodox.

Until the end of the sixteenth century, the Catholic Church had generally ignored the heretical implications of Copernicanism, and been satisfied to treat it as a mere mathematical hypothesis, useful for calculation, as in the case of the reform of the calendar so successfully carried through in 1582. There was good tradition for this : Oresme in the fourteenth century and Nicholas of Cusa in the fifteenth had both discussed arguments in favour of a moving Earth, and both had shown that the apparent contradictions of Scripture could be dealt with harmlessly.* The fundamentalist position was not a Catholic one, and there was good authority for treating Scripture allegorically : had not St. Augustine declared that it was only when he learned that the Old Testament could be so treated that he had been able to accept the tenets of Christianity ? In 1576 a Spanish theologian Diego de Zuñiga (Didacus à Stunica) had treated this problem admirably : in a *Commentary on Job* (published in 1584) he used the text " who shaketh the earth out of her place, and the pillars thereof tremble," (Job, ix : 6) to show that though the immobility of the Earth was commonly spoken of in Scripture, yet there was, as here,

* In fact, Oresme believed that the Earth stood still, and therefore the relevant Scriptural passages were astronomically valid.

also authority for its mobility. And it was a well-established scholastic tradition that when Scriptural passages appeared to contradict each other, reason might be applied to the resolution. Hence, the author concluded that the Pythagorean doctrine was not contrary to Scripture—a conclusion not specifically refuted by the Church until 1616.

Various new factors influenced the attitude of the Church after 1600, among which must be included Bruno's adoption of certain Copernican doctrines ; this certainly suggested, what had not been apparent before, the philosophical dangers inherent in the Pythagorean hypothesis. It was not for his espousal of Copernicanism that Bruno, rashly returning to Italy in 1591, was imprisoned first by the Venetian and then by the Roman Inquisition. There were plenty of charges against him : he was an apostate monk ; he had espoused atheistic Epicurean doctrines ; he appeared to have taken an Arian stand on the nature of the Trinity ; he was a magician of sorts. When pressed to recant he was obdurate, insisting that there was nothing to recant, and trying instead to show his judges the beauties of his mystic pantheism. The only strange element in the whole case was the reluctance of the Inquisition to judge that he was " an impenitent and pertinacious heretic " ; it took eight years before he was finally condemned and burned. In all the indictment there is no mention of Copernicanism, nor did it occur to anyone that there should be. Once Bruno was dead, however, it was difficult to forget that this astronomical hypothesis in particular could be used for dangerous purposes if its physical truth were upheld. And astronomers were soon to insist on its physical validity more strongly, and more publicly.

Protestants, especially Lutherans, had been quicker to condemn Copernicanism ; they did not see it as an astronomical *hypothesis*, in spite of Osiander's preface, but as a *system* fatal to the truth of the Bible. This was not only because of their insistence on the literal truth of Scripture, but, ironically, because they were well informed. Luther's disciple, Melanchthon, was connected with the University of Wittenberg, and must have heard of the new theory from Rheticus, even before Rheticus went to Frauenburg. At least Luther knew enough of the theory in 1539 to denounce it; in one of his "Table Talks" he is said to have castigated

the new astronomer who wants to prove that the Earth goes round, and not the heavens, the Sun and the Moon; just as if someone sitting in a moving waggon or ship were to suppose that he was at rest, and that the Earth and the trees were moving past him. But that is the way nowadays; whoever wants to be clever must needs produce something of his own, which is bound to be the best since *he* has produced it! The fool will turn the whole science of astronomy upside down. But, as Holy Writ declares, it was the Sun and not the Earth which Joshua commanded to stand still.[29]

Melanchthon, writing after the publication of *De Revolutionibus*, in his own *Elements of Physics* (1549), was more detailed in his rebuttal, but the essence of his argument was the same. Only fools, seized with a love of novelty, try to insist that the Earth moves; "it is a want of honesty and decency to assert such notions publicly, and the example is pernicious. It is the part of a good mind to accept the truth as revealed by God and to acquiesce in it." [30] Calvin never even mentioned Copernicus; but his belief in the literal truth of Scripture was no less absolute.[31] With all this it is no wonder that some, like Tycho, found the motion of the Earth too hostile to religious faith for serious contemplation.

Yet though the gradual recognition by the various Christian sects of the dangers inherent for dogma in the new astronomy seemed to make inevitable a conflict between science and religion, the issue was not often faced publicly. Many scientists accepted the Copernican system privately, and discussed it with friends, while avoiding public commitment. Others salved their consciences by a partial acceptance. And many more boldly asserted that the Churches were in error, and there was no need to insist that Scripture and astronomy conflict. Paradoxically, the Protestant restriction was strongest in the early years, when there was least evidence for the truth of the Copernican system ; whereas the Catholic attack was fiercest when, for the first time, it began to appear that there might be physical, as well as mathematical and aesthetic grounds for adopting the heliostatic system.

Whatever the reservations of individual scientists, Copernicanism, modified during the course of sixty years, was in far more flourishing state after 1600 than it was in the 1540's. This is the more strange, since throughout the later years of the sixteenth century there had been no great new discoveries to render the Copernican system one bit more probable than it had been in 1543. Indeed, its later modifications had been of a nature to repel rather than attract rational minds—the extension of the sphere of fixed stars towards infinity, the abolition of crystalline spheres, the introduction of mysterious forces to account for the motion of planets—all these tended to suggest that Copernicanism belonged to the mystics. Tycho, the greatest practical astronomer of the age, was an anti-Copernican : his work was of no immediate benefit to Copernicanism, and the theoretical discoveries of Kepler were required to pull together all the significant advances in astronomy of the sixteenth century into a form which supported the Copernican system. But nevertheless there were more Copernicans among serious astronomers than is usually indicated ; although the later sixteenth century was not an age readily receptive of new ideas. And in the sixty years since the publication of *De Revolutionibus*, Copernicanism had been so thoroughly debated and so widely discussed that even laymen knew the arguments for and against it, and a casual reference was intelligible to an ordinary literate audience. These years of discussion rendered the system familiar, and reduced its novelty ; this in turn helped to make it more acceptable when new arguments in its favour were forthcoming. And they served as well to nullify the force of anti-Copernican arguments, which were stultified by repetition. In spite of the insistence by anti-Copernicans that nobody but a fool could fail to perceive the incontrovertible force of their arguments, such fools continued to become astronomers, and to win converts. The debate had been long and public yet conducted with remarkable mildness in a violent age ; it was not to end without passion and drama.

CHAPTER V

THE FRAME OF MAN AND ITS ILLS

Anatomical study has one application for the man of science who loves knowledge for its own sake, another for him who values it only to demonstrate that Nature does nothing in vain, a third for one who provides himself from anatomy with data for investigating a function, physical or mental, and yet another for the practitioner who has to remove splinters and missiles efficiently, to excise parts properly, or to treat ulcers, fistulae and abscesses.[1]

In 1542 Andreas Vesalius (1514–64) wrote with characteristic Renaissance smugness that " those who are now dedicated to the ancient study of medicine, almost restored to its pristine splendour in many schools, are beginning to learn to their satisfaction how little and how feebly men have laboured in the field of Anatomy from the time of Galen to the present day."[2] It was his own belief that his great treatise *On the Fabric of the Human Body* (1543) was the first real step forward from Galen, no small boast in view of the high esteem in which Vesalius, like his contemporaries, held the great Greek physician of the second century A.D. Modern criticism has tended to agree with Vesalius in thinking both that a revival of anatomy was a necessary preliminary to the improvement of medicine, and that the work of Vesalius himself is a landmark in that revival. The hazards of a date—1543—have brought together two diverse figures, Vesalius and Copernicus, who shared a respect for the ancients and a desire to raise modern science at least to the level of ancient science.

Progress in anatomy before the sixteenth century is as mysteriously slow as its development after 1500 is startlingly rapid. One cannot say that it was because anatomy was a forbidden subject, for the old myth that human dissection was prohibited throughout the Middle Ages has long since been dispelled. It is true that Islamic writers laid little stress on anatomy, in spite of their knowledge of the magnificent work of Galen in this field ; they emphasised rather the identification of disease and the compounding of drugs, and this bent was transmitted to Western Europe through the writings of Avicenna (979-1037). The Moslem lack of interest in anatomy seems to have stemmed from religious prohibition ; but there was no such prohibition by the Church in Christian Europe.* Indeed, it appears that distaste for opening the human body after death was a relatively late development (perhaps even appearing after the revival of anatomy), for a fifteenth-century Florentine physician, Antonio Benivieni, habitually performed post-mortem examinations and commented with surprise when, after he had treated an obscure but interesting incurable disease, the man's " relations refused through some superstition or other " to allow him to open the body and investigate the cause of death.[3] Post-mortems were frequently performed in the fourteenth century, both privately and publicly, and members of the university faculties were commonly called in as consultants in legal cases when it was desirable to ascertain

* What the Church did forbid was the boiling up of bodies to produce skeletons. The edict (1300) was the result of what threatened to become an over-popular practice because of the desire of rich Crusaders and pilgrims to have their bones laid to rest at home. The edict was responsible for the many subterfuges such as robbing gallows and charnel houses to which later anatomists resorted in order to acquire bones when bodies were readily available for dissection.

if death were due to natural or unnatural causes. (One wonders how they were able to decide.)

Nevertheless, anatomy as such was little practised. One obvious reason was the lack of a guide. Surprisingly, Galen's anatomical treatises escaped the first great wave of translation in the twelfth and thirteenth centuries, when so much of his purely medical work was translated. All that was available of his brilliant anatomical investigations was a short treatise called *De Juvamentis Membrorum* (*On the Function sof the Members*), a truncated version of his physiological treatise *On the Use of the Parts*. It was a highly abbreviated paraphrase of little more than half of the original, dealing cursorily with the function of the limbs and digestive organs, and retaining Arabic nomenclature. This could suggest a reason for studying the body, and provide a list of the principal organs, but it was of little help in directing men to a clear picture of the correct approach to anatomy. There was thus little to stimulate investigation, and even less to help if investigation was attempted. In fact, medical men at first found they had quite enough to do in mastering the immense mass of material presented to them in books. Besides, they not unnaturally tended to accept the Moslem view, that medicine should deal with disease and its causes rather than try to fathom the structure of man. Even the surgeon had little need to know anything more than surface anatomy and the articulation of the limbs, the latter useful in case of dislocations.

The first step towards a rediscovery of human anatomy was a revival of interest : the first indication that this had taken place is the appearance of an *Anatomy* by Mondino de' Luzzi, written in 1316. Mondino (*c.* 1275–1326), a professor at Bologna, was perhaps influenced by the animal dissection undertaken at Salerno in the previous century, perhaps by the growing demands of the surgeons, certainly by his reading of Galen's *De Juvamentis* mentioned in the proemium. Judging by the use which Mondino made of *De Juvamentis*, it was only imperfectly Latinised, for Mondino's terms are nearly all Arabic in origin : he was, in fact, one of those Arabicised physicians whom the poet Petrarch was to attack so vehemently in the next generation. Mondino's approach was simple : without preamble, he plunged into a brief and crude description of the parts of the body, beginning with those of the abdominal cavity and proceeding via the thorax to the head and extremities. This order became traditional in anatomical study, partly from the example of Mondino, partly from the need to examine first the parts most subject to decay. Mondino's intention does not seem to have been to write a detailed textbook, but rather to provide a rough outline of procedure for dissectors ; here there are no precise directions to follow in dissecting, and no attempt at exact nomenclature. Mondino has clearly dissected a body in the way he describes, but he could not, even if he had wished to do so, delineate the position and nature of every organ. Yet the work is thoroughly professional, and Mondino is not wholly subservient to his authorities.

Because of its succinctness and utility, Mondino's *Anatomy* became the standard textbook of the medical schools ; for about this time most universities incorporated into their statutes the provision that all medical students should see one or even two anatomies (always and naturally performed in the winter) ; and these same statutes usually specify Mondino as a text. Indeed there was no other ; and references to Galen are more usually to the supplementary text of *De Juvamentis*. This remained true for another century, even though Niccolo da Reggio in 1322, six years after Mondino had finished his *Anatomy*, completed a translation of Galen's *On the Use of the Parts*, a book which Mondino would certainly have used had it been available to him. In fact, Galen was relatively neglected, because Mondino had replaced him.

By 1400, anatomical dissections were established as a regular part of the curriculum in most medical schools,* and a standard procedure had been developed. The cadaver was laid on a table, around which the students clustered closely ; the actual dissection was performed by a demonstrator (often a surgeon) while the professor on his high lecture platform read the prescribed text which was Mondino or sometimes, later, Galen's *Use of the Parts*. This is the famous scene depicted in the frontispiece to a 1493 Italian edition of Mondino (and many other woodcuts of the period) and doubtless represents official practice, though there are other fifteenth-century depictions of anatomical scenes where a less formally pedagogic dissection is in progress. Presumably students were also able to attend post-mortem dissections when their own professors were engaged for the purpose, and the records indicate that these were fairly frequent. (One can perhaps account for the seemingly exaggerated claims of later anatomists as to the number of bodies they dissected by assuming that they lump true dissections and post-mortems together.) The printing-press helped to establish Mondino's as the official text ; the first printed edition appeared in 1476, after which there were at least eight more editions in the fifteenth century, and over twenty in the sixteenth. At the same time, commentaries on Mondino were naturally being produced by the professors who lectured on anatomy, and it was in the form of commentaries on Mondino, rather than on Galen, that new anatomical treatises were presented. Of these a typical example is that of Alessandro Achillini (1463–1512), who was alternately a professor of philosophy and of medicine : his *Anatomical Annotations* (published posthumously in

---

* Some universities were slower : Tübingen only introduced anatomical studies in 1485, and the statutes stipulated that dissections were to be conducted every three or four years ! Even in 1538, when the use of Mondino was forbidden, they were still infrequent. But the Tübingen medical faculty achieved little fame—the better faculties were much more insistent on dissections.

1520) reveal that his lectures did not go much beyond Mondino. Yet he, like Mondino, clearly had performed dissections, and tradition assigns to him a number of minor anatomical discoveries. Achillini's work is mainly of interest in showing how anatomical study and original anatomical investigation was slowly taking root among professors of medicine.

In the early years of the sixteenth century anatomy was undoubtedly regarded as far more important than had been the case before, and anatomical studies were pursued with great vigour, and in a new way. The chief stimulus in this direction came, rather improbably, from humanism which, soon after denouncing the Arabic tradition represented by Mondino, made available the superior Greek tradition of Galen. Just as fifteenth-century astronomy rebelled against mediaeval texts and tried to return to the pure fount of Greek tradition with an intensive study of the works of Ptolemy, so in anatomy and medicine there was an attempt to restore medicine by a reconsideration of the works of Galen. First, naturally, came new editions of the texts known to the Middle Ages ; among the more famous new translations are those by Thomas Linacre (?1460–1524), humanist, physician and founder of the College of Physicians, who concerned himself with medical texts as well as with Galen's great physiological treatise *On the Natural Faculties* (1523). The most influential of Galen's works in the early years of the sixteenth century was *On the Use of Parts*, available by 1500 in a number of versions direct from the Greek, which set the style for having a discussion of the function of each organ in conjunction with anatomical dissection. It had a further curious advantage in having been unknown to Mondino, which gave it extra prestige in the anti-mediaeval and anti-Arabic climate of the period. Every attempt was made to get these Galenic works into the hands of medical students : thus in 1528 there was published in Paris a series of four handy texts in pocket size, including *On the Use of the Parts* (in the fourteenth-century translation of Niccolo da Reggio), *On the Motion of Muscles* (newly translated) and Linacre's five-year-old version of the *Natural Faculties*. The rise in importance of the medical school of the University of Paris dates from the renewed interest in Galen indicated by these publications and by the activities of the Paris faculty. It was Johannes Guinther of Andernach (1487–1574) (in spite of his name, a professor at Paris) who first published a Latin translation of a newly discovered and most important Galenic text, *On Anatomical Procedures* (*De Anatomicis Administrationibus*, 1531). Guinther was a medical humanist, rather than a practising anatomist, but his contributions to the advance of anatomy are none the less great, and in spite of the later criticism of his pupil Vesalius, Guinther did perform anatomical dissections, as well as make translations. Vesalius as a student assisted Guinther in preparing the professor's own textbook, *Anatomical Institutions according to the opinion of Galen for Students of Medicine* (1536).

The real worship of Galen begins with the rediscovery of the *Anatomical Procedures*, and its commentary by Guinther. (His Latin version of 1531 was followed in 1538 by a Greek text, prepared for the press by a group of scholars which included the botanist Fuchs. There were many editions of both the Latin and the Greek versions through the course of the sixteenth century.) It was a Galenic treatise wholly new to the Renaissance, whose superiority to Mondino, and even to the commentators on Mondino, was conspicuous. Its immediate impact is clearly indicated in the rearrangement of procedure that was now adopted. Galen had begun, not with the viscera like Mondino, but with the skeleton, for, as he insisted, " as poles to tents and walls to houses, so are bones to living creatures, for other features naturally take form from them and change with them."⁴ This was a much needed injunction, for the skeleton was poorly known. Here too Galen indicated clearly the nature of his anatomical material ; lamenting the impossibility of studying human anatomy at Rome, he explained why he had chosen apes and other animals, while insisting that one should procure human cadavers whenever possible. (Unfortunately, this warning was not always heeded.) After the bones, Galen proceeded to the study of the muscles of the arms, hands and legs, followed by the nerves, veins and arteries of the same limbs ; then the muscles of the head. Only then did he proceed to the internal organs of the body, which he classed by function—alimentary, respiratory (including the heart) and the brain. This is a totally different method of procedure from that of Mondino, both in the order in which the organs are treated and the manner in which they are discussed ; the immediate influence of the work is indicated by those treatises (including that of Vesalius) which follow the Galenic procedure.

When one considers both the novelty and the intrinsic value of the Galenic texts, it is not surprising that sixteenth-century anatomists eagerly seized upon them, at the same time denouncing the established tradition of the medical schools, and those in particular who claimed that Mondino and his fifteenth- and sixteenth-century commentators were preferable to Galen. Galen's work was really so immeasurably superior to what had been done in the intervening period that admiration and adulation was inevitable and desirable. For until anatomists learned what Galen had to teach there was little chance they would ever learn more about anatomy than he had known. It is not surprising that adulation sometimes turned into worship, and the conviction that Galen could do no wrong ; nor that the critics who opposed Galen because they believed that mediaeval anatomists were better became confused with those who, following Galen's precepts, and exploring the problems of human anatomy, found that Galen erred. So John Caius was content to devote a major part of his life to the editing of Galen's works, and regarded dissent from Galen as an indication of academic flightiness and irresponsibility. He found Galen a perfectly adequate guide when he lectured on anatomy to surgeons, and thought others should do so too.

The astonishing thing is that contemporary with the rise of Galen worship there actually were anatomists bent on following Galen's example and admonitions, who did dissect with a fresh eye (even though the other was usually fixed on the text of the *Anatomical Procedures*). Galen certainly would have envied those, like Vesalius, who had access to human cadavers, and would have had only scorn for the fact that such men, with advantages he himself lacked, were often reluctant to accept the evidence of their own eyes, and preferred to believe that Galen was describing human anatomy when, as he himself had carefully pointed out, he knew only animal anatomy. But what scientific apprentice has not, many times since the sixteenth century, preferred to trust the authoritative text rather than his own unskilled eyes? It took time to create an independent school of anatomy, even as it takes time to make an individual anatomist. And in spite of the comparative abundance of human dissection material, it was not quite as abundant as anatomists boastfully made out; much preliminary dissection was performed on animals, and the lessons of this early training often persisted in spite of later experience.

At about the same time that humanism was influencing anatomy through the rediscovery of Galen, artistic circles were influencing anatomy in quite a different way. Every studio manifested an interest in surface and muscle anatomy as part of the attempt at naturalistic portraiture. The greatest exemplar of this tradition is Leonardo da Vinci (1452–1519), but he was merely the best of a large group which includes Dürer and Michelangelo, as well as many lesser artists, some of whom turned their hands to anatomical illustration. Leonardo was introduced to anatomy in the studio of Verrochio (1435–88) who insisted that his pupils learn anatomy thoroughly: he taught them to observe surface anatomy, and also had them study flayed bodies, so that they could learn enough about the play of muscles to represent them accurately in action. Artists of the late fifteenth century commonly tried their hands at the dissection of human and animal subjects in pursuit of artistic anatomy; they could also attend anatomy demonstrations, either the public dissections which took place every winter in Italian universities, or the private lessons which were also widely available.

The earliest anatomical drawings of Leonardo, made about 1497–9, show only slight knowledge of dissection, though already profound understanding of surface anatomy. He began about this time to plan a great book *On the Human Figure*, intended to portray living, artistic anatomy rather than structural and physiological anatomy. Soon after 1503, however, Leonardo's approach began to change. First, he had access to more dissecting material (though never, apparently, to as much as he claimed). Then, perhaps about 1506, he read Galen's *On the Use of the Parts*, which stimulated him to further studies on bones and muscles, taught him much about anatomical fact and procedure, and interested him in physiological functioning. (He is often as scathing about the statements of Mondino as any medical humanist.) It is to this period that his greatest work belongs, much of it based upon the centenarian whose superficial anatomy he studied during visits to the hospital, and whose body he later dissected and compared with that of a seven-month foetus. Side by side with his studies on man were studies on animals : partly because, with Galen, he assumed that animal and human anatomy was basically identical, and partly from the demands of art. Leonardo studied the anatomy of the horse for the great projected equestrian statue of Ludovico Sforza, and he was as interested in the proportions of animals as in those of human bodies.

Some of Leonardo's work is extraordinary : with the advantage of an eye trained to observation he saw as clearly as any professional anatomist the correct relationships and forms of bones, muscles and organs, and his mechanical ability suggested to him a number of ingenious techniques for studying individual organs. Some of his work is poor : he either had not really observed what he drew, or had observed it wrongly. But the whole—and the level of his competence is generally high—is transformed and illuminated by the drawings which fill every page of his notes, for Leonardo was a peerless anatomical illustrator. He had, of course, the great advantage of being both observer and draughts-man. There is almost no page of his manuscript notebooks which is not a thing of beauty in its own right, and in all that he did, Leonardo looked for the hidden beauty which he believed to lie behind all (or almost all) the body's frame and structure. Leonardo stands in a class by himself : a great artist, he made of his anatomy a work of art.

Leonardo stands apart for another reason : he worked in secret and published nothing. He was known to be working on anatomy and a few artists saw some of his illustrations. In fact, his influence may have been real on anatomical illustration, though his influence on anatomy was nil. Anatomical illustra-tion developed amazingly during the first years of the sixteenth century, to such an extent, indeed, that it is tempting to judge the worth of every work chiefly by its illustrations. This would surely be wrong. Whether the illustrations have independent artistic merit is really irrelevant to their purpose ; there was as much luck as judgement involved in whether an anatomist could secure the services of a good artist or not. Even the accuracy of the drawings may reflect the artist rather than the anatomist, for, as in the case of herbals, it is not at all clear how closely anatomists were able to work with their artists. Anatomical illustration appears rather suddenly in the early sixteenth century, for though the first books on anatomy are illustrated, there are no anatomical drawings. Usually the illustrations are of dissection scenes or of surgical operations, though there were also the "wound men" indicating the probable location of difficult sword cuts, and the crude figures indicating the astrological significance of various regions of the body. The attractively illustrated *Anatomical Bundle* (*Fasciculo di Medicinae*, 1493), which includes the text of Mondino, has an interesting seated female figure with the body opened to show the reproductive organs ; though the drawing is naturalistic the anatomy emphatically is not.

The first anatomist to take advantage of the possibilities of anatomical illustration was Berengario da Carpi (*c.* 1460–*c.* 1530), who was associated with the University of Bologna which had a good anatomical tradition. Berengario published a commentary on Mondino in 1521, followed in 1522 by a short book with a long title : *A Short but very Clear and Fruitful Introduction to the Anatomy of the Human Body, Published by Request of his Students,*[5] and both books were illustrated with true anatomical drawings. The second work contains a number of plates designed to illustrate the muscles : the artist gives a spirited view, rendering the scene arresting by drawing the body with normal facial expression, the figure in each case cheerfully holding back flaps of skin to display muscular structure. This method of demonstrating living anatomy was further developed later to give complete "muscle men" and skeletons. The figures in Berengario are all set in a bare landscape, which developed into the almost blighted and ruined background of the figures in the illustrations to the works of Vesalius, the climax of anatomical illustration in this genre.

Anatomical illustration was undoubtedly valuable, especially in the absence of a competent technical vocabulary. And it produced some wonderful picture-books. But it also had its disadvantages. Most noticeable is that it tended to draw attention away from the text, which it did not necessarily represent accurately. This was particularly undesirable in books like those of Vesalius which are more than a mere outline of anatomy. In the sixteenth century, some anatomists complained that illustrations even drew students away from dissection ; having a picture, they felt less necessity to observe for themselves. As Vesalius put it, " I am convinced that it is very hard—nay, futile and impossible—to obtain real anatomical or therapeutic knowledge from mere figures or formulae, though no one will deny them to be capital aids to memory." [6]

This difficulty still exists, especially when appraising the work of sixteenth-century anatomists : in looking at the pictures one is all too apt to forget the text, which is a far better measure of scientific achievement. And the text by itself is always interesting. Every anatomist of the period shares a certain common attitude. Thus each declares that anatomy needs clarification, because the professors are such blockheads ; and each claims to have learned this need through his own dissection of innumerable cadavers. Equally, each betrays the fact that his anatomical investigations were in reality based on relatively few cadavers, supplemented by autopsies on the one hand, and numerous animal dissections on the other. This last fact explains many anomalies. It has always been a puzzle to understand why sixteenth-century anatomists " saw " in the human body what Galen described for animals, and it has been assumed that they were wilfully blind or stupid. Aside from the fact that it is often quite easy to " see " what a textbook or manual says should be seen, very often sixteenth-century anatomists used the same animal material as Galen, partly because it was readily available, partly because it did more closely resemble what Galen described. Hence the "five-lobed" liver, found in dogs and apes but not in man, yet commonly shown in anatomical illustrations including the early drawings of Vesalius. Hence, too, the insistence that the *rete mirabile* was present in man, though it was known to be difficult to detect in a body long dead.* Hence, too, the universal habit of representing the right kidney as higher than the left, though in man it is lower : clearly the ideas of most anatomists (even of Leonardo and Vesalius) were so firmly fixed by early dissection of animals that they never rearranged their vision when they dissected man, an indication of how difficult it may be even for practised eyes to see aright.†

Beginning about 1520 there was a great rush of anatomical works, one after the other, of various degrees of originality. All are, naturally, more or less influenced by Galen, either physiologically or anatomically. Each of these books has its own merits and its own discoveries; all together represent the "new anatomy." It is difficult to distinguish them chronologically for books were often years in preparation ; one of the distinctions of Vesalius was the way in which he rushed into print. Among the earliest of the new kind of anatomical treatise are those of Berengario da Carpi : the *Commentary on Mondino*, and the *Brief Introduction*. As befitted a commentator, Berengario organised his work on lines laid down by his authority, but his was immeasurably superior to Mondino's. He explained his position in the dedication :

There are many books which discuss anatomy, but they are not well arranged for the reader's comfort. The authors seem to have borrowed fables from other volumes instead

* The *rete mirabile* is a network of vessels at the base of the brain, found in cattle but not in man. The difficulties connected with it are indicated by the comment of Niccolo Massa (*c.* 1489–1569) in his *Introduction to Anatomy* (1536) : "some dare to say that this rete is a figment of Galen . . . but I myself have often seen the rete, and have demonstrated it to the bystanders so that no one could possibly deny it, though sometimes I have found it very small." 7 Vesalius used to keep the head of an ox or lamb handy when dissecting a human head, in order to demonstrate the *rete* plainly.

† Galen says, "The right kidney lies higher in all animals," 8 but says nothing about man. The rest of the paragraph makes it plain that he is referring to apes and cattle.

of writing genuine anatomy. For this reason there are few
or none at all who now understand the purpose of this
necessary and important art.[9]

And he proved his own understanding by demonstrating his own
achievements in actual dissection. He studied "the reader's
comfort" too ; for, in contrast to Mondino, Berengario was
clear, direct, careful to explain both the names and positions of
organs, indicating how they were to be handled for the most
effective dissection, what precautions were necessary. Reading
his account, anyone would feel almost capable of picking up a
dissecting knife and going to work. Berengario was not startlingly original, but he did observe with a fair degree of accuracy.
He was amusingly contemptuous of the "common opinion" that
the *rete mirabile* is found in man, for " I have never seen this net,
and I believe that nature does not accomplish by many means
that which she can accomplish by few means " : [10] since it is not
necessary, there is no need to imagine it. There are many other
anatomical works in this period which have merit : the *Introduction to Anatomy* of Niccolo Massa (1536) ; *On the Dissection of the
Parts of the Human Body*, by the French printer Charles Estienne
(published in 1545, though begun about 1530 ; Estienne (1504–
1564) ingeniously provided illustrations by taking figures from
contemporary artists and having anatomical details inserted) ;
the *Anatomy of Mondino* by Johannes Dryander (1541) ; each
introduced some new names and new facts worthy of note.
But none is significantly superior to any other.

It is the distinction of Vesalius to have produced a work
far superior to all others, anatomically, pictorially and physiologically, so far superior as almost to eclipse the work of his
contemporaries. Vesalius had certain advantages, especially that
of having been educated into the new anatomy. Born in 1514,
he studied first at Louvain, where he learned Latin and Greek,
absorbed the humanist love of languages, and found pleasure in

dissecting animals. In 1533 he went to Paris for formal medical training ; though he stayed only three years, and though he was later to characterise his teachers as ignorant of practical anatomy, in fact the years at Paris formed his anatomical outlook. Here, under Guinther of Andernach, he was introduced to Galen's *Anatomical Procedures* ; he assisted Guinther in the preparation of his *Anatomical Institutes* ; and he was profoundly influenced by the tremendous interest in Galen displayed by the medical faculty and the printers of Paris. The scorn that he later heaped on his teachers is at least partly a measure of how much they taught him : for they and Galen combined to teach him to approach anatomy, not as a textbook subject, but as a subject for research. The value of seeing for oneself, the intimate connection between anatomy and physiology—these were Galenic precepts, and Vesalius followed the path his education fitted him to follow, though he left his teachers far behind.

Vesalius left Paris for Louvain in 1536 when war forced the closing of the medical school ; for a year he lectured and demonstrated with *éclat*, published his thesis, and then departed for Italy. Here, at Padua, he secured his M.D. and immediately, in spite of his youth, appointment as Lecturer in Surgery. He lectured on anatomy and as a result of his first experiences published the six sheets known as the *Tabulae sex* in 1538. These are large, perhaps so that they could be pinned on a wall, and combine illustration and text on each sheet. Characteristically, the first sheet carries a dedication which explains that it was at the demand of the students and other professors that Vesalius produced this work. The first three sheets (with drawings by Vesalius himself) represent the liver and associated blood vessels, together with the male and female reproductive organs, the venous system and the arterial system ; the drawings are adequate but contain traditional errors with respect to the shape of the liver and uterus, and the relative positions of right and left kidney.

The last three sheets, drawn by Jan Stephen van Calcar (a pupil of Titian) represent the three aspects of the skeleton, in living posture, with a text naming the bones. These sheets seem to have started a fashion, and after their appearance many anatomical sheets were produced for student use.

For the next few years Vesalius lectured and dissected furiously, until he felt satisfied that he had solved the major problems in anatomy and was competent to present the results of his work to the public. He presented his achievements not in one book, but in two, both published in 1543 : *On the Fabric of the Human Body* (*De Humani Corporis Fabrica*) and a brief handbook, about the size of Berengario's, the *Epitome*. This is a fantastic achievement in the time available, the more so as Vesalius was in this period concerned with editing Guinther's *Anatomical Institutes* and contributing to the 1541 Latin edition of Galen (the Giunta edition). In 1543 Vesalius left Padua for Basle, to see his books through the press ; when copies were available, he took them to the Emperor's court in Germany, hoping to secure a court position. He was successful ; after his appointment as Imperial Physician to Charles V, Vesalius had little time for dissection, and his anatomical activity nearly ceased, though the second edition of the *Fabrica* in 1555 includes a fair amount of revised material. This second edition immediately preceded his appointment as physician to Philip II of Spain, just as the first had preceded his appointment as physician to Charles V. He appears to have been less successful in Spain than he was in Germany and the Low Countries, and about 1562 he gave up his post ; his activities are then obscure, but he died on a pilgrimage in 1564, intending to return to teaching at Padua.

What makes the *Fabrica* superior to all other anatomical books of the period (apart from its dramatic and artistic illustrations) *

* These were attributed to Jan Stephen van Calcar (1499–*c*. 1550) by the sixteenth-century art historian, Vasari. Modern students have doubted this,

is its plan and its scope. As the title indicates, it is more than an account of structural anatomy ; its size shows at once that it is no mere handbook. The influence of Galen was still strong on Vesalius : the content of the sections follows the plan of Galen, not of Mondino ; and Vesalius included in the last book many of the vivisection experiments described earlier by Galen on the effect of cutting and tying various nerves. The first book treats the skeleton ; the second, myology, carefully showing all the muscles and their relations ; the third and fourth books the venous, arterial and nervous systems ; the fifth and sixth the organs of the abdominal and thoracic cavities and the brain.

Vesalius was in part writing an anti-Galenic polemic ; at least he was ever eager to attack the Galenists, even his own masters. He enjoyed disagreeing with Galen, as when he argued that the vena cava has its origin in the heart, not the liver, an argument he pursued in some detail. But in fact he could not have written his great work without Galen ; there is a real sense in which Vesalius began with Galen rather than the human body, in the same way in which Copernicus began with Ptolemy rather than with the physical world. Neither Copernicus nor Vesalius was any the less original for that. Vesalius kept one eye on Galen, but the other was quick to look for possible discovery : for no anatomist of the sixteenth century felt that he had really established himself as an independent worker unless he found something that had escaped Galen, which he was the first to dis-

---

because the figures are as superior to those in the *Tabulae Sex* as the text of the *Fabrica* is to that of the earlier work—though it is possible that the artist had learned as rapidly as the author. In place of Jan Stephen van Calcar, the only candidate is an unknown, also a member of Titian's studio. It seems difficult to believe that so spirited a draughtsman as the artist who drew the pictures for the *Fabrica* should be otherwise unknown ; though it is odd that Vesalius, who had given Jan Stephen credit for his work in the *Tabulae Sex*, did not mention the name of the artist of the *Fabrica*.

cover. Vesalius is no exception. Nor was he an exception in persisting in error in spite of many dissections " with his own hand." (His insistence that the right kidney is higher than the left is a case in point.) Vesalius was exceptional in the amount of new material that he saw, and in his detailed and lively comments. It is almost a pity that the illustrations are so fine, for they are not as accurate as the text. Occasionally the figures include both animal and human anatomy telescoped together, not, generally, through confusion, but because Vesalius was, in the text, discussing comparative anatomy, and permitted the artist to make a combined figure, either for simplicity or to save his time.

Perhaps the most striking aspect of Vesalius' work is the pains he took to deal with the relation between individual organs and the body as a whole. What starts as a complete skeleton ends as a few bones ; what starts as a flayed (but active) human figure displaying its surface muscles is dissected layer by layer until only a few individual muscles are left ; the bodily cavity is considered as a whole before its individual parts are discussed. This is different from the standard method of procedure as much as the aim of the book is different : for Vesalius was not writing an elementary text and handbook, but a great monograph designed as a replacement for Galen.

Not unnaturally, Vesalius was as much concerned with the use of the parts of the body as with their structure, with physiology as well as anatomy ; indeed he made, like Galen, little distinction between the two. Structure is, where possible, related to function ; thus Vesalius considers very carefully the difference in fibre structure between veins and muscles as these are related to their action and purpose. The main function of the veins is to serve the body in conveying nourishment, so their structure is adapted for this purpose :

Nature gave straight fibres to the vein ; by means of these
it draws blood into its cavity. Then since it has to propel

the blood into the next part of the vein, as though through
a water-course, she gave it transverse fibres. Lest the whole
blood should be taken at once into the next part of the vein
from the first without any pause, and be propelled, she also
wrapped the body of the vein with oblique fibres.* [11]

For, " The Creator of all things instituted the veins for the prime
reason that they may carry the blood to the individual parts of
the body, and be just like canals or channels, from which all
parts suck their food." [12]

Following normal Galenic physiology (all he knew) Vesalius
assumed that the veins take their origin from the liver (a belief
presumably engendered by the striking size of the vena cava)
and that their function is to carry the nutritive blood to various
parts of the body, while at the same time removing waste products. Similarly, the arteries are presumed to distribute the vital
spirit to all parts of the body. Very noticeable is the strongly
mechanical concept of bodily function : for attraction and repulsion represent inhalation and expulsion, and the whole venous
system is compared to a water supply, an analogy that was to
serve Harvey in good stead seventy-five years later. The insistence upon the importance of fibre structure (especially detailed in
the discussion of the lungs, in Book VI) was to be maintained
continuously after Vesalius, and became emphatically mechanistic
in eighteenth-century physiology.

Nutrition Vesalius discussed at great length in connection
with the anatomy of the abdominal cavity. He had nothing very
original to say, but he expressed clearly the common conclusions
of sixteenth-century anatomists and of Galen :

Thus food and drink are taken from the mouth through the
stomach into the belly, as into a certain common workshop
or storehouse, that squeezes everything enclosed within it,

* According to Vesalius, straight fibres are responsible for attraction,
oblique fibres for retention, and transverse fibres for expulsion.

mixes it and concocts it, and protrudes what is concocted into the intestine. Thence, the branches of the vena porta suck away what is best of that concocted juice, and most suitable for making blood, together with the moister remnant of this concoction, carrying it to the hollow of the liver. . . . However, the liver, after admitting the thick juice and fluid, adds an embellishment to it necessary for the production of perfect blood. It expels a double waste, that is, the yellow bile, the lighter and more tenuous waste, then the atrabilious or muddy juice, thick and earthy. . . . But the blood is led through the vena cava propagated in a very numerous series of branches to the parts of the body : and what in it is similar and appropriate to the individual parts they attract to themselves, assimilate, and place in position. What is superfluous and what waste arises in this concoction they exclude from themselves through their own ducts.[13]

To Vesalius, as to Galen, the arterial system was both less important and less interesting than the venous system. The venous system derived its importance both from its responsibility for nutrition, and from the necessity of knowing the exact position of each vein for successful phlebotomy. Besides, the structure of the arterial system was less controversial than that of the venous system, though there were plenty of questions to ask about the structure of the heart, from which the arterial system arose :

The dissension among medical men and philosophers concerning the great artery is much less than that about the origins of the veins and nerves. For Hippocrates, Plato, Aristotle and Galen lay down that the heart is the fount and origin of the arteries, as is reasonable . . . But if philosophers and leading physicians have decided that the heart is the fount of the arteries, nevertheless they do differ not a little about the sinus of the heart from which the great artery springs, since some contend that it arises from the middle sinus of the heart, others from the left one. But as this controversy turns rather upon the ventricles and sinuses of the heart, than upon the origin of the artery, and since there are only two ventricles in the heart, we shall confirm the origin of the great artery in the grander left sinus of the heart.[14]

A more difficult problem was the question of the nature of the septum, the thick wall dividing the right side of the heart from the left. The surface of the septum is covered with little pits which Galen, not unreasonably, concluded to be very small pores ; he therefore assumed that they existed to allow a small amount of blood to percolate from the right side of the heart to the left. The importance of this became greater in the sixteenth century as the interest in detailed physiology increased. Vesalius was predisposed to accept the idea that these pits went through the septum, though after careful examination he could not detect any passage. He could only conclude that there was no certainty in this matter :

Conspicuous as these pits are, none (as far as can be detected by the senses) permeate from the right ventricle into the left through the septum between the ventricles ; nor did any passages, even the most obscure, appear to my eyes, by which the septum would be made pervious, although these are described by the professors of dissection because they have a most strong persuasion that the blood is carried from the right ventricle to the left. Whence also it is (as I shall advise more plainly elsewhere) I am not a little doubtful of the heart's action in this respect.[15]

Inevitably, having considered the structure of the heart, Vesalius next considered "the function and use of the heart and of its parts so far described and the reason for their structure." [16] This seems natural, but there was a problem troubling Vesalius : if one considers the natural faculties of the heart and lungs one must become involved in the theological question of the nature of the soul. But, properly speaking, Vesalius argued, this is medical as well as theological, and is, therefore, a fit topic for a work on anatomy :

Furthermore, lest I should here meet with any charge of heresy, I shall straightway abstain from this discussion about the species of the soul, and of their seats. For today, and particularly among our countrymen (Italians) you may find many judges of our most true religion, who if they hear anyone murmur something about the opinions on the soul of Plato, or of Aristotle and his interpreters, or of Galen (perhaps because we are dealing with the dissection of the body, and ought to examine things of this kind at the beginning) they straight away imagine he's wandering from the faith, and having I don't know what doubts about the immortality of the soul. Not bothering about that, doctors must (if they don't wish to approach the art rashly, nor to prescribe and apply remedies for ailing members improperly) consider those faculties that govern us, how many kinds of them there are, and what is the character by which each is known, and in what member of the animal the individual ones are constituted, and what medication they receive. And especially, besides all this, (if our minds can attain it) what is the substance and essence of the soul.[17]

Having thus proclaimed his right to discuss such sensitive questions, Vesalius proceeded to a detailed discussion of the functions of the heart and certain associated functions of the liver and brain. He concluded that :

Just as the substance of the heart is endowed with the force of the vital soul, and the unique flesh of the liver with the faculty of the natural soul, in order that the liver may make the thicker blood and natural spirit and the heart may make the blood which rushes through the body with the vital spirit, and thus these organs may bring materials to all parts of the body through channels reserved for them, so . . . the brain . . . prepares the animal spirit.[18]

This is perhaps a rather lamely orthodox result of his bold proclamation of the rights of free medical inquiry, for it was a conclusion to which Galenists could readily subscribe, but at least Vesalius had the advantage of proclaiming his independence, however little use he made of it.

The work of Vesalius is so imposing—partly because it so often transcends anatomy—that the work of his contemporaries appears somewhat tame by comparison. But he was a member of a fertile and original generation, and when he left Padua the university found no difficulty in finding worthy successors, for there were many Italian anatomists, each of whom made his own contributions. One of the most interesting is Eustachio (1520–74), a practising physician in Rome who, alone among the great anatomists, was not associated with a university. His work was often more accurate than that of Vesalius, but he published little. In 1563 he produced a small book (*Opuscula Anatomica*) in which he compared organs in man and in animals, pointing out that Vesalius and many others had discussed animal kidneys, not human kidneys, and noting the difference between, for example, the venous system in the arm of a man, an ape and a dog ; he also published on the anatomy of the ear. He planned a detailed and comprehensive survey, but only the plates (published in the eighteenth century) now survive. Among the professors of anatomy at Padua was Fallopius (1523–62) ; his *Anatomical Observations* are especially good on the female reproductive system, whose organs he described in detail. Other Paduan anatomists, like Realdus Columbus * (d. 1559) and Fabricius of Aquapendente were concerned with the physiological anatomy of the venous system. Anatomy and physiology were slowly

* Cf. ch. IX, below.

separating and becoming specialised, though each remained a necessary adjunct to the other.

There was, in the sixteenth century, a divergent, though less fruitful, physiological tradition, best represented by the French physician Jean Fernel (d. 1557). His *Natural Parts of Medicine* (1542) was modelled on Galen's *On the Use of the Parts*, just as the *Fabrica* of Vesalius was modelled on Galen's *Anatomical Procedures*. But Fernel was far less original, and he underestimated the importance of anatomy. Though he devoted the first section of his book to an adequate anatomical discussion, he apparently felt that physicians were paying too much attention to anatomy, and too little to medical practice. As a contemporary reported :

Often too I have heard him declare the absurdity of going on toiling, even into old age, in turning over books of anatomy and reading the properties of simples, without ever looking at a sick man or actually seeing the things which the ancients have described in the sick. He argued that it is better, after perusing once and then once again, some skilled and well-written but brief compendium of anatomy, and going through it attentively, to pass straight to the things themselves as they can be seen and observed in number of sick persons, and not lose time, and the years life has to offer, in reconciling a multitude of authors whose statements disagree. Today, he said, books on anatomy are almost more numerous than are the sick, and there are more writers of herbals than there are herbs to describe.[19]

This slightly muddled plea for empiricism and attention to the needs of the sick by no means deterred Fernel from pursuing a thoroughly unoriginal concept of the workings of the human body. Though he did emphasise the importance of distinguishing between the functioning of the body in health and the body in disease, he was unable to do more than to present a " modern " version of Galen. In this capacity his book was to prove acceptable to the fiercely Galenic medical faculty of the University of Paris.

It was physicians who began the new study of anatomy in the sixteenth century, but it was far more useful to the surgeons. Though the surgeon was both less highly educated and lower on the social scale than the physician, the latter did try to keep the surgeon informed of the latest anatomical advances, in spite of the disputes that continually sprang up between the two kinds of practitioners. This was especially true in England, where surgeons achieved an assured position in the mid-sixteenth century. There the old battle between barbers and surgeons had ended in 1540 with the creation of a united company, entitled to supervise surgical practice. This they did by keeping a check on the candidates for apprenticeship (who were expected to have attended a grammar school long enough to acquire a little Latin), and setting examinations. They further provided for the education of present and prospective surgeons by establishing a Readership in Anatomy, for which they secured the right to the bodies of four criminals yearly. These readers were able men, who conscientiously fulfilled the duties of their office. The first, Thomas Vicary (d. 1561), published a respectable treatise on anatomy ; the second, John Caius, established the tradition of appointing university graduates. The English surgeons of the later sixteenth century therefore had advantages of opportunity, education and respectability, in strong contrast to the position in France. There the surgeons proper formed the College of Saint-Cosmas, but their more active and numerous inferiors, the barber-surgeons, were unorganised, and their training varied with the master under whom each served an apprenticeship. English and French surgeons shared the advantages of hospital appointments, where they usually really began to learn their trade, for among the poor there was much opportunity for observation and practice. From there, most surgeons went on to serve in war, and the battlefield proved an even better school than the hospital for training distinguished men.

Properly, there was a clear division between medicine and surgery, though the two professions inevitably overlapped. Surgeons were supposed to deal with external medicine, physicians with internal ; surgeons dealt with wounds, fractures, and childbirth, cut for the stone, performed amputations and (at the physician's orders) let blood.* The surgeons were supposed to appeal to the physicians when medicines were required, but they often prescribed drugs on their own account, arguing that the fever deriving from wounds lay within the province of the surgeon, or that the surgeon was entitled to administer the purges which were a necessary preliminary to operations. There was, in the sixteenth century, one disease which was nearly always left to the surgeons, partly because they had first treated it, partly because its outward manifestations were lesions of the skin, which the surgeon was clearly entitled to treat. This was " the new sickness of the armed forces," the "French disease," *lues venerea*, syphilis.† Whether brought from the New World or not (and the sixteenth century nearly always thought that it had been) this undeniably became epidemic in the armies at the siege of Naples in 1495, and spread from thence all over Europe with frightening rapidity. Every surgeon who wrote at all wrote on the new disease and its cure, as surely as he also wrote on the treatment of gunshot wounds. Remedies varied : mercury (already widely used for skin diseases, and a great expeller of

* In France, the surgeon proper applied bandages and external remedies ; he did not perform operations nor let blood. The barber-surgeon, who did all these, gradually replaced the more restricted surgeon, and all the notable French surgeons of the period were barber-surgeons.

† This last name was given the disease by Fracastoro in an allegorical poem (1530) which both accounts for its origin in mythological terms, and gives an accurate clinical picture of it.

peccant humours through salivation) and guaiacum, a wood from South America, were certainly the favourites. Controversies on which was the better raged furiously and mingled with the later controversy over chemical versus herbal remedies in medicine.

One of the earliest writers on the surgical problems of the army was Giovanni da Vigo (1460–1525), an Italian who became surgeon to Pope Julius II. Vigo has a poor reputation because in his work *On the Art of Surgery* (1514) he advocated the cauterisation of gunshot wounds, which he believed must be poisoned by the lead of the bullets. (Perhaps he was influenced by the high incidence of tetanus.) He also wrote on the ligaturing of arteries under certain conditions (a technique forgotten since antiquity); on new surgical instruments; and on syphilis, for which he advocated the ingestion of mercurial drugs. His book with its brief anatomical introduction was rapidly translated into all the major vernaculars, and was highly influential. Indeed, the chief claim to fame of Ambroise Paré (1510–90) is his denunciation of Vigo's practice of cautery, which he replaced with mild dressings. Paré, after an apprenticeship as a barber-surgeon, and a couple of years' service as house-surgeon at the Hôtel-Dieu in Paris, went on the campaign of 1536 as private surgeon to the general in command of infantry. Here he began to acquire experience and to devise new methods of treatment; after a number of campaigns (and passing the barber-surgeons' examinations in 1541) he became surgeon to a succession of French kings. Paré wrote voluminously on gunshot wounds, dislocations, amputations (where he extended the use of the ligature), obstetrics (he devised new procedures and instruments), and burns; he discussed specific case histories, and took good care to proclaim the manifest superiority of his methods over those of his contemporaries and opponents. Very similar was the slightly later career of the English surgeon William Clowes (1544–1604) whose surgical casebook (*A Proved Practice*, 1587, revised and enlarged as *A Profitable and Necessary Book of Observations*, 1596) is a delightful account of the difficulties and achievements of a successful surgeon, for Clowes wrote with zest and skill, and he had a wide practice.

Surgical casebooks make depressing enough reading, for wounds were slow to heal and excessively painful ; but at least the surgeon was, usually, dealing with an ailment which he was competent to cure. The practice of the contemporary physician was dismal in the extreme : he could, sometimes, diagnose the disease, but there was little or nothing he could do to alleviate or cure it, and most of his methods of treatment seemed rather to aggravate than mitigate discomfort. Yet every physician boasted spectacular cures, and certainly physicians kept their patients' confidence. Purges, bloodletting and a variety of complex and nauseous drugs were the inevitable prescription, whether the nature of the disease was known or not. Almost the only positive treatment (from a modern point of view) was the isolation of the contagiously sick, the quarantine developed in the later Middle Ages and applied to leprosy and plague.

One of the few treatises of this period dealing with a purely medical (as distinct from physiological, anatomical, surgical or pharmaceutical) problem is the work of the humanist physician and astronomer Fracastoro, *Contagion, Contagious Diseases and their Treatment* (1546). Fracastoro was familiar with numerous contagious diseases, old and new, from consumption to typhus ; he tried to classify them by degree and method of contagion.

There are, it seems, three fundamentally different types of contagion. The first infects by direct contact only. The second does the same, but in addition leaves *fomes*, and this contagion may spread by means of that *fomes*, for instance scabies, phthisis, bald spots, elephantiasis and the like, (by *fomes* I mean clothes, wooden objects, and things of that sort, which though not themselves corrupted can, nevertheless, preserve the original germs of the contagion and infect by means of these). Thirdly, there is a kind of contagion which not only is transmitted by direct contact or by *fomes* as an intermediary, but also infects at a distance ; for example, pestilent fevers, phthisis, certain kinds of ophthalmia, exanthemata of the kind called variolae [typhus], and the like. These different contagions seem to obey a certain law ; for those which carry contagion to a distant object infect both by direct contact and by *fomes* ; those that are contagious by means of *fomes* are equally so by direct contact ; not all of them are contagious at a distance, but all are contagious by direct contact.[20]

Fracastoro believed in " seeds of contagion," imperceptible particles somehow mechanically transmitted from the sick to the healthy. This essentially atomic theory of disease seemed to conform to experience when carefully examined ; but it was not really useful, and Fracastoro's investigation is an example of pure medical research similar to the case histories of Hippocrates. However meticulous his examination of the problem and however clear his vision, he could not apply his theory in any constructive way except to confirm the advantages of isolation. Perhaps this is why, though he was a physician, he fled from Verona during the plague of 1510 to live in the safety of his country estate. Knowledge can make cowards.

Another development in the recognition and classification of disease was consideration of occupational and regional diseases. Paracelsus (*c.* 1493–1541) wrote a book on miners' diseases (1533–4) in which he commented on mercurial and arsenical poisoning, though he somewhat diminished the value of his observations by regarding all diseases associated with metals as " mercurial " in origin. Some years later Agricola (1490–1555), physician in a mining town, included in his great work on metals (*De Re Metallica*, 1556) a short section on the diseases peculiar to miners. Similarly, travellers were peculiarly subject to certain diseases like scurvy, and had as well the risk of meeting entirely new diseases, especially in the tropics. In 1598 there appeared a work on tropical medicine entitled *The Cures of the Diseased In Remote Regions : Preventing Mortalitie, incident in Forraine Attempts, of the English Nation* (by George Wateson) ; Hakluyt thought it ought to be included in his *Voyages*, but he was dissuaded by William Gilbert to whom he submitted it for approval : Gilbert denounced the views of Wateson, and promised instead to provide Hakluyt with a better and more complete work on tropical and arctic medicine. He failed to do so ; and this is the only glimpse we have of Gilbert's medical interests.

Although from the modern point of view there was little the sixteenth-century physician could do to alleviate suffering and cure disease, he naturally was unaware of this fact and had full faith in his chosen remedies. Most of these were traditional, many even Greek in origin, but there were new and exciting developments in drugs in the late fifteenth and early sixteenth centuries. These resulted from the markedly increased use of chemical remedies in spite of the criticisms of herbalists who adhered to tradition. Controversy about the advantages of chemical drugs usually centred around the name of Paracelsus, but the use of non-herbal remedies antedates that strange mystic figure, whose ideas are so much more magical than medical ; chemical drugs had always been used to a certain extent, especially externally. Chaucer's physician knew that " gold in physick is a cordial " ; and his " cordial waters," improved by the addition of flecks of gold, are still in existence, though rather as an after-dinner liqueur than as a specific against the plague.

Cordials or strong waters—distilled liquors—were introduced into Europe in the course of the fourteenth century, originally as

...icines. The German physician (Michael Puff von Schrick) who is the author of the first printed book on distillation (1478) wrote hopefully, " Anyone who drinks half a spoonful of brandy every morning will never be ill," and assured his readers that brandy would revive even the dying.[21] Cordials were widely used in time of epidemics. They were distilled from a variety of substances, ranging from wine and fermented grains to fruits and herbs, and a cult of "elixirs" and "essences" developed in the fifteenth century. This is well displayed in the books of the fifteenth-century surgeon Hieronymus Brunschwygk (c. 1450–1512), *The Small Book of Distillation* (*Liber de Arte Distillandi de Simplicibus*, 1500) and *The Big Book of Distillation* (*Liber de Arte Distillandi de Compositis*, 1512) being among the first of a long series of distillation books.* Brunschwygk describes how to achieve the "essence" of a substance by macerating it with water or spirit of wine (alcohol) and then distilling ; and also how to construct a still. As his books were soon translated into various vernaculars, the art of treating herbal remedies chemically rapidly spread throughout Europe ; it was already partially established by 1500, at least in Strasbourg, for Brunschwygk described the practice of the surgeons and physicians he knew.

Although the official pharmacopoeia issued at Augsburg in 1564 lists chemical remedies only for external use, in fact many substances totally unrelated to herbs were prescribed at this time. As early as 1514 Giovanni da Vigo had advocated giving mercury internally in the treatment of syphilis, not merely externally as a salve, claiming infallible success. There was good reason not to follow Vigo's practice : mercury vapour was, of course, known to be poisonous, and there was a possibility that liquid mercury was too. In fact it is not, but taken internally in any considerable

---

* An interesting example is a work by Gesner, published pseudonymously in Latin in 1552, and in English in several different translations and under various titles, including the picturesque one of *The Newe Jewell of Health* (1576).

amount it produces uncomfortable symptoms, including excessive salivation. And doses were massive ; it was said that some surgeons prescribed so much mercury that their patients' bones were found to be filled with mercury instead of marrow! It was also expensive. No wonder that many physicians and their patients preferred to trust guaiacum, the Holy Wood of the New World, arguing that the remedy for a disease of the New World must be found there. There were many attested cures ; as guaiacum is totally without medical value these cures were either imaginary, or else the initial diagnosis of syphilis was wrong ; but there were many failures as well. It was fear of the misuse of mercury that caused the Augsburg Senate to warn apothecaries in 1582 not to prepare " substances which are known to be detrimental or poisonous, such as . . . Turpethum minerale and other purging mercurials." [22]

Paracelsus leapt gleefully into the mercury-guaiacum debate on the side of mercury, partly because it was the side of novelty against tradition, partly because he believed in a homoeopathic principle that like must combat like, and violent illnesses demanded violent treatments. He enthusiastically adopted also the use of antimony, a newer and even more violent drug than mercury. Antimony is an emetic ; various antimonial compounds were used (though tartar emetic is a slightly later preparation), but the simplest and most common method was to fill a cup made of antimony with wine, allow it to stand overnight, and drink the contents on rising. (No more eloquent testimony to the power of the drug is needed.) Along with pure antimony * and mercury went their compounds, usually prepared by the action of mineral acids (newly discovered in the fifteenth century); these rejoiced in such names as *Mercurius vitae* (an antimonial preparation which in fact contains no mercury), balm and regulus

---

* Antimony in this period always meant the sulphide ore ; the metal was called *regulus* of antimony.

of antimony, and spirit of quicksilver. Mineral acids, especially oil of vitriol (sulphuric acid), were also enthusiastically adopted in medical use by Paracelsus ; since his patients appear to have survived, perhaps he prescribed these in the form of metallic salts. The most famous of Paracelsan remedies was his Laudanum, praised by himself and his followers as an almost miraculously curative substance ; this, rather oddly, was a mild herbal mixture, similar to drugs known to Galen, and containing no opium ; it was later Paracelsans who prepared their laudanum from opium.

Perhaps the violent nature of the new chemical remedies appealed to an age in which new and extremely violent epidemic diseases—syphilis, typhus—appeared to match the violence of war and religious controversy. Perhaps their novelty had an attraction for an age prepared for newness ; perhaps it was merely desperation in the face of recurring disease. Certainly their use spread rapidly, only partly under the aegis of the mystic medical theory of Paracelsus. In spite of all official restrictions chemical remedies were increasingly in demand by both patients and physicians. The so-called iatrochemists (medical chemists) and chemical physicians were usually regarded as followers of Paracelsus, though most of them were lukewarm partisans, accepting the idea of chemical medicine but rejecting the extremes of medical mysticism demanded by Paracelsus. One of the most fervent Paracelsans was Joseph Du Chesne, generally known as Quercetanus ; writing a defence of his own doctrines in 1575 he said cautiously :

> As touching *Paracelsus* I have not taken upon mee the defence of his divinitie, neither did I ever thinke to agree with him in all points, as though I were sworn to his doctrine : but . . . I dare be bold to say and defend, that he teacheth many things, almost divinely, in Phisicke, which the thankful posteritie can never commend and praise sufficientlie.[23]

What Quercetanus most commended in Paracelsus was his use

of metals and salts as drugs ; this he made plain in his polemic *On the Truth of the Hermetic Medicine against Hippocrates and the Ancients* (1603), a panegyric on chemical remedies ; the English translator thoughtfully added recipes for the use of converts. Another Paracelsan, Oswald Croll (1580–1609), professor of medicine at the University of Marburg, in his *Royal Chemistry* (*Basilica Chymica*, 1608) devoted his energies to propaganda for the chemical approach to medicine, declaring " Without this chemical philosophy, all Physick is but lifeless." [24] His work achieved extensive popularity, especially after it began to appear with practical advice on the use of chemical drugs, added by another Marburg physician, Johann Hartman.

One of the oddest works on this subject was *The Triumphal Chariot of Antimony* of Basil Valentine, supposedly a fifteenth-century Benedictine monk, most probably in fact the German "editor," Johann Thölde ; this was published in German in 1604 and subsequently in Latin, French and English. Like its title, the book is presented in a misleadingly antiquated and alchemical style ; in fact it is a thorough investigation of the chemical and medical properties of antimony, and an ardent defence of its use in medicine. The author gives recipes for the preparation and administration of a wide variety of antimonial drugs ; he admits that chemical medicines are nearly all deadly poisons, but he argues that chemists prepare them so that their poisonous nature is removed, and that, besides, their derivation from poisons permits them to act as antidotes to all poisonous ills, by their very nature.* The *Triumphal Chariot of Antimony* had a profound influence on the acceptance of antimonial compounds in medicine, and on the chemical study of metals.

---

* This continued to be a fruitful defence in the seventeenth century. About 1650 the young Robert Boyle wrote " An Essay of Turning Poisons into Medicines," instancing this phenomenon as an example of the infinite goodness and diversity of the ways of God.

At the end of the century, as at the beginning, there was a new medicine and an old ; the new medicine was not now distinguished by its appreciation of anatomy, but of chemistry. The old-fashioned physician defended " Galenicals " (herbal preparations) ; the new physician championed "Spagyric" (chemical) drugs, and took a generally anti-Galenist position. Some moderates tried to combine the two points of view ; the German Daniel Sennert (1572–1637) tried to show that every conceivable scientific position was reconcilable in medicine, as the title of his most famous book—*On the Agreement and Disagreement of the Chemists with Aristotle and Galen* (*De Chymicorum cum Aristotelicis et Galenicis Consensu ac Dissensu*, 1619)—indicates. But the controversy continued to rage, and conservative forces only slowly gave ground. The French king might issue edicts against antimony, but its use was so extensive in Paris in the early seventeenth century that the edict was soon rescinded. The English College of Physicians was among the more conservative bodies in this respect, and as late as 1665 a group of London physicians set up a " Society of Chymical Physicians " in opposition to the official body, contending that only violent chemical medicine was adequate to deal with the Great Plague ; the fact that no distinguished names appear in the new society indicates that it was perhaps as much recognition of the mischief played by violent remedies as old-fashioned adherence to tradition that produced reluctance to pin all dependence to chemical drugs.

Chemical medicine had a doubtful influence on medicine : it had a wholly benevolent effect upon the practice of chemistry. As the demand for the new drugs grew, there was a corresponding demand for recipes. Apothecaries needed instruction, and a new kind of chemist appeared to satisfy this demand. The chemical teacher was also the first writer of chemical textbooks. One of the earliest to establish a course of lectures for apothecaries was the Frenchman Jean Beguin (*c.* 1550–1620), who in 1610 published a book entitled *Chemistry for Beginners* (*Tyrocinium Chymicum*) for the use of his pupils ; though Beguin wrote in Latin, the work was soon available in vernacular versions, and continued popular for nearly the whole of the century. Beguin's aim was to teach " the Art of dissolving natural mixed Bodies, and of coagulating the same when dissolved, and of reducing them into salubrious, safe, and grateful Medicaments " ; [25] to this end he wrote a work of instruction, simple, precise and detailed. There was no room for alchemical secretiveness in the works of the new chemical teachers. The new pharmacy soon received official sanction ; in France, a royal decree ordered the establishment of a chemical chair at the Jardin du Roi, the official botanic gardens, in 1626, and a number of useful textbooks appeared as a result. Chemistry was already a university subject in some German medical faculties ; it soon became a commonplace adjunct. Whatever reservations conservative physicians might have, chemical medicine was accepted and chemical drugs were indispensable. This was a revolution in medical practice ; but it was a rational one, rather than the mystic revolution proclaimed by Paracelsus. Iatrochemistry triumphed, but Paracelsan chemico-medicine was even less acceptable than it had ever been. Medical chemistry joined metallurgy and pyrotechnics as a practical craft.

CHAPTER VI

RAVISHED BY MAGIC

The sciences themselves which have had better intelligence and confederacy with the imagination of man than with his reason, are three in number : Astrology, Natural Magic and Alchemy ; of which sciences nevertheless the ends or pretences are noble.[1]

To the layman, the scientist has always seemed something of a magician, seeing further into the mysteries of nature than other men, and by means to be understood only by initiates. The line separating Copernicus or Vesalius, ordering the stars and planets in their courses and penetrating to the innermost secrets of construction of the human body, from Faust, selling his soul to the devil for the knowledge that is power, was narrow indeed to the popular mind of the sixteenth century. Physician, alchemist, professor all then wore the same long robe, which might mark either the scholar or the magician. And when so much of what was new in science was concerned with the very frontiers of knowledge, and dealt with almost unimaginable problems of the organisation, complexity and harmony of Nature, scientists themselves were puzzled to know certainly where natural philosophy stopped and mystic science began. When problems failed to yield to traditional methods they were tempted to cry with Faust,

> Philosophy is odious and obscure ;
> Both Law and Physick are for petty wits ;

> Divinity is basest of the three,
> Unpleasant, harsh, contemptible and vile :
> 'Tis magic, magic that hath ravished me.[2]

The difficulty was not that there was no difference between natural philosophy and mystic science ; but rather that men saw that each rational science had its magical, occult or supernatural counterpart. Applied astronomy might be either navigation or astrology ; applied chemistry either metallurgy or the search for the philosophers' stone. Yet at the same time even the most ardent practitioners of the mystic branches knew that their form of science was not as intellectually or morally reputable as the more normal forms. Astrologers were better paid than instrument makers, but even non-scientists in the sixteenth century knew in their hearts that astrologers, like magicians, gazed on things forbidden, though they also knew that not many of them were really in league with the devil.

This aspect of the mystic sciences even lent them a certain glamour, because forbidden learning was almost certain to be more exciting and more important than mere licit knowledge. Certainly astrology flourished outrageously in the sixteenth century, along with alchemy, natural magic (the occult form of the not yet invented experimental physics), even spiritualism. As later in the nineteenth century, the natural philosopher was attracted to this last, the most occult of all the magical arts ; yet John Dee, hiring a medium to gaze at a crystal ball, became thereby no less an enlightened applied mathematician than Sir Oliver Lodge, three centuries later, nullified his contributions to physics by his addiction to *séances* and table rappings of a not very different sort. In each case it was only essential for the scientist to be aware of the difference between his two fields of interest ; for in each case he accepted mysticism as the road only to that knowledge inaccessible by ordinary scientific means of investigation. The area to which magic could be and was

applied in the sixteenth century was still very great ; it is fascinating to observe the way in which, out of the muddled mysticism
of sixteenth-century thought and practice, the scientifically valid
problems were gradually sifted out to leave only the dry chaff of
superstition.

The attack on astrology by fifteenth-century humanists like
Pico della Mirandola had appealed particularly to literary men,
already developing that cool and rational scepticism so characteristic of Montaigne ; it had far less influence on astronomers.
Yet though there was no agonising re-appraisal in the early
sixteenth century, astrology was already noticeably on the
defensive, and was to remain so throughout the century. Society,
while it continued to pay the astrologer better than the astronomer,
increasingly expressed its disapproval of the "mathematician"
who dabbled in the occult wisdom of the stars. And "judicial
astrology " (the casting of personal horoscopes, as distinct from
general prognostication) was legally forbidden, though commonly practised. It was a most tempting pursuit, especially in
its medical aspects, for what physician and what patient would
not want to try all means to determine an accurate prognosis ?

Indeed, many were introduced to astrology by medicine.
Jean Fernel, educated as a physician, neglected to practise his
profession for some years because he found astrology so much
more interesting ; as his sixteenth-century biographer explained,

> contemplation of the stars and heavenly bodies excites such
> wonder and charm in the human mind that, once fascinated
> by it, we are caught in the toils of an enduring and delighted
> slavery, which holds us in bondage and serfdom.[3]

Fernel spent his own and his wife's fortune on the construction
of "mathematical" (i.e. astrological) instruments, and only de-
sisted under the stern admonitions of his father-in-law and the
necessity of earning a living. This was about 1537, when the
Paris medical faculty was enforcing the edict against the practice

of astrology quite strictly. Indeed, only a year later the Parlement of Paris, at the request of the faculty, publicly condemned Michael Servetus (1509–53), then a pupil of Guinther von Andernach, later more famous as a radical theologian, for giving public lectures on judicial astrology ; legal proceedings being slow, he managed to forestall more severe punishment by hastily publishing *A Discourse in Favour of Astrology*, a denunciation of Pico and others and a defence of astrology on the grounds that it had been accepted by both Plato and Galen. The Italian physician and mathematician Cardan (1501–76) was, about the same time, an ardent practitioner of astrology, who cast horoscopes of himself, his patients, and even, it was said, of Jesus Christ ; his patients might approve, but his colleagues regarded this as so highly dubious a practice as almost to amount to malpractice.

Professional astronomers tended to take astrology for granted ; there is no doubt that most of them agreed with the common people that it was a legitimate application of astronomical knowledge, as well as a useful way of gaining a livelihood. They tended to avoid judicial astrology in the ordinary way, preferring general prediction and calculation, though they all upon occasion cast horoscopes for men of importance whose personal fortunes might affect the well-being of nations. So Regiomontanus produced his *Ephemerides*, the astrological almanacs which allowed others to calculate horoscopes and to be aware of such significant events as eclipses and planetary conjunctions ; a common form of publication in the sixteenth century. There were general, non-technical works, like the *Prognostication Everlasting*, of the English practical mathematician Leonard Digges, from which a literate man could derive useful predictions. Every eclipse and every comet produced its spate of ephemeral literature, asserting that these heavenly events foretold famine, war and pestilence for mankind ; this omnipresent trio never failed to oblige. (The astronomer Kepler, in his first formal attempt at astrology, predicted famine, a peasant uprising and war with the Turks for the year 1595, three events which duly occurred.) Even Galileo cast horoscopes for his patron, the Grand Duke of Tuscany, coolly rational as he might be on other occasions.

The greatest observational astronomer of the sixteenth century, Tycho Brahe, had begun his astronomical career from astrological interest, and he never lost either his preoccupation with astrology or his conviction that it was true applied astronomy. The nova of 1572 provided him with a splendid opportunity for prediction, of which he took full advantage whenever he wrote about its scientific significance. As, so he thought, this was only the second time in the history of the world that a nova had appeared, he was not hampered by precedent, and was able to offer, for once, a cheerful prognostication. The nova, he concluded after some years' consideration, predicted a New Age—because it followed the conjunction of Saturn and Jupiter by nine years and was reinforced five years later by a comet—and foretold the future for the whole world—because it lay in the eighth sphere. The New Age would be one of peace and plenty ; it would begin in Russia in 1632, sixty years after the nova's appearance, and thence would spread all over the world. An enterprising London printer published an English version in 1632 (entitled *Learned Tico Brahe His Astronomicall Coniectur of the New and Much Admired* * *Which Appeared in the Year 1572*) quite undeterred by the fact that Tycho's original astronomical prediction was patently inexact. But as Tycho had said, " These Prognostic matters are grounded only upon conjectural probabilitie," [4] for the matter was difficult ; or, as he was to write later, " it will hardly be possible to find in this field a perfectly accurate theory that can come up to mathematical and astronomical truth." [5] Yet it was always worth trying, for " we ought not to imagine that God and Nature doth vainly mock us, with such new formed bodies, which do presage nothing to the world." [6]   Indeed, after his study of planetary orbits, he

took Astrology up again from time to time, and . . . arrived at the conclusion that this science, although it is considered idle and meaningless not only by laymen but also by most scholars, among whom are even several astronomers, is really more reliable than one would think ; and this is true not only with regard to meteorological influences and predictions of the weather, but also concerning the predictions by nativities, provided that the times are determined correctly, and that the courses of the stars and their entrances into definite sections of the sky are utilized in accordance with the actual sky, and that their directions of motion and revolutions are correctly worked out.

But, though he was sure his astrological method was correct, he would not make it public. "For it is not given to everybody to know how to use it on his own, without superstition or excessive confidence, which it is not wise to show towards created things." [7]

This tendency to keep esoteric knowledge secret, because only the initiated can be trusted with it, is the chief reason why the magical sciences are so obscure ; their practitioners wrote long, passionate apologies and defences, but seldom revealed their methods. Astrology was, for a mystical science, singularly open in its methods ; alchemy by contrast was the most secretive. Tycho, who practised alchemy along with astrology, defended himself again on the grounds that not everyone was to be trusted with such powerful knowledge. As he explained :

I also made with much care alchemical investigations or chemical experiments . . . the substances treated are somewhat analogous to the celestial bodies and their influences, for which reason I usually call this science terrestrial Astronomy.   I have been occupied by this subject as much as by celestial studies from my twenty-third year, trying to gain knowledge . . . and up to now I have with much labour and at great expense made a great many discoveries with regard to metals and minerals as well as precious stones and plants and similar substances. I shall be willing to discuss these questions frankly with princes and noblemen and other distinguished and learned people, who are interested in the subject and know something about it, and I shall occasionally give them information, so long as I feel sure of their good intentions and their secrecy. For it serves no useful purpose, and is unreasonable, to make such things generally known, since although many people pretend to understand them, it is not given to everybody to treat these mysteries properly according to the demands of nature and in an honest and beneficial way.[8]

Most alchemists shared Tycho's arrogant certainty that only the alchemist could judge whom to initiate into the " subtle science of holy alchemy," and that only initiates could be trusted. Certainly they were successful in keeping their meaning secret, and few who are not alchemists can pretend to understand what they wrote on the subject.

By the fifteenth century alchemy, a relative late-comer to Europe, was firmly established, and laymen had learned that most alchemists were cheats. Yet there is no doubt that many alchemists shared Tycho's conviction that the toilsome life of the alchemical laboratory could lead to something higher than the making of gold. An instructive example is the English *Ordinall of Alchimy* (1477) of Thomas Norton of Bristol, a longish work in rather doggerel verses of great if naïve charm. Norton had the highest aims, and his alchemy is thoroughly Christianised : only the upright and pure can succeed in the work and only a devout apprentice can hope to learn it from his equally devout master.

Masterfully marvellous and Archemastry
Is the tincture of holy Alchemy
A wonderful Science, secret Philosophy,
A singular grace and gift of th'almighty :
Which never was found by labour of Man,
But it by Teaching or Revelation began.
It was never for Money sold nor bought,
By any Man which for it hath sought :
But given to an able Man by grace.[9]

Norton described his own apprenticeship, and catalogued some of the technical processes, like the degrees of fire and the hierarchy of colour ; though he professed to be writing clearly and simply so that all could understand, and though his mysticism was comparatively simple, one cannot learn much about alchemical processes from him. Most alchemical treatises are similar : fairly clear on processes like calcination, distillation, sublimation, digestion, rather vague on recipes, shot through with an ancient and complex symbolism that readily lent itself to illustration in printed books, but did not so readily lend itself to comprehension.

By 1500 alchemy was becoming even more mystic in the face of competition from the technical chemical processes which provided so many instructive and detailed books in the sixteenth century, delightfully illustrated and written in clear, simple, layman's language. These covered a multitude of subjects and varied from Brunschwygk's books on distillation (1500 and 1512) ; and the little German *Bergbüchlein* and *Probierbüchlein*, Agricola's great Latin treatise *De Re Metallica* (1556), all on mining and assaying ; Biringuccio's Italian treatise on metals and minerals (*Pirotechnia*, 1540) ; to Neri's *Art of Glass* (1612). All were more informative and instructive than any alchemical work, and made the alchemist seem by comparison excessively secretive, obscure and occult. These authors despised alchemy ; Biringuccio remarked scornfully :

The more I look into this art of theirs, so highly praised and so greatly desired by men, the more it seems a vain wish and fanciful dream that it is impossible to realize unless someone should find some angelic spirit as patron or should operate through its own divinity. Granted the obscurity of its beginnings and the infinite processes and concordances that it needs in order to reach its destined maturity, I do not understand how anyone can reasonably believe that such artists can ever do what they say and promise.[10]

Biringuccio distinguished two kinds of alchemy. One " takes its enlightenment from the words of wise philosophers . . . This they call the just, holy and good way, and they say that in this they are but imitators and assistants to Nature." This is at once the better branch, and the more fascinating ; " it is indeed so ingenious a thing and one so delightful to students of natural things that they cannot forego the expenditure of all possible time, labour, or expense," because it continually produces something new, especially such useful results as medicines, colours for painting, and perfumes. Even though somewhat uncertain and suspect, this was a science to be tolerated. But there was another form of alchemy, a " sister or illegitimate daughter " to enlightened alchemy which was so evil that " usually only criminals or practisers of fraud exercise it. It is an art founded only on appearance and show . . . it has the power of deceiving the judgement as well as the eye . . . it contains only vice, fraud, loss, fear and shameful infamy." [11] This corrupting form was the tricky and cheating alchemy the rational man laughed at, and it was practised by the alchemist of derisory literature from Chaucer's *Canon's Yeoman's Tale* of the late fourteenth century to Ben Jonson's *Alchemist* of the early seventeenth. Few ever believed the claims of the alchemist to have made real gold, though the alchemist maintained his claims undeterred.

The late sixteenth century saw a number of notorious claims to transmutation ; in a sense these men truly succeeded in making gold, for they all acquired vast sums from credulous patrons, at least for a while. (Whether they believed in their own claims it is often difficult to tell.) Edward Kelly, whom Jonson mentions by name, was apparently as much of a fraud in alchemy as he was in the crystal-gazing which led John Dee to follow him half over Europe ; Kelly's imposture proved too obvious even for the credulous Emperor Rudolf, and Kelly ended his life trying to escape from the prisons of Prague. A more ambiguous figure is the Scotsman Alexander Seton (d. 1604), known as the Cosmopolite, who was reported to have performed prodigies of transmutation on his way through Holland in 1602. After wandering through Germany, always receptive to mystic science and where he had great success, transmuting even iron by means of a mysterious red (or perhaps yellow) powder, he finally arrived at the court of the Elector of Saxony, where his success was also remarkable. Too remarkable, because the Elector decided he needed to know what the powder was ; when Seton refused to divulge his secret, he was tortured and imprisoned, and only rescued by a sympathetic and clever fellow practitioner, Michael Sendivogius. Seton died soon after his rescue, bequeathing his remaining supply of powder to Sendivogius, but not the secret of its composition. Both Seton and Sendivogius were convinced of the realities of transmutation ; yet at the same time, and while writing with an excessive amount of symbolism, they did have some idea of chemical theory. Indeed Seton was one of the first to discuss the theory that there are nitre or nitrous particles in the air which play an essential role in combustion. The ideas of Seton and Sendivogius (it is impossible to separate them) were set out at large in a very popular work, *New Light on Chemistry* (*Novum Lumen Chymicum*, 1604), which purported to be by Sendivogius, but is probably really by Seton. The book is a fair enough example of the new theoretical alchemy that developed in the early seventeenth century; it has far more chemical content than the older works, but was still deliberately obscure and semi-mystic.

The most famous—and most baffling—figure of sixteenth-century alchemy, as well as the best known, is Paracelsus (1493–1541). Paracelsus had a great fascination for his contemporaries, and has often charmed those who have made his acquaintance since: there are many who, painstakingly searching through his enormous collection of writings, claim to have found much of philosophical and scientific import; he has been called the first systematiser in chemistry, a great naturalist and a great mystic philosopher.* Others have been violently repelled, and found his writings obscure, unduly superstitious, and scientifically worthless. Everything about him is so complex and difficult that almost any interpretation is justified. Even his name is surrounded with obscurity. By the end of his life he was known as Philippus Aureolus Theophrastus Bombastus von Hohenheim Paracelsus, though at the beginning Bombast von Hohenheim only appears. His parentage was dubious; his father, William von Hohenheim, was apparently an illegitimate son of the German noble whose name he bore, and his mother was (probably) of peasant origin. Though he grew up in a small Swiss mining village, and nearly always wrote in the local dialect, with many Latin words mingled where required, he seems to have been well educated in the traditional subjects, though whether he ever formally studied medicine is uncertain. His whole life is obscurely full of wandering; whenever he settled down his violent actions and prejudices soon aroused equally violent opposition—as when he burned the books of Galen and Avicenna in a students' bonfire at Basle on St. John's Eve, 1527—and

* Most translations of his works have been made by those holding this view; this has tended to make them clearer than Paracelsus left them.

he was forced to move on. Like the philosopher Giordano Bruno, he was equally adept at attracting disciples and repelling colleagues.

Paracelsus wrote enormously, and not always consistently, on all sorts of subjects; for he shared to the full the Renaissance passion for novelty and universality. He combined iconoclasm with appeal to 'experience,' primarily mystic experience. He attacked reason because it was opposed to magic, and magic was to him the best key to experience:

Magic has power to experience and fathom things which are inaccessible to human reason. For magic is a great secret wisdom, just as reason is a great public folly. Therefore it would be desirable and good for doctors of theology to know something about it, and to understand what it actually is, and cease unjustly and unfoundedly to call it witch-craft.[12]

His attitude to alchemy is complex: he certainly believed that the alchemist should be more concerned with the preparation of drugs than with transmutation—partly because he was aware of the alchemist's dubious reputation—but at the same time he insisted that alchemy could best be described as "an adequate explanation of the properties of all the four elements—that is to say of the whole cosmos—and an introduction into the art of their transformation." [13]

Both alchemy and medicine were, he thought, controlled by an *archeus* and a *vulcanus* and both were associated with the arcana. *Vulcanus* and *archeus* together effect all chemical and medical operations; they work together, but separately, and the *archeus* is an inner *vulcanus*. Thus in the *Labyrinth of Errant Physicians* he wrote:

This is the way that nature proceeds with us in God's creatures, and as follows from what I have said before, nothing is fully made, that is, nothing is made in the form of ultimate matter. Instead all things are made as prime matter and subsequently the *vulcanus* goes over it and makes it into ultimate matter through the art of alchemy. The *archeus*, the inner *vulcanus*, proceeds in the same way, for he knows how to circulate and prepare according to the pieces and the distribution, as the art itself [alchemy] does with sublimation, distillation, reverberation, etc. For all these arts are in men just as they are in the outer alchemy, which is the figure of them. Thus the *vulcanus* and the *archeus* separate each other. That is alchemy, which brings to its end that which has not come to its end, which extracts lead from its ore and works it up to lead, that is the task of alchemy. Thus there are alchemists of the metals, and likewise alchemists who treat minerals, who make antimony from antimony, sulphur from sulphur, vitriol from vitriol, and salt into salt. Learn thus to recognize what alchemy is, that it alone is that which prepares the impure through fire and makes it pure.[14]

This might conceivably mean that Paracelsus thought of alchemy as practical metallurgy, but it is more likely, in view of his other writings, that he never clearly understood the nature of technical processes ; he certainly did not understand what happens when a metal was chemically extracted from its ore.

His most important purely chemical work is the *Archidoxis* written about 1525 and published posthumously in 1569) ; even here the microcosm, man, was always discussed in connection with the mysteries of the macrocosm. Paracelsus was much influenced by the work on distillation of the late fifteenth century, combining its technique with the alchemical theory of elements. The result was conviction that to attain the quintessence one must separate the elements by fire, " for all things must go through fire in order to attain to a new birth, in which they are useful to man."[15] Confusingly, Paracelsus spoke both of the

Aristotelian elements—earth, air, fire and water—and the three chemical principles—salt (solidity), sulphur (inflammability, malleability, yellowness) and mercury (fluidity, density and metallic nature). Metals he took to have a special importance, an importance enhanced by the ease with which, so he said, the elements could be separated from them. The processes he described, some comprehensible and some not, are usually those of metallurgical preparation and alchemical endeavour to make the perfect metal, combined in a peculiarly obscure fashion and designed for obscure ends. Paracelsus seems to have been genuinely, if not necessarily intimately, familiar with the processes he described, but not in the least original in devising new ones ; there is not one single chemical discovery ascribable to him.

His real achievement was to excite disciples to turn to a modification of traditional alchemy, replacing mediaeval symbolism with Renaissance symbolism, and traditional alchemical interest in metals with a new medical interest. Paracelsan alchemists laid stress on alchemy as the key to medicine, either through the preparation of drugs or the chemical interpretation of physiology (the chemistry of the *archeus* being already vitalistic, this was the easier to do). There was still a great interest in metallic reactions and processes in an attempt to separate out the elements : this involved the conversion of metals (and metallic minerals like antimony, always in the sixteenth and seventeenth centuries the sulphide ore, not the metal) from their natural forms by calcination and the action of fire into " passive " forms in which they might more easily be acted upon by the *archeus*.

This kind of alchemy was more talked about than practised by Paracelsus himself ; the real Paracelsans in this case are his later followers. One of the most interesting and influential works of this sort is the *Triumphal Chariot of Antimony* of "Basil Valentine" ; this is often obscurely worded and its aim is certainly difficult to understand, but it contains a wealth of information about antimony and its compounds, and many new processes. Though it displays much alchemical symbolism of a fairly obvious sort, there is none of the rhodomontade mysticism of Paracelsus, and much of the book is recognisably about chemistry. Even the consideration of elements reads more clearly than Paracelsus could have phrased it :

Mercury is both inwardly and outwardly pure fire ; therefore no fire can destroy it, no fire can change its essence. It flees from the fire, and resolves itself spiritually into an incombustible oil ; but when it is once fixed no cunning of man can volatilize it again.[16]

The generation of metals is here considered a natural process :

The first principle is a mere vapour extracted from the elementary earth through the heavenly planets, and, as it were, divided by the sidereal distillation of the Macrocosmus. This sidereal infusion, descending from on high into those things which are below, with the aero-sulphureous property, so acts and works as to engraft on them in a spiritual and invisible manner a certain strength and virtue. This vapour afterwards resolves itself in the earth into a kind of water, and out of this mineral water all metals are generated and perfected.[17]

In spite of its cosmic beginning, this is essentially closer in spirit to the theories of those who wrote about mining and smelting metals than to those who wrote about transmuting them. Most of the *Triumphal Chariot of Antimony* is devoted to describing the preparation of innumerable antimony compounds—glass, liver, regulus, flowers, arcanum, mercury, and best of all, the star—each readily identifiable. Rather disappointingly, the arcanum turns out to be merely sal ammoniac (ammonium chloride) distilled with spirit of vitriol (sulphuric acid) and spirit of wine (alcohol), the original antimony having been thrown away in the first " operation"; the final result was (probably) ether. (This is the first record of its preparation.) The "star" of antimony was made by smelting the regulus of antimony (the metal) and allowing it to cool slowly, especially in the presence of iron, to form a characteristic crystalline structure ; this was a phenomenon which, it was felt, must indicate that the alchemist was well set on the road to discovery, and many methods of preparing it were discussed all through the seventeenth century. As the author of the *Triumphal Chariot* expressed it, " Antimony is a mineral made of a terrestrial vapour changed into water, which sidereal change is the true Star of Antimony." [18] No doubt he knew. All such operations upon metals as here described involved prodigious quantities of strong mineral acids, and for the first time these became essential reagents in chemical practice.

The introduction of alchemy into medicine was not so strange as it might appear, for medicine also was, in some aspects, an occult science. Certainly astrological prediction played an important role in prognosis ; just as true magic played a role in therapeutics. Fernel, in his physiological treatise *On the Hidden Causes of Things*, found it well to explain that

No magic can create the real thing as it is ; it can only produce a semblance or ghost of a thing, which deceives the mind as does a conjurer's trick. Hence magic does not cure ; it is never sure or safe ; it is always capricious and perilous. I have seen a jaundice of the whole body removed in a single night by a scrap of writing on a paper hung round the neck. But the malady soon returns, and may be worse than before. The cure is plainly fictitious and merely makeshift. [19]

In spite of his (moderate) scepticism of the methods of his fellow practitioners, it never occurred to Fernel that many of the ingredients of the drugs he prescribed—powdered skull or filings of loadstone—were equally magical, as were many of the more noxious drugs. Paracelsus was not being any more mystic than his contemporaries when he explained that the emerald

is a green transparent stone. It does good to the eyes and memory. It defends chastity ; and if this be violated by him who carries it, the stone itself does not remain perfect.[20]

Most sixteenth-century physicians would have agreed. But Paracelsus' acceptance of the doctrine of signatures—the theory that each and every natural object has stamped upon it some sign of its utility to the relief of man's ills—was far more wholehearted than was the case with most physicians, and more literally insisted upon. Thus he wrote :

Behold the Satyrion root, is it not formed like the male privy parts ? Accordingly magic discovered it and revealed that it can restore a man's virility and passion. And then we have the thistle : do not its leaves prick like needles ? Thanks to this sign, the art of magic discovered that there is no better herb against internal pricking. The *Siegwurz* root is wrapped in an envelope like armour, and this is a magic sign showing that like armour it gives protection against weapons. And the *Syderica* bears the image and form of a snake on each of its leaves, and thus, according to magic, it gives protection against any kind of poisoning.[21]

Paracelsus was stating explicitly the theory behind the practice ; such herbs continued to be used well into the seventeenth century (and beyond in popular medicine) though fewer knew the magical reason for their efficacy as time went on.

Thomas Norton, explaining that the word " alchemy " derives from a King Alchimus, a noble " clerk " who laboured long to find the " work," added that

King Hermes also he did the same
Being a Clerk of Excellent fame ;

> In his *Quadripartite* made of Astrology,
> Of Physic [medicine] and of this Art of Alchemy,
> And also of Magic natural
> As of four sciences in nature passing all.[22]

Of all the mystical and occult sciences, natural magic was, ultimately, the most fruitful. Superficially, natural magic was to natural philosophy what astrology was to astronomy, but in fact it dealt with problems with which natural philosophy had failed to cope. It was, besides, a far more truly empirical art than astrology or even alchemy. When sixteenth-century writers on the occult tried to defend magic as the key to nature—as they did by deriving the word "mage" from Persian, in which it was supposed to mean "a wise man, or a Philosopher, so that Magic contained both Natural Magic and the Mathematicks"[23]—it was natural magic which supported their contention. The German writer Agrippa von Nettesheim (*c.* 1486–*c.* 1534), author of *The Vanity of Arts and Sciences* (1530) as well as numerous books on the occult, took

> Natural Magick . . . to be nothing else, but the chief power of all the natural sciences ; which therefore they call the top and perfection of Natural Philosophy, and which is indeed the active part of the same, which by the assistance of natural forces and faculties, through their mutual and opportune application, performs those things that are above Humane Reason.[24]

Natural magic, as Agrippa here implied, was the study of the occult and mysterious forces of nature by natural means rather than by supernatural ones ; it thus differed from pure magic both in its means and in its ends, which were benign rather than demonic. As the Italian G. B. della Porta (d. 1615) wrote in 1589 :

> There are two sorts of Magic : the one is infamous, and unhappy, because it hath to do with foul spirits, and consists

of Enchantments and wicked Curiosity ; and this is called Sorcery. . . . The other Magic is natural ; which all excellent wise men do admit and embrace, and worship with great applause ; neither is there any more highly esteemed, or better thought of, by men of learning.[25]

The forces which natural magic undertook to investigate were sympathies and antipathies, signatures, magnetic attractions, the virtues of stones and herbs, mechanical arts, optical illusion—all strange by-lanes of nature over which the natural magician alone had control. Hence his contemporary English name of Archemaster. The scope of his domain was ably outlined by the mathematician John Dee who in his preface to the first English translation of Euclid (1570) wrote not only on the dignity and importance of mathematics, pure and applied, but on the study of nature in all its aspects. He was particularly eloquent about the potentialities of Archemastry :

This Art, teacheth to bring to actual experience sensible, all worthy conclusions by all the Arts Mathematical proposed, and by true Natural Philosophy concluded : and both addeth to them a farther scope, in the terms of the same Arts, and also by his proper Method, and in peculiar terms, proceedeth with help of the foresaid Arts, to the performance of complete Experiences, which of no particular Art, are able (formally) to be challenged. . . . Science I may call it, rather, than an Art : for the excellency and Mastership it hath, over so many, and so mighty Arts and Sciences. And because it proceedeth by Experiences, and searcheth forth the causes of Conclusions, by Experiences : and also putteth the Conclusions themselves, in Experience, it is named of some *Scientia Experimentalis*.[26]

Dee thought that archemastry exceeded observational sciences like astronomy and optics in the attention it gave to " doctrine Experimentall." One must, of course, be wary of the word "experience" in the sixteenth century : to Paracelsus and other mystics it meant all too often occult experience, just as seeing with one's own eyes meant denying orthodox doctrine. But Dee, though a mystic himself on other occasions, here truly meant genuine observation of nature ; for Dee, as for others, natural magic was near to becoming experimental science. On occasion indeed, sixteenth-century natural magic was indistinguishable from true experimental science in its investigation of the effects of mysterious forces by means of observation and experiment. Natural magic and experimental science finally parted company when the latter was allied to that particular form of natural philosophy known as the mechanical which endeavoured to understand both the effects of such mysterious forces, and their cause, in truly rational terms. So, for example, when magnetism could be successfully explained by a consideration of the orientation of the particles of the loadstone, or the rainbow by a consideration of reflection and refraction of light, natural magic lost its importance and became not archemastry, but conjuring. But this was not finally to happen until the middle of the seventeenth century.

The best natural magic, by understanding nature, sought to master and control it ; it thus shared the power ascribed later by Francis Bacon (1561–1626) * to experimental science. Bacon indeed was well aware of the close relationship between the two, though he saw that natural magic had a side characterised by charlatanry, which was not the case with experimental science. Bacon accurately described natural magic as

> The science which applies the knowledge of hidden forms
> to the production of wonderful operations ; and by uniting
> (as they say) actives with passives, displays the wonderful
> works of nature.[27]

This was its worthy side ; there was also that " popular and degenerate " side which

* See below, ch. VIII.

flutters about so many books, embracing certain credulous and superstitious traditions and observations concerning sympathies and antipathies, and hidden and specific properties, with experiments for the most part frivolous, and wonderful rather for the skill with which the theory is concealed and masked than for the thing itself.

The two aspects of natural magic, serious and popular, experimental and mystic, noted by Bacon in the 1620's are characteristic of the subject throughout the sixteenth century. Most works on natural magic were written with an eye to laymen, who even then enjoyed reading about and perhaps performing simple and ingenious experiments. A good example is *On Subtlety* (1551 in Latin, and then translated into French) by the Italian physician and mathematician Cardan. Cardan combined mathematics, mysticism and medicine in a successful amalgamation, though even he could not always quite keep his various interests separate. The subjects of *On Subtlety* are those things which are perceived by the senses, but which the intellect comprehends with difficulty if at all : the realm of natural magic, in fact. When Cardan discussed such things as the nature of the four elements, his real interest was in mechanical contrivances. He described a gimbal-mounting for keeping a chair stable on the deck of a ship at sea (the so-called Cardan's suspension), fireworks (introduced in a consideration of fire), pumps (introduced in a consideration of water), meteorology (the phenomena associated with air), geology (the phenomena associated with the element earth), optics, the burning glass, gems and their influence and activities (including the electrical action of rubbed amber), the virtues of plants, the generation of animals, the character and temperament of man, the phenomena associated with the senses of hearing and smell, the nature of the intellect and the soul of man, arts and crafts, marvels, and demons. The result is an experimentally oriented version of a mediaeval *Summa*, with the emphasis on the forces of sympathy and antipathy, rather than on the more orthodox explanations of scholastic natural philosophy.

In vitalising such forces as sympathy and antipathy lay the key to natural magic, for the natural magician regarded them as the dominant forces of nature. Cardan was by no means the first to treat them in detail ; the physician Fracastoro had written a book on the subject in 1546. Here he discussed such diverse questions as why the magnet turns to the north, how the remora or sucking fish stops a ship (he believed that the fish lived near magnetic mountains, which were responsible for the effect) and how the seeds of contagion of a disease happened to affect one person and not another. Fracastoro's work is more theoretical and less experimental than later works on natural magic, but he shared with their authors a preoccupation with such mystic forces as were involved in the attraction of iron to the loadstone or of chaff to rubbed amber.

The famous treatise on *Natural Magic* of della Porta (published in 1558, revised and enlarged in 1589 ; both editions were often reprinted) was frankly a work of popularisation even though written in Latin. Porta made a parade of learning, but his interest was really that of the party conjurer who deceives the eye by the quickness of his hand or mind before demonstrating how the thing was done. In effect, Porta treats the same subjects as Cardan, but without the interest in natural philosophy which made Cardan discuss the phenomena associated with each element in turn ; Porta was more interested in tricks, such as how to make near things seem far, how to make a plain woman appear beautiful (or, cruelly, to detect the methods she has used to try to appear beautiful). Biology for Porta was the production of strange new plants and animals ; metallurgy the study of apparent changes in metals ; the chapter on gems deals with counterfeiting. There is, on the one hand, a miscellany of information about practical affairs—housekeeping, cooking, medicine, distillation, hunting and fishing—on the other hand there are chapters on magnetism, optics, hydraulics, statics which are superior to those of Cardan. Underlying the whole is the belief that by observing the sympathies and antipathies associated with natural objects, one can arrive at an understanding of their essential nature and a control of their virtues, since it is by means of the hidden virtues (or forces) of things that apparent wonders are performed.

Porta wished to explain how apparent wonders actually came about, just as he was genuinely interested in trying and testing the marvels he wrote of; indeed, he had some real comprehension of the role of experiment in investigation. His chapter on the loadstone is much concerned with such trivia as how to move figures of men and animals placed on a table without touching them, effected by making them of iron, and moving a loadstone held under the table. Yet at the same time he concerned himself with more important and complex properties of the loadstone which he tried, often ingeniously, to test experimentally. And he showed a most healthy scepticism about some of the more surprising powers attributed to the loadstone. Whereas Agricola solemnly repeated the story that one could destroy the virtue of a magnetic compass by rubbing it with garlic, or even by breathing on it after eating garlic, Porta reported that

It is a common Opinion amongst Sea-men, That Onyons and Garlick are at odds with the Loadstone : and Steersmen, and such as tend the Mariners Card are forbid to eat Onyons or Garlick, lest they make the Index of the Poles drunk. But when I tried all these things, I found them to be false : for not onely breathing and belching upon the Loadstone after eating of Garlick, did not stop its vertues : but when it was all anoynted over with the juice of Garlick, it did perform its office as well as if it had never been touched with it.[28]

Nor did Porta concern himself with the ever-popular notions of magnetic mountains or the magnetic island described so vividly by Olaus Magnus in his *History of the Northern Nations* (*Historia de Gentibus Septentrionalibus*, 1555), a great compendium of wonders (which even described the sticks of wood on which one could slide over the surface of snow).

Porta's optical marvels are more often than not perfectly respectable experiments presented in a context of wonder such as the use of a lens as a burning glass, the use of magnifying lenses, the building of a camera obscura. His pneumatical experiments are also perfectly sound examples of simple engineering, mainly derived from Hero of Alexandria's *Pneumatics*, a work of the first century A.D. dealing with the physical nature of air, and the ways in which its elastic powers could be used to work mechanical contrivances. The late fifteenth century gave it considerable attention, and it received a Latin translation at the hands of the mathematician Commandino (1509–75) in 1575. It was subsequently translated into Italian by several authors. Cardan had known something of Hero ; Porta had access to the Latin version. Later the book became the model for a host of others dealing with mechanical toys and wonders,* like Jean Leurechon's *Récréations Mathématiques* (1624) and Gaspar Ens' *Mathematical Thaumaturge* (1628), which are natural magic without any serious element of mystic belief. Its influence is also reflected in serious sixteenth-century engineering works, like those of Jacques Besson, and—at the other extreme—in the work of avowed mystics like Robert Fludd. Here humanism and natural

---

* Hero's book was one of those which Regiomontanus intended to print ; it is perhaps significant that Regiomontanus was supposed to have built many mechanical animals, including a fly and an eagle, both of which were reputed to fly as if living.

magic combined ; all discussions of air, atmosphere, the vacuum, and of contrivances worked by air and water pressure, derive equally from Hero and from the basic content of natural magic.

Clearer proofs, in the discovery of secrets, and in the investigations of the hidden causes of things, being afforded by trustworthy experiments and by demonstrated arguments, than by the probable guesses and opinions of the ordinary professors of philosophy : so, therefore, that the noble substance of that great magnet, our common mother (the earth), hitherto quite unknown, and the conspicuous and exalted powers of this our globe, may be the better understood, we have proposed to begin with the common magnetick, stony, and iron material, and with magnetical bodies, and with the nearer parts of the earth which we can reach with our hands and perceive with our senses ; then to proceed with demonstrable magnetick experiments ; and so penetrate, for the first time, into the innermost parts of the earth.[29]

So William Gilbert (1540–1603) addressed " The Candid Reader, studious of the Magnetick Philosophy " whom he hoped to attract to his great work *On the Magnet, Magnetick Bodies also, and on the great magnet the Earth ; a new Physiology, demonstrated by many arguments and experiments* (*De Magnete*, 1600). This, the first full-length treatise on the magnet, is often taken to be the first great work in modern experimental science, and its claim is supported by more than the title. But it is also, and more certainly, the last important work in natural magic, for Gilbert's work is closer to Dee's archemastry than any other sixteenth-century attempt to study the hidden forces of nature. Gilbert showed how successful the experimental investigation of such forces could be ; but he was not outside the tradition. Most of the "magnetical" writers whose ideas Gilbert quoted (not always, though usually, in derision) are those who had regarded magnetism as occult; indeed, Gilbert's first chapter is almost a bibliography of works on natural magic. Gilbert's method was not very different from that of Porta; many of his experiments were very like those of earlier writers; and he was still, as befitted a physician, equally interested in the medicinal properties of the loadstone. He was more ingenious, more thorough, more curious, and possessed of a better power of evaluating experimental results. Hence, though his aim was that of previous writers, his conclusions are more striking and more important.

Gilbert was interested in more sweeping problems than the mere behaviour of objects under the force of attraction, though he investigated such behaviour most thoroughly. This is amply illustrated by his first great conclusion : that the Earth itself is a magnet. As a consequence, loadstones must be nothing but iron ore magnetised by lying in the correct orientation to the magnetic poles of the Earth; and when smiths produced magnetism in the bars of iron they hammered, this was because the iron lay in a North–South line, and like natural iron ore, was affected by the magnetic poles of the Earth. The compass needle must point to the North because it was attracted by the terrestrial pole (not the celestial, as earlier writers had thought). From this it followed, Gilbert believed, that the variation of the compass needle from true North was caused " by the inequality of the projecting parts of the Earth," [30] which is an engagingly rationalistic explanation.

When he came to examine the phenomenon of attraction, Gilbert for the first time clearly distinguished between *magnetism* —the attraction of the loadstone for iron—and *electricity* *—the

* Though Gilbert did not use the word electricity, he did coin the term *electrics* for those bodies which acquire the power of attracting light objects when subjected to friction.

attraction after rubbing of hard translucent objects like amber, jet, some gems, glass, resin, sulphur for any small, light bodies. Gilbert designed the first electroscope (versorium, he called it), a lightly poised metal needle, to assist his careful examination of electrics and his attempt to find out what changes in their physical state, such as those caused by heating and chilling, would do to their powers of attraction. Even in his brief investigation of electricity, Gilbert proceeded with a thoroughness of experimental detail which is utterly captivating. And in those sections which treat of the magnetic nature of the Earth, the nature of the loadstone, the methods by which its attractive force can be strengthened, and the investigation of its directive force, Gilbert provides a wonderful array of interesting and ingenious experiment mostly original, as he grandiosely indicated by the guidebook device of small and large stars in the margin. Many of these experiments involved the use of his favourite " terrella " or spherical loadstone, which he truly imagined to be, in its behaviour, the Earth in miniature.

Gilbert also, like his contemporaries, attempted to find out how to use the properties of terrestrial magnetism to aid navigation. Variation from true North had been known for a couple of centuries, and Columbus had been fascinated by the discovery that it changed from East in the Mediterranean, to zero off the Azores, to West as he proceeded westward across the Atlantic. By 1599 there had been a long period of collection of data, which many hoped could be used for longitude determination if the data were plotted on a global grid. This method was advocated by the Dutch engineer and mathematician Simon Stevin (1548–1620) in his aptly entitled *Havenfinding Art, Or the Way to Find Any Haven Or Place Appointed at Sea* (published in Dutch and English in 1599). Gilbert was doubtful : he thought variation was affected by many forces and changed from place to place in too complicated a fashion to make it accessible in navigation.

Gilbert preferred to think that position could be determined from the recently discovered dip (inclination). Dip had first been described by a contemporary of Gilbert, Robert Norman, a London instrument-maker. He was accustomed to magnetise his needles after they were mounted, and found difficulties with long needles, which dipped so violently towards the compass card that their ends touched before the needle was stable. Faced with this problem, Norman studied the matter seriously, devised a declinometer to measure the amount of dip, and published his ideas in a little book quaintly entitled *The Newe Attractive* (1581). Gilbert extended Norman's work with the help of the terrella, and discussed the possibility of devising a very elaborate grid, together with a special quadrant, whereby measurement of dip could be used to establish latitude at sea. The advantage of Gilbert's method of latitude determination was that it could be used in cloudy weather, so, though difficult, it had merit. It is the first contribution to the art of navigation by experimental science, as distinct from mathematical science, and Gilbert had some reason to be proud of his suggestion ; as he boasted,

We may see how far from unproductive magnetick philosophy is, how agreeable, how helpful, how divine! Sailors when tossed about on the waves with continuous cloudy weather, and unable by means of the celestial luminaries to learn anything about the place or region in which they are, with a very slight effort and with a small instrument are comforted, and learn the latitude of the place.[31]

He never knew that dip, like variation, changes with time as well as place, and that neither latitude nor longitude can be determined magnetically.

In spite of occasional encomiums on the " magnetick philosophy," and digressions on the feeble reasoning powers of his predecessors, Gilbert maintained a steady experimental outlook through nearly three-quarters of his book. He had steadfastly interested himself in experiments to illustrate the magnetic abilities of the loadstone, and had given little attention to discussing the possible cause of magnetism. There is but one brief chapter on the causes of magnetic and electric attraction, where Gilbert derived electricity from *matter* and magnetism from *form*; though this is conceived in terms of Aristotelian philosophy, experiment finds its place even here, and the occult is firmly rejected. Gilbert was certain that the magnetic power of the Earth is a natural force and " is neither derived nor produced from the whole heaven by sympathy or influence or more occult qualities, nor from any particular star." [32] Yet Gilbert was, after all, a natural magician, not a natural philosopher, and speculation about forces would creep in. So, at the end of his careful discussion of dip and navigation, there is suddenly a chapter entitled " Magnetick force is animate, or imitates life; and in many things surpasses human life, while this is bound up in the organick body," a title suitable to Cardan. Gilbert was partly trying to ascribe an *anima*—a moving soul—to the Earth, instead of merely to the celestial sphere, to support his belief that the Earth moves. But there is more than that, and Gilbert demonstrated a profound belief in a totally animate universe. Aristotle, he insisted, erred in denying a moving soul to the Earth; he did not indeed properly appreciate what Gilbert knew, that " the whole universe is animated, and that all the globes, all the stars, and also the noble Earth have been governed since the beginning by their own appointed souls and have the motives of self-conservation." [33]

This is not very like a purely experimental investigator, nor was his anti-Aristotelian mysticism wholly a defence of the possibility that the Earth moved. Gilbert genuinely believed that magnetism, however subject to control by the natural magician who thoroughly understood the ways in which it manifested itself, was truly an occult force. When, as in the *New Philosophy of the Sublunary World* (written soon after *De Magnete*, but published posthumously in 1651), Gilbert concerned himself with the heavens, the Earth and the Moon, it was this slightly magical strain of anti-Aristotelianism which predominated, to make a work so non-experimental that it seems to differ more in tone from *De Magnete* than is, in fact, the case. It is here, as Francis Bacon was to complain later, that Gilbert " made a philosophy out of the loadstone," attempting to explain gravity by the magnetic attraction exerted by the Earth on all bodies, perhaps even on the Moon. Here, too, Gilbert discussed a host of questions lying on the borderline between scholastic philosophy and natural magic : the possibility of an interplanetary vacuum, the composition of the heavenly bodies, the actions of the tides (related to magnetism again), and various " meteorological " (atmospheric) phenomena, like winds, the formation of rainbows, and fountains. It is actually a continuation of the last ten chapters of *De Magnete* ; and just as Gilbert regarded his conclusions about the Earth's magnetic and animate nature as directly derived from his earlier experiments on magnetic bodies, so, one must presume, he thought the later work also derived from magnetic experiments.

Once again Gilbert appears to be more nearly a natural magician than a seventeenth-century experimental scientist. His failure to treat more adequately of theoretical subjects reveals more clearly than anything else the gap that separated natural magic from the new experimental learning to be developed in the seventeenth century. Even Gilbert's experimental genius was cramped by his attempts to restrain it within the bounds of a method which, however much it endeavoured rationally to understand the way in which the forces of nature worked, and to control them, yet assumed these forces to be impervious to rational comprehension, because it knew them to be occult and essentially unknowable. Magic was a delight to the inquiring mind ; but even when white and natural, it led men all too often astray. No wonder that there was to come a period of revolt and rejection of the occult, a revolt so violent that it led to a veritable excess of rationalism.

CHAPTER VII

THE USES OF MATHEMATICS

Perspective, Astronomy, Musike, Cosmographie, Astrologie, Statike, Anthropographie, Trochilike, Helioscopie, Pneumatithmie, Menadrie, Hypogeodie, Hydragogie, Horometrie, Zographie, Architecture, Navigation, Thaumaturgike and Archemastrie.[1]

When Henry Billingsley, university graduate and successful London merchant, published the first English edition of Euclid in 1570, he invited John Dee, England's leading mathematician, to write a preface dealing with the virtues and advantages of mathematical learning. He and Dee hoped to " stir the imagination mathematical : and to inform the practiser mechanical " how necessary the study of mathematics was for all manner of useful arts, as well as for the study of nature. It was not just that mathematics was used in those applied sciences which, in exuberant Greek derivatives, Dee listed with Renaissance gusto; for mathematics, in the eyes of its fifteenth- and sixteenth-century practitioners, was not merely an abstract art for the specialist. To them, as to the Greeks, the term meant all the sciences of magnitude and number, and their practical applications.

Though geometry was the branch of mathematics which had been most esteemed by the Greeks, they had not neglected other branches. The Pythagoreans had judged mathematics to consist of four divisions : geometry, arithmetic (number theory), astronomy and music ; for they regarded astronomy as being applied geometry, and music as applied arithmetic. This classification had persisted, to reappear in the quadrivium of the mediaeval university. Plato, influenced by the Pythagoreans, had emphasised the role of mathematics in science as well as in philosophy. Pure mathematics, in Platonic doctrine, because it dealt with the world of perfect, unchanging, abstract ideas, was the best possible training for the philosopher who wished to study the nature of ideas, forms and essences. Mathematics reflected the unchanging reality behind the flux and uncertainty of the world of the senses ; hence for the Platonist to study nature was to search for the mathematical laws which govern the world. Though Aristotle had protested that magnitude and body were different things, and natural philosophy and mathematics could not be the same, the Platonic tradition continued to appeal to many minds. The fifteenth century's intensification of interest in Platonic and neo-Platonic doctrine helped to encourage the view that mathematics was not only the key to science, but included within its competence the greater part of what the seventeenth century was to call natural philosophy. One has only to recall that Copernicus wrote for mathematicians, and might well have called his book " Mathematical Principles of Celestial Revolution " to realise how the anti-Aristotelian tendency of the age was apt to express itself by the attempt to treat mathematically what Aristotle had treated qualitatively.

The Platonic tradition was of enormous consequence for Renaissance mathematics. Most obviously, it encouraged the study of pure mathematics and the search for previously neglected Greek mathematical texts. It stimulated the founding of chairs of mathematics in the new humanist schools, like the Collège Royale in France, though these were intended as linguistic centres. It helped the revival of professorships of mathematics in the established universities, though it did not raise the professors' salaries. It suggested that mathematics was better training for the mind than dialectic. It offered a number of useful varieties of mathematics, suitable for non-academic education : fortification for the gentleman-soldier, surveying for the landed proprietor, practical astronomy and some knowledge of the use of maps for all. On a less rational plane, Platonism and neo-Platonism encouraged so much number mysticism and astrology that to the layman " mathematicus " and " astrologer " were identical. (Indeed, they were when the mathematician was a Cardan or Dee, though the latter protested that he only dealt with " marvellous Acts and Feats, Naturally, Mathematically and Mechanically wrought," and it was unfair to call him a conjurer for that.) Many a young man must, like Fernel, have progressed from elementary geometry and the doctrine of the sphere to the delights of astrological prediction ; no wonder that careful fathers like Vicenzo Galilei warned their sons of the dangers of a subject at once dubious in reputation and poor in remuneration.

The popularisation of science and the new awareness of the needs of the technical man affected mathematics strongly. In the mediaeval university, all students attended lectures on Euclid ; now they expected the professor of mathematics to cover a wider range by dealing with practical mathematics : everything from the doctrine of the sphere to the use of mathematics in war, navigation or engineering. Mathematicians were eager to exploit the host of newly discovered ways in which they could aid the unlearned, from teaching the merchant how to reckon his profits to showing the instrument-maker how to draw the scales on the brass plates of his wares. So great was the demand that there sprang up a new profession of semi-learned mathematical practitioners, men skilled in the practical aspects of mathematics, who knew how to apply geometry and trigonometry to the problems of scientific measuring devices. Many of these gave mathematical lectures in the vernacular, a practice especially common in London in the second half of the sixteenth century, and wrote books of elementary instruction in plain, simple and easy language.

A fair example is *A Booke Named Tectonicon* by Leonard Digges, published in 1556 and often reprinted. Digges said he had planned a " volume, containing the flowers of the Sciences Mathematicall, largely applied to our outward practise, profitably pleasant to all manner men in this Realme " ; while waiting to complete it he produced this smaller work, whose subtitle declares it to be a book

briefly shewing the exact measuring, and speedie reckoning all manner of Land, Squares, Timber, Stone, Steeples, Pillers, Globes, &c. Further declaring the perfect making and large use of the Carpenters Ruler, containing a Quadrant Geometricall. Comprehending also the rare use of the Square. And in the end a little Treatise adjoyning, opening the composition and appliancy of an Instrument called the Profitable Staffe. With other things pleasant and necessary, most conducible for surveyers, Land-meaters, Joyners, Carpenters and Masons.

Truly an indispensable mathematical handbook, suitable for learned and unlearned alike.

An earlier attempt to apply mathematics for the use of the craftsman was exemplified in the *Course in the Art of Measurement with Compass and Ruler* (1525) by the artist Albrecht Dürer (1471–1528). This is an example of Dee's " Zographie," the application of mathematics to art. Painters had not long since solved the problems of perspective and methods of creating the illusion of three dimensions on a two-dimensional canvas ; the results were exhibited in Jean Pélerin's *On Artificial Perspective* (1505) embodying the developments of over half a century. But this was empirical knowledge ; more sophisticated now, more learned in academic subjects, many painters wanted to know the mathematics and theory of the art of " false perspective." Not that mathematics could teach them how to paint, but that many were filled with curiosity to know why the tricks of the trade worked.

Hence Leonardo's studies on the mathematics of proportion, or the elaborate vernacular treatises of Dürer, which made Latin and Italian knowledge available to Germans ; for, he thought, " geometry is the right foundation of all painting " [2]—and of all arts as well as the building crafts—and should be available to all.

Not all applied mathematics was dedicated to elucidating the practices of the craftsman ; an immense amount of interest centred on the mathematical background of the theoretical sciences. The work of fifteenth-century astronomers had amply demonstrated the need for detailed mathematical analysis of astronomical problems. That was for the specialist ; on a more elementary level there was the geometry of the sphere, which helped to make elementary astronomy also mathematical. Some tutors began with the ancients ; Linacre used his own translation of Proclus (1499) to introduce the English royal children to the beginnings of astronomy ; others preferred to write new treatises on the sphere, treatises which stressed the terrestrial rather than the celestial spheres, and became geographical rather than astronomical. Cosmography became a common subject ; treatises ranged from the learned and thoroughly mathematical works of such men as Peter Apian, professor of mathematics at Ingolstadt, or Oronce Finé (1494–1555), professor of mathematics at the Collège Royale, to intentionally popular works like that of Sebastian Münster (1489–1552), who had been educated in mathematics at Heidelberg before he turned to lecturing on Hebrew at Basle. They all helped spread an understanding of the importance of mathematics in geographical exposition.

Navigational problems such as fifteenth-century astronomers had tried to solve were still in the domain of the applied mathematician. More erudite now, as well as better aware of the needs of the seaman, the professors of mathematics were as eager as their predecessors to devise new methods to aid sailors ; and even more ingenious. Boldly they tackled the problem of longitude, untouched in the fifteenth century. Peter Apian and Oronce Finé suggested that longitude might be determined by the "method of lunar distances," which involved the measurement of the angular distance of the Moon from certain stars; this involved further study of lunar motions and accurate tables, but it was more promising than the timing of lunar eclipses, not frequent enough to be useful. Gemma Frisius (1508–55), Apian's pupil and a professor of mathematics at Louvain, suggested the use of clocks for longitude determination; this was a fantastically optimistic proposal in view of the current inaccuracy of time-pieces. Jacques Besson, professor of mathematics at Orléans, invented a universal instrument for navigation, timekeeping and astronomy, which he described in *Le Cosmolabe ou Instrument Universel concernant toutes Observations qui se peuvent faire Par les Sciences Mathematiques, Tant au Ciel, en la Terre, comme en la Mer* (1567) and thoughtfully included a fine picture of an observer sitting in an improbably large chair, mounted on gimbals to minimise disturbances from the ship's roll and pitch; he did not say how the sailors would find room on deck for their work.

These methods, though possible for shore-based mathematicians, were too complex and uncertain for use at sea. No wonder that practical men, like Robert Norman and Simon Stevin, thought the mathematical professor an uncertain guide even though their own methods were not necessarily better. Those who had been at sea were extremely critical; Robert Hues (1553–1632), Oxford graduate and professional mathematician, had some right to speak from experience after accompanying Thomas Cavendish on his circumnavigation of the world (1586–8); he was full of scorn for mathematicians who thought to calculate longitude from lunar motion:

this is an uncertaine and ticklish way, and subject to many difficulties. Others have gone other ways to worke; as, namely, by observing the space of the Aequinoctial hours betwixt the Meridians of two places, which they conceive may be taken by the help of sundials, or clocks, or hour glasses, either with water or sand or the like. But these conceits long since devised, having been more strictly and accurately examined, have been disallowed and rejected by all learned men (at least those of riper judgements) as being altogether unable to perform that which is required of them.[3]

But Hues had little to offer instead beyond cartographic assistance ; and the practical man, joining forces with the natural magician, was as fallible as the mathematician ; for the compass needle proved to change its variation and declination with time, and to be no help in the problem.

In spite of ingenious suggestions, better tables and improved instruments (like the backstaff * described by John Davis in *Seamen's Secrets*, 1594) sailors at the end of the sixteenth century, as at the beginning, preferred to depend primarily on dead-reckoning, with astronomical assistance where this proved readily practicable. Even here the mathematician had good advice to offer, not all of which was accepted. The learned man knew that a great-circle route was the shortest distance between two points ; but the sailor preferred the method of parallel sailing or "running down the latitude," whereby the ship made its way to the required latitude as directly as wind and current permitted, and then sailed East or West until land was sighted. This kind of navigation was facilitated by the invention of the " log " for measuring the ship's speed, from which its day's run could be calculated. An English invention, it was long an English monopoly, though it was described at length by William Bourne in the popular *A Regiment for the Sea* (1573). Formerly the sailor

* This was a quadrant, modified so that the navigator turned his back to the Sun, whose altitude he measured by observing the shadow cast by a movable vane.

had estimated his ship's speed by throwing overboard a chip of wood and watching it travel past the length of the ship while he paced the deck to determine the time it took to do so. Now the sailor heaved astern a log of wood fastened to a line knotted at equal, fixed intervals, and counted the number of knots that ran out during a length of time measured by a sand glass. (Hence the practice of giving a ship's speed in knots, since the distance between each knot was designed to measure a speed of one nautical mile per hour.) For accuracy, the knots needed to be properly spaced and the sand glass accurately calibrated, two measurements hardly ever systematically undertaken. But when the length of a degree of terrestrial arc (which determined the nautical mile) was far from settled, sailors were justified in refusing to worry as long as they erred on the safe side. As they said, it was better to be a day's sail behind their calculated position than even a cannon shot ahead of it.

Here the mathematician was ready with advice, and sometimes achievement. Edward Wright, the Cambridge-trained mathematician who had learned about practical navigation on an expedition to the Azores in 1589, was the first to note the desirability of measuring the Earth's surface to determine the length of a terrestrial degree with some accuracy, and made an astronomically based improvement. The first actual English measurement was that by Richard Norwood (1590–1675), seaman, mathematical teacher and surveyor : he paced out the distance between London and York when he had occasion to go from one city to the other, and published the result in *The Seaman's Practice* (1637). Great improvements in tables, methods of calculation and instrumental aids appeared in the early seventeenth century, notably the use of Gunter's sector (first described in 1607), a calculating instrument which greatly reduced the amount of tedious computation required in dead-reckoning.

Whether by dead-reckoning or astronomical methods (increasingly sophisticated now, as more and more mathematicians compiled tables, developed simplified methods and published books) all navigation involved the use of maps and charts. By the beginning of the sixteenth century nearly all land maps were based on some form of projection, but the " plane chart " * still held supremacy at sea. In the plane chart distances between meridians were the same at all latitudes, whether near the equator or the poles, and large errors were thereby introduced at high latitudes. The Portuguese mathematician Pedro Nuñez (1502–1578), a successor to Zacuto in his interest in applying mathematics to the improvement of navigational methods and techniques, tried to analyse the problem mathematically in his *Tracts* (1537) ; his analysis became better known when a Latin version appeared in 1566 under the title *On the Art of Sailing*. Nuñez discovered that on a sphere a rhumb line or loxodrome (a line of constant compass heading) is not a straight line, as it is on a plane, but a spiral terminating at the pole. He also noted that since the meridians on a globe converge, a true sea chart should not have its meridians everywhere equally spaced. Nuñez designed a quadrant which would enable one to find the number of leagues in a degree along each parallel, but he was unable to solve the much more important mathematical problem of finding a projection which would give the required convergence and make rhumbs straight lines.

Many references to the problem are to be found in subsequent books on mathematical navigation ; the next real step towards its solution was made by Gerard Mercator (1512–94). Mercator studied mathematics under Gemma Frisius and lectured at Louvain until his Protestant faith made it necessary for him to leave the Low Countries for Germany. There he became a

* So called because it treated the (spherical) Earth as if it could be mapped on a flat plane.

mathematical instrument-maker and a globe and map designer and publisher. His globes reflect both his mathematical ingenuity and his knowledge of the work of Nuñez, whose loxodromic spiral he engraved on some of them. He also worked out the proper relation between the length and width of the gores which made up the map on the globe (printed paper slips pasted on the globular core) : he divided his map into twelve gores, cutting off each twenty degrees from the pole, and providing two extra circular gores for the poles, a procedure which ensured a higher degree of accuracy than previous methods. His world map of 1569, not a true sea chart though ostensibly " for the use of mariners," further utilised the notions of Nuñez : here Mercator spaced out the meridians towards the poles, apparently by guess-work, though he may have used trigonometric methods. He never explained how he derived his figures ; others, though they might admire, could not duplicate his work, and Mercator never made another such map.

The next map-maker to publish a map on " Mercator's " projection was the Dutchman, Jodocus Hondius (1563–1611), who made use of the work of the English mathematicians whom he had met while a refugee in London between 1584 and 1595. The English mathematicians proved better at solving the problem than Mercator. The inspirer of their work was John Dee, who had travelled to the Low Countries in 1547, " to speak and confir with some learned men, and chiefly Mathematicians," [4] among them Gemma Frisius and Mercator. (Dee brought back some of Mercator's globes, which he gave to his College.) A year later Dee was back on the Continent, first, briefly, as a student at Louvain, then as a teacher of mathematics at Paris ; here he met Finé, Fernel and others, acquired a reputation as an ingenious mathematician, and established correspondence with a number of continental workers, including Nuñez. Dee was thus in close touch with the work being done in navigation and cartography. Two of his colleagues, Thomas Hariot (1560–1621) and Edward Wright, claimed success in the matter of a loxodromic chart. Hariot discussed the matter briefly in the fifth part of Hues' *Treatise on the Globes* (1594), but he gave no precise data or method. The first real discussion was in Edward Wright's *Certaine Errors in Navigation, Arising either of the ordinarie erroneous making of the Sea Chart, Compasse, Crosse staff, and Tables of declination of the Sunne, and fixed Starres detected and corrected* (1599). Wright had been in no hurry to publish ; perhaps he agreed with Dee, that mathematical knowledge was sufficiently esoteric to warrant secrecy, though he did not share Dee's addiction to the magical sciences. Wright's work circulated for some time in manuscript before it was published ; he claimed at last to make it public only to forestall a pirated edition under another's name. Indeed, as he knew, Hondius had already made use of his work, and without acknowledgement, though he had shown his tables to Hondius only under the promise of secrecy.* If his work were to become common property, it might as well be accurately done, and he might as well claim the credit.

Wright's intention was to analyse all the errors commonly associated with the usual methods of dead-reckoning : in particular he treated the errors inherent in the use of the plane chart, showing their geometrical and physical sources and the ways of avoiding them. Wright supplied tables of rhumbs, showed how to use these tables and the new charts based upon them ; how to find the distance from one place to another on the new charts, given latitude and longitude, and how best to plot a course. In fact everything the practical man needed to know, and with

* When accused, Hondius admitted his fault, but tried to explain that he had failed to acknowledge his debt to Wright in print—as he claimed to have done verbally—only because the Latin translation was too poor to publish under Wright's name ! The true reason was more probably what he wrote to Wright's friend Henry Briggs, professor of geometry at Gresham College : " the profit thereof moved me." [5]

the tediums of calculation and computation removed as far as possible. No wonder that Hondius—not a mathematician, and not even a skilled cartographer—had been able to draw a map on the new projection.

Wright's description of the geometrical problem involved in this new projection illustrates the clarity of his thought and his style : he wrote,

Suppose a spherical superficies with meridians, parallels and the whole hydrographical description drawn thereupon to be inscribed into a concave cylinder, their axes agreeing in one.

Let this spherical superficies swell like a bladder (whilst it is in blowing) equally always in every part thereof (that is as much in longitude as in latitude) till it apply, and join itself (round about and all along also towards either pole) unto the concave superficies of the cylinder : each parallel upon this spherical superficies increasing successively from the equinoctial towards either pole, until it come to be of equal diameter with the cylinder, and consequently the meridians still widening themselves, till they come to be so far distant every where each from the other as they are at the equinoctial. Thus it may most easily be understood, how a spherical superficies may (by extension) be made a cylindrical, and consequently a plain parallelogram superficies.[6]

It was, of course, not enough to see that the problem could be simplified if a cylinder (which can be unrolled to form a flat surface) were used instead of a sphere, though this was more than either Nuñez or Mercator had seen. It was necessary to work out tables to permit the construction of maps upon this projection. Wright did both ; and after the publication of his work any map-maker could draw a map on the now familiar Mercator projection, so particularly suited to the sea chart, since now a rhumb line is a straight line, and a constant compass course can be laid out with a ruler. That a great circle route is not so simple was still, obviously, of no concern to seamen not interested in finding the shortest distance between two points, since wind and current would never permit them to sail it even if it had been easily plotted.

The new projection did not become instantly popular, though it was fairly common within a generation. If it did not make its way more rapidly, this was partly because maps were so popular that any botch could be sure of a good sale, and many map publishers blindly pirated their predecessors. Most maps were not for seamen at all, but for gentlemen—" to beautifie their Halls, Parlers, Chambers, Galeries, Studies, or Libraries with," as Dee observed [7]—for diplomats, for travellers and for scholars. Sailors, English and Dutch especially, pinned their faith to the "Waggoner," the useful *Mariner's Mirror* compiled by Lucas Janszoon Waghenaer (published in Dutch in 1583, and subsequently in many editions in many vernaculars). This was a handy and simple manual of elementary navigational methods, complete with tables, astronomical rules and old-fashioned charts of European waters. The printers who published and republished it saw no need to improve on it, even when better charts and navigational rules were available and it became more and more out of date.

It was perhaps to remedy this situation that the States General of the Netherlands in 1605 commissioned Willem Blaeu (1571–1638) to write a new seaman's guide. Blaeu belonged to the scientific school of cartographers : he was no mere publisher of maps, but a competent and highly trained mathematical instrument-maker who had spent two years under the supervision of Tycho Brahe at Uraniborg, studying astronomy, geography and the construction of precision instruments. The result of his work, published as *The Light of Navigation* in 1612, was a much improved manual, complete with new and corrected astronomical and nautical tables and a new set of sea charts, all drawn on Mercator's projection. It was the first of many works which embodied the advances of the later sixteenth century, and in which English mathematicians now repaid the debts contracted earlier to Portuguese and Dutch applied mathematics. The log, the backstaff and Wright's elucidation of Mercator's projection all came into use throughout European practice in the first half of the seventeenth century.

The instrument-maker was not the only craftsman who needed guidance from the mathematician : the engineer found mathematics equally essential. Civil as well as military engineering was a thriving profession throughout Europe in the sixteenth century, especially in Italy, but north of the Alps as well. There was a general interest in machinery, as the many beautiful picture-books of the period testify. There was Jacques Besson's *Theatre des Instrumens Mathematiques & Mechaniques* (1579) ; Ramelli's *Le Diverse et Artificiose Machine* (1588, published bilingually in French and Italian) ; Faust Veranzio's *Machinae Novae* (c. 1595) ; Zonca's *Novo Theatro di Machini et Edificii* (1607) ; Branca's *Le Machine* (1629) ; and many more, all describing power machines, pumps, mills, cranes, bridges, fountains, war machines, pneumatic and hydraulic devices.

There was some humanist influence, derived from Greek and Roman works, but mostly the interest and the novelty came from the flourishing practical technology of the age, such as that illustrated in Biringuccio's *Pirotechnia* or Agricola's *De Re Metallica*. The " practical men " who built these machines were by no means always ignorant of mathematics, and there were a host of mathematically trained inventors equipped to design lathes for cutting cylinders and cones and similar refinements. Ramelli (1531–90) was eager to insist on the advantages of mathematical knowledge : his book carries a preface entitled *De l'excellence des mathematiques, ov il est demontré combien elles sont necessaires pour acquerir tous les arts liberaux.* Besson described himself as " docte Mathematicien ", and characterised mechanics and engineering as the true goals of mathematics : " the contemplation of the proportions of numbers, points, and measures of artificial things is useless unless related to action, so that it follows that mechanics is the fruit of geometry, and consequently its goal." [8] This, of course, was the ideal ; but certainly the sixteenth century thought that the building of machines was a mathematical art.

The science behind this art was mechanics, or mathematical physics : the study on the one hand of the laws of simple and complex machines, and on the other hand of the behaviour of bodies on which these machines were based, that is to say, of statics and dynamics. The fifteenth century had been little interested in such problems. The sixteenth century enjoyed the advantage of a twofold stimulus : the printing of mediaeval works on physics, and the collection and editing of the works of Archimedes. The treatises of Archimedes had been well known to fourteenth-century scholars ; but the mediaeval approach to statics derived less from the method of Archimedes than from that of the pseudo-Aristotelian *Mechanical Problems*. This, the earliest theoretical discussion of the theory of simple machines, embodied a dynamical approach, treating all cases of rest as similar to the equilibrium of a balance.* Archimedes, on the contrary, dealt with rest only, and treated statics as a branch of mathematics, concerning himself with the precise handling of magnitudes. His works were too complex to attract publishers

* The author of the *Mechanical Problems*, who followed the Aristotelian tradition quite closely, reduced all simple machines to the lever, which he related in turn to the circle through the balance and wheel. He gave the first statement of the law of the lever in qualitative terms, and was critical of Aristotelian dynamics.

in the fifteenth century ; the first fairly complete Latin text (excerpts were printed earlier) was the version drawn from various sources and edited—badly, his enemies said—by Niccolo Tartaglia (1500–57), for publication in 1543. A more accurate translation, with a Greek text (though oddly, not the one from which the translation had been made), was published a year later.*

The combination of the ready availability of Archimedes with the publication of mediaeval texts started two different sorts of investigation. The interesting comments on statics by Leonardo da Vinci, early in the century, derive exclusively from the mediaeval tradition. In contrast, Simon Stevin at the end of the century was motivated solely by Archimedean considerations, and a rigid insistence on a statical approach both to equilibrium problems and to fluid mechanics. Reflecting on the old problem of how it is that objects at the bottom of a lake or the sea are not crushed by the weight of the water above them, Stevin arrived at an enunciation of the hydrostatical paradox that the pressure of a fluid upon a solid body immersed in it is proportional to the height of the column of fluid immediately above it, and not to the total volume of fluid in which it is immersed. His logical, quasi-mathematical approach was similar to that later employed by Pascal.

Stevin himself was most proud of his elucidation of the equilibrium conditions of bodies on an inclined plane, which he illustrated on the title page of *The Elements of the Art of Weighing* (published in Dutch in 1586) 9 with a motto intended to show that he had taken the wonder out of an apparent marvel (see figure 6). He imagined a triangular surface ABC, with base AC parallel to the horizon, and side AB twice side BC ; over this he imagined hung an endless chain on which fourteen spheres had

---

* The translation was that made by Regiomontanus, a correction of an early fifteenth-century version. The complete mathematical text had to wait for translation (by Federigo Commandino) until 1558.

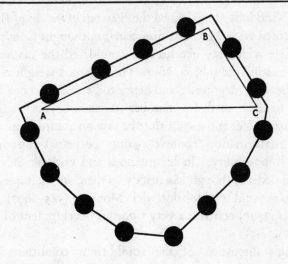

FIG. 6. STEVIN'S DEMONSTRATION OF EQUILIBRIUM ON AN
INCLINED PLANE

been fastened at equal intervals, all the spheres being of the same size and weight. Unless there is to be a perpetual motion of the chain about the triangle, which Stevin regarded as absurd and impossible, it must rest in equilibrium with two spheres on BC and four on AB : because otherwise there will be a perpetual motion of the chain about the triangle. Since the chain is in equilibrium, the lower portion may be removed without disturbing the equilibrium of the rest. Hence the length of the inclined planes will be directly proportional to the "apparent weight"—the component along the direction of the plane—supported along the plane, which is equivalent to saying that on inclined planes of equal height a given force will sustain a weight proportional to the length of the plane. Note that Stevin here used a "triangle" (though he sometimes preferred to call it a

213

prism) ; and indeed he defined the *Elements of the Art of Weighing* as concerned with " gravity, dissociated in thought from physical matter," [10] which he, like his age, considered the mathematical way of treating the subject. In fact he regarded weight as similar to number and magnitude, and hence to be discussed in a manner similar to that used for number (arithmetic) or magnitude (geometry). Yet at the same time he saw no absurdity in arguing in this mathematical context against perpetual motion as a *physical* impossibility. In his methods and outlook Stevin was an Archimedean, though less strictly so than, for example, Commandino's pupil Guidobaldo del Monte (1545-1607) whose *Mechanics* (1577) contains a very rigorous development of statical principles.

Stevin's discussion of the equilibrium conditions on an inclined plane was ingenious and original, but it was by no means the only possible approach to the problem. Another approach had been considered by Jordanus Nemorarius in the thirteenth century, one based in turn on that in the *Mechanical Problems*, and this tradition flourished at the same time as the Archimedean. Indeed, the two could be combined, as they were by Galileo (1564-1642) * in the treatise *On Mechanics* which he wrote about 1600 for his private pupils in Padua. It is an elementary analysis of the five simple machines (inclined plane, lever, windlass, pulley and screw) with a brief discussion of the elements common to all of them. Although Galileo thought little of his contributions, and did not regard them as sufficiently original to merit inclusion in the *Discorsi* of 1638, modern writers have noted that, in fact, he was the first to see that simple machines could not create work, but merely transformed its method of application. Galileo always equated input and output of a machine, either in terms of power and distance or of force and speed. His analysis

* His early work on physics is described below, pp. 221 f. ; his life and astronomical work are described in ch. XI.

was at the same time suggestive of further problems to be investigated mathematically, and clearly related to the actual physical world. His approach is revealed by his conclusion to the discussion of the steelyard and lever :

> And to sum up, the advantage acquired from the length of the lever is nothing but the ability to move all at once that heavy body which could be moved only in pieces by the same force, during the same time, and with an equal motion, without the benefit of the lever.[11]

As the *Mechanics* was widely read in Italy (although until 1649 only in manuscript versions) and in France (in a translation by Mersenne published in 1634), it had a wide influence.

The Aristotelian elements detectable in *On Mechanics* by no means indicate that Galileo was, at this period, in any sense an Aristotelian, however thoroughly he may have been grounded in the Peripatetic doctrine during his student days. He was already both an anti-Aristotelian and a devout disciple of " the superhuman Archimedes, whose name I never mention without a feeling of awe." [12] Indeed he had already, in a treatise *On Motion* (*De Motu, c.* 1590), used Archimedean physics as a weapon against Aristotelian dynamical principles ; in this approach he was influenced by the work of Niccolo Tartaglia and G. B. Benedetti. Many mathematical writers—Leonardo da Vinci in his manuscripts, Tartaglia, Benedetti, Galileo's Pisan teacher Bonamico—had already tried to mathematise the impetus theory of dynamics. This theory, thoroughly explored in late mediaeval physics, experienced a new lease of life in the sixteenth century when the experience of gunners and the growing anti-Peripatetic spirit of the age combined to show the glaring errors inherent in Aristotle's discussions of motion. The sixteenth-century attempts to make impetus dynamics rigorously mathematical were doomed to failure, as Galileo was to realise after the completion of *De Motu*, for impetus was a qualitative, not a quantitative force. But the very impossibility of the attempt made Galileo realise the necessity for a new dynamics which should somehow satisfy both the Archimedean demand for an expression appropriate to abstract magnitudes moving through geometrical space (an approach adopted by Benedetti and further pursued by the young Galileo) and the exigencies of real bodies rolling down physical inclined planes.

No one in the sixteenth century could write about the physics or mathematics of moving bodies without reflecting the ideas of Aristotle. Aristotle had related all motion to the medium in which a body moved, and also to its position in the universe ; anyone who wrote against Aristotle—like Benedetti in his *Book of Divers Speculations on Mathematics and Physics*—had always to remember that Aristotle had satisfactorily explained how and why bodies fall and projectiles move, and his theory had to offer explanations of the same kind. " Natural " motion, including the motion of falling bodies, had for Aristotle required no cause other than previous displacement ; for natural motion was the result of a body's intrinsic tendency to seek its natural place in the universe. A " heavy " body was one which tended to fall " down " (towards the centre) ; a light body one which tended to rise " up " ; both down and up being determined absolutely with respect to the centre of the universe. Absolutely heavy bodies and absolutely light bodies had only one tendency ; relatively light and heavy bodies were those which could either rise up or fall down, depending on where they found themselves. The body displaced "knows" that it is so, and hence "knows" its goal, so it moves faster (accelerates) as it approaches its destination.

One other factor is involved in natural motion : the medium. Recognising that the denser the medium through which a body moves, the slower the motion, Aristotle argued that the speed is inversely proportional to the density of the medium. Hence in a vacuum, where no medium exists, the speed of a falling body would be infinite. This was to Aristotle a manifest absurdity, and a solid argument against the possibility of the existence of a vacuum. Again, the heavier the body, the greater the ability to overcome the resistance of the medium, so the swifter the fall ; hence, the speed of a falling body is also directly proportional to the weight of the body. Projectile motion did, in Aristotle's view, require a force, not only to initiate it, but to ensure its continuance, because it was forced, not natural, motion. For Aristotle (as for Descartes later) all such motion had to proceed by impact, and he imagined that again the medium played the essential role, maintaining the push initially imparted by hand or sling. But the push of the medium gradually grew less and less with time, until it was finally worn out ; at this point gravity, previously inoperative, took over, and the body dropped under natural motion. Since forced motion and natural motion did not mix, all projectiles were regarded as having straight-line trajectories, not curved ones.

Now these theories, though they appeared to offer answers to all possible questions concerning bodies in motion, were not wholly satisfactory, and a certain amount of criticism began very early, especially in regard to the question of whether, in fact, bodies fall at speeds exactly proportional to their weights. Alternative answers were slow to develop, and it was only at the end of the classical period, among sixth-century commentators on Aristotle, that the impetus theory was first adumbrated. This theory preserved the outlines of Aristotelian thought, the doctrine of natural places and the incompatibility of mixed motion, while at the same time its proponents rejected Aristotle's view that a body continues to move after an initial application of force because the air pushes it along. This they did on two grounds : first, because air (as Aristotle himself had said) naturally resists motion ; and second, because the motion of heavy bodies lasts longer than that of light bodies, although air moves light objects more easily than heavy ones. These logical arguments they supported by examples drawn from experience, familiar facts to be continually repeated for centuries. In place of Aristotle's theory, they supposed that the moving force gave to the body moved an *impetus* (also known as an "impressed force" or "moving virtue"). Just as *heat* was the name given to the quality possessed by a body which is hot, so *impetus* was the name given to the quality possessed by an object that moves; and just as the heat gradually wears off after the fire is removed, so impetus must do when the moving force is removed.

The impetus theory reached its height of sophistication in the fourteenth century, in the hands of English mathematicians at Merton College, Oxford, and others at the University of Paris. They used it to explain how a falling body increases its speed (because at every instant the tendency to fall is added to the existing impetus to move) and thereby dispensed with the notion that speed increases as the goal is approached. They even used impetus to account for the unchanging and eternal revolution of the celestial spheres. More important, they recognised that speed itself (and not merely impetus) could be treated as a quality of the moving body. They had devoted great ingenuity to the development of both geometric and arithmetic expressions for the *variation* of qualities in general, working on the assumption that the "intension" of any quality (like heat, or whiteness) could be denoted numerically. Thus, they argued, a body of heat 8 would be hotter than one of heat 4; and a speed 8 would be faster than speed 4. (Of course, these numbers are purely arbitrary, and had no physical meaning.) One of the important questions treated was the comparison of a quality that varied (say from 9 to 1) with a quality that remained constant, a process known as the calculus of qualities. The most fertile such calculus of qualities was that developed for the discussion of the "latitude"

or variation of forms and qualities by Nicole Oresme, the great
mathematical philosopher of the fourteenth-century University
of Paris. Essentially it was a method of plotting the "intension"
of a quality geometrically against something else—its "extension"—which was constant (a period of time, for example).
If the variation were linear, Oresme called it "uniform"; if

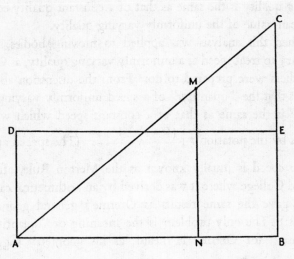

FIG. 7. THE LATITUDE OF FORMS: GEOMETRICAL ANALYSIS

non-linear, "difform." He therefore represented uniform variation by a sloping straight line, difform variation by a curve. Thus
in figure 7, the intensity of any uniformly varying quality
is represented by the length of the vertical line MN, increasing
uniformly as N moves from A to B. As Oresme expressed it:

> The quantity of any linear quality at all is to be imagined by
> a surface whose longitude or base is a line protracted in
> some way in the subject . . . [i.e. the extension] and
> whose latitude or altitude is designated by a line erected

perpendicularly upon the protracted base line [i.e. the intension].[13]

In the figure, therefore, the quantity of the uniformly varying quality MN is the area of the triangle ABC ; and this is, obviously, equal to the area of the rectangle ABED, when E is the mid-point of BC. Hence, Oresme concluded, the quantity of a uniformly varying quality is the same as that of a constant quality equal to the mean value of the uniformly varying quality.

When this analysis was applied to moving bodies, it was necessary to treat speed as a uniformly varying quality, as Oresme and others were prepared to do. From the discussion above, it follows that the " quantity " of a speed uniformly varying from v to V is the same as that of a constant speed which we may express by the notation $v + \dfrac{(V-v)}{2} = \dfrac{V+v}{2}$. (The special application to speed is usually known as the Merton Rule, after the Oxford College where it was derived by an arithmetical calculus, which gave the same results as Oresme's general geometrical analysis.) The only problem is the meaning of " quantity " in this case : for Oresme it meant, as his geometry suggested, distance.

There were many other problems involved, however ; and these were not to be cleared up until Galileo took them in hand. Though the Merton Rule provided a way of treating accelerated motion, it was not applied to falling bodies before the sixteenth century, because no one was bold enough to assume that such bodies are uniformly accelerated. And the mathematicians who discussed the intension and remission of such qualities as speed did not relate this directly to impetus, which remained a useful explanation of why bodies moved, without being necessarily involved in the purely mathematical discussion of *how* mathematical magnitudes moved.

Impetus theory in the sixteenth century was a muddled

subject, for it had had no consistent development. It was used to attack Aristotelian theory quite as much as to endeavour to understand the actual problems of moving bodies ; and the tacit belief that impetus theory could be treated in Archimedean fashion (which it could not) inevitably introduced confusion. Besides, each mathematician was interested in some special aspect of the problem, and few considered kinematics as a whole. Thus Tartaglia interested himself in the motion of bodies almost entirely from the point of view of ballistics, and his task was not made easier by his attempt to reconcile Aristotelian physics with the observations of gunners. (In spite of what might seem obvious, this was not in fact a case of reconciling traditional and out-of-date theory with the discoveries of clear-sighted empirics. Gunners made as many mistakes as Aristotle : they knew for a fact that a cannon ball increased its speed after leaving the gun for some little time, so that muzzle velocity was not maximum velocity.) Tartaglia regarded the imposition of impetus as responsible for forced motion, but for long he believed with Aristotle that natural and forced motion could not mix. Hence the trajectory of a projectile must consist of two straight lines ; later, perhaps in the face of observation, he decided that gravity must act continuously, always drawing the projectile a little away from a straight path into a slightly curved one ; or, as he put it, " there is always some part of gravity drawing the shot out of its line of motion." [14] He hesitated whether to describe the acceleration of a falling body in terms of its distance from its starting point or its approach to its terminal point, but he could not make up his mind. It was Benedetti, even more anti-Aristotelian than Tartaglia, who first liberated himself from the concept of a "goal" and began to consider only the past history of the falling body, without trying to anticipate its future, in attempting to establish the speed of the body at any given point.

Galileo's work *On Motion* (*De Motu*) belongs in the general tradition of Tartaglia and Benedetti. Though it is far superior to their work, it shows that even so penetrating a mind as Galileo's could not render the problem of falling bodies and projectiles clear and simple as long as it was considered within the framework of impetus physics. Galileo tried to write an elementary but exhaustive account. So his first chapters are concerned with the nature of heavy and light ; here he broke with Aristotle by denying the existence of light bodies. Lightness, he said, is relative ; apparently light bodies move upwards because heavy ones fall down below them, but in reality all bodies are more or less heavy. This notion he seems to have derived from a consideration of floating bodies ; and indeed much of this part of Galileo's mechanics is derived from Archimedes' hydrostatics. He is as much concerned with the rise of light bodies in water as with the fall of heavy ones in air, and so he regards the resistance of the medium (whether air or water) as a kind of buoyancy which supports less dense bodies more effectively than it does more dense ones. His theory can, in fact, be reduced to the view that bodies fall at speeds proportional to their densities (not their weights, as Aristotle had supposed) less the density of the medium. Or, as he said, speed will be " measured by the difference between the weight of a volume of the medium equal to the volume of the body, and the weight of the body itself." [15] Hence in air, for example, objects made of the same material, having the same density, would all fall at the same speed, irrespective of their weights. If one has two objects of the same *weight*, however, the denser would fall the faster.* If the density, or buoyancy, of the medium were to be progressively decreased, then the objects

* As this treatise was written while Galileo was at Pisa, it would have been this theory which was to be tested by the Leaning Tower of Pisa experiment—if indeed it took place. Others, like Stevin, had described dropping balls of different weights and the same material, which duly hit the ground together.[16] It proved Aristotle wrong, without establishing an alternative theory.

would both fall progressively faster until in the limit (i.e. a vacuum) their speeds would be proportional to their densities. (Galileo says to their weights, meaning their relative weights.) Thus motion in the vacuum is possible, Aristotle notwithstanding, even though objects of different materials still fall at different speeds in it.

Using weight as the determining factor, Galileo derived some rather peculiar notions about acceleration in free fall. According to his reasoning, a falling body has first to overcome the force which placed it in position, so its initial motion is accelerated motion. Once its characteristic speed of fall is attained, there is no further acceleration ; indeed, there can be none because, so Galileo argued, a constant force must produce a constant speed. Since heavy bodies have a greater force to overcome, they attain their characteristic speed more slowly than light ones. By this reasoning, Galileo was able to deny Aristotle's contention that unopposed natural motion would be infinitely swift, as in a vacuum, and opened the way for later consideration of the speed of bodies falling with no resisting medium. At the same time, Galileo was forced to conclude that true inertia is impossible, although he had some inkling of its practical existence. From a consideration of inclined planes (from which he later drew the conclusion that there *was* inertial motion) he here remarked that if one takes the case of a perfectly smooth body and a frictionless surface, one can conclude that " any body on a plane parallel to the horizon will be moved by the very smallest forces, indeed, by a force less than any given force." [17] This is, obviously, very close to the concept of inertia, still denying true inertial motion. Benedetti, indeed, had seemed to state the case for inertial motion more clearly—but only for abstract bodies moving through geo-metrical space.

How Galileo progressed from this world of involved Archi-medean-Aristotelian-impetus physics to a totally new dynamics is by no means clear. He wrote relatively little on mechanics between *De Motu* and the *Dialogue on the Two Chief Systems of the World* (1632) which embodies many of his conclusions on the subject.* He was mainly concerned, at least after 1604, with astronomy and polemic. But there are a few glimpses of his laborious progress from one system to another. The most famous is contained in a letter to Paolo Sarpi, and is dated 16 October 1604.[18] He wrote :

Reflecting on the problems of motion for which, in order to demonstrate the accidents which I have observed, I needed an absolutely certain principle which I could take as an axiom, I arrived at a proposition which seemed reasonably natural and self-evident : which being supposed, I demonstrate everything else, namely that *the spaces passed over in natural motion are as the squares of the time* and that consequently, the *spaces passed over in equal times are as the series of odd numbers.* And the principle is this : That the speed of a body falling in natural motion is proportional to its distance from its point of origin.

Now this is most curious ; for what Galileo has proved is the familiar law of free fall, namely $s = 1/2at^2$; but the self-evident natural axiom by which he claims to have derived this conclusion, that the instantaneous velocity is proportional to the distance traversed, is quite wrong. That speed was related to distance traversed (rather than to elapsed time) was an eminently natural assumption ; speed had been regarded as proportional to distance by, for example, Leonardo and Benedetti, and was still to be so regarded by Descartes, who never was able to correct this erroneous view. It was the almost inevitable result of trying to deal mathematically with falling bodies ; for as long as mathematics was primarily geometry, space rather than time is the most

* For a discussion of his later work in dynamics, see *Galileo to Newton* (vol. III in this series).

obvious dimension to consider. Only much later did Galileo
come to see that, although a constant cause must produce a
constant effect, this constant effect may be a rate of change, not
a fixed value; that is, it may be uniform acceleration rather than
constant speed. From this, ultimately, derives the law of inertia.
But in a very real sense a more intense degree of mathematisation
was required before the mathematical point of view could show
itself really consonant with the empirical test which Galileo
perhaps tried at this time, the rolling of balls down an inclined
plane in the manner so graphically described in the *Discourses on
Two New Sciences* of 1638. Yet his conclusions of 1604 amply
justified his faith in the mathematical approach, even though it
was to be some years before his mathematical reasoning could be
perfected. Galileo's early work shows both the strengths and the
weaknesses of sixteenth-century applied mathematics in the world
of physical bodies.

The mathematics used by the applied mathematician was not of
a notably high order; indeed, initially he used little which had
not been available to earlier centuries. The simplified computation possible with Hindu-Arabic numerals had been known to
the learned since 1200, as had algebraic methods of solving simple
equations, while the geometry used in surveying, navigation,
perspective and mechanics went little beyond Euclid. The trigonometrical requirements of navigation and astronomy were
mathematically complex, so that trigonometry mad esignificant
advances in the hands of astronomers and (later) pure mathematicians, as did advanced computational methods. Most of the
men who wrote on pure mathematics in this period wrote on
applied mathematics as well, so that theory and practice went
more serenely hand in hand than is usually the case. By the early
seventeenth century pure mathematics had reached a stage of
complexity far exceeding its state of a century earlier; for the

stimulus to this advance one can look both to the practical demands and the influence of humanism. For the later sixteenth century was powerfully influenced by mathematicians of the age after Euclid ; and it must be remembered that before 1550 even Archimedes had been better known for his mechanical than for his mathematical work. Later Greek mathematicians had been almost unknown until Regiomontanus and other mathematical humanists rescued their work from near oblivion and called attention to its importance.

It was not, in fact, until the second half of the sixteenth century that advanced mathematics received much attention from translators. One of the most important contributors to this work was Federigo Commandino (1509–75), mathematician to the Duke of Urbino, whose humanist court rejected astrology, leaving Commandino free to devote himself to a study of Greek mathematics. He was an indefatigable and able translator, with complete command of both Greek and mathematics. He was responsible for the first reasonably complete text of the mathematical work of Archimedes (a text which made available the *Sand-Reckoner*, in which the heliocentric system of Aristarchus is described) ; and he himself did sound work on the centres of gravity of solids, using Archimedean methods. He also made a translation of the *Conics* of Apollonios (1566), a text superior to those of Regiomontanus and of J. B. Memus (published 1537) ; yet it was only in the last quarter of the century that mathematicians began to study conic sections seriously. Commandino also translated the valuable *Mathematical Collections* of Pappus, and a number of other treatises on pure and applied mathematics. The algebra of Diophantos had been known only to mathematicians like John Dee who could read Greek ; it appeared in Latin in 1575, and subsequently suggested a host of new problems to the already flourishing algebraists.

Geometry was undoubtedly at once the most useful and the

most advanced branch of mathematics ; perhaps for that reason it received relatively less attention than other branches. Necessarily, much time and effort was expended in assimilating the work of the ancients, who had gone so far, and only advanced mathematicians could hope to succeed in developing novel forms. There was much Archimedean geometrical analysis of solid and plane surfaces, work that was only to show its worth in the next century. Francesco Maurolyco (1494–1575), considered one of the best of the sixteenth-century geometers, and an important writer on geometrical optics, wrote on conic sections, treating them, as Apollonios had not, as actual plane sections of the cone. There was a continued interest in the regular Platonic solids and, for the first time, an interest in skewed solids, first pictured by Luca Pacioli (d. *c.* 1510) in his *Divine Proportion* (1509), and often discussed thereafter. Kepler, the astronomer, published *Wine-Vat Stereometry* in 1615 ; in the course of trying to ascertain the proper method of judging the cubic contents of a wine cask he treated the determination of areas and volumes by means of infinitesimals, rather than by the more normal method of exhaustion, and discussed a wide variety of solids produced by the rotation of a conic section about any straight line lying in its plane, this investigation produced ninety-two differently shaped solids. Kepler, like Maurolyco, contributed to the development of mathematical optics, and more influentially. Indeed, the most important geometrical works in the later sixteenth century were concerned with the application of geometry to optics, astronomy and mechanics.

The most widely pursued form of mathematics in the fifteenth and sixteenth centuries was what we should now regard as the most elementary : the art of reckoning with Hindu-Arabic numerals, and the solution of numerical problems which tacitly required quadratic and cubic equations for their presentation. These two types of mathematics were usually subsumed under the general term Arithmetic, which by this time had lost its original Greek meaning of number-theory and had begun to replace the mediaeval term algorism. (*Algorism*—reckoning with Arabic numerals—is a corruption of the name of the ninth-century Islamic mathematician al-Kwarizmi ; the word *algebra* is a corruption of the title of the treatise in which he described the art of solving problems arithmetically rather than geometrically.)

The use of Arabic numerals had been known to specialists for centuries ; al-Kwarizmi's treatise on the subject was one of the first Arabic texts translated in the twelfth century, and the thirteenth-century treatise of Leonard of Pisa (misleadingly called the *Book of the Abacus* : in fact it made the abacus unnecessary) was a clear, concise and useful summary of the principal methods required. But the Arabic numerals were slow to replace the use of the abacus. This was not really so peculiar as it seems ; even the sixteenth century found the rules of simple arithmetic very difficult to comprehend, and long division was truly long in the time required to accomplish it. At the same time, quick and easy methods for the simpler operations of arithmetic were much in demand, especially in the merchant cities of Italy and Germany, and the later fifteenth century saw the appearance of numerous vernacular treatises designed to satisfy the demand. These surveyed the field from numeration to double-entry bookkeeping ; from simple addition to the solution of complex problems involving quadratic equations ; and from multiplication to the extraction of roots. Indeed the most complete and detailed fifteenth-century treatise, the *Summa* of Luca Pacioli (written in 1487 and published in 1494), includes arithmetic, algebra and (relatively briefly) practical geometry, making a manual of useful mathematics.

In both arithmetical and algebraic operations, some form of abbreviation was desirable, and indeed to print without contractions was an unheard of idea to fifteenth-century type-setters, still influenced by manuscript style. The earliest arithmetical signs were short forms of plus and minus, mere contractions ; our modern forms for these two operations first appear as a commercial symbol to indicate overweight and underweight bales or boxes of merchandise. Most sixteenth-century algebraic symbolism was also abbreviative rather than symbolic of operation. This is confirmed by the use of the term "cossist" for a writer on algebra ; it derives from the Italian use of the word *cosa* (thing, equivalent to Latin *res*) to designate the unknown quantity in a problem. Separate terms were used for powers to avoid writing the whole series of words ; it was only slowly that the advantages of a symbolism which clearly displayed the relationships between powers was recognised. (Even at the end of the seventeenth century, mathematicians wrote a^2 and aa indifferently.) Because we are familiar with a system of arithmetical and algebraic symbolism which has been standard for over two centuries, it is tempting to assume that each symbol has inherent merit, and hence to hail the early adoption of any one of these symbols as a great achievement. Sixteenth-century developments show the fallacy in this reasoning ; most modern symbols owe their survival more to luck than merit, and many equally useful and valid symbols appeared only to be lost, as mathematicians slowly worked from abbreviated (often called syncopated) algebra to true symbolic algebra.★

Each writer developed his own symbolism, and drew exclusively on those predecessors who wrote in his own vernacular —for algebra and arithmetic were popular arts—so that national schools of algebraic notation tended to emerge. It is equally true that there is almost no sixteenth-century writer in the field

★ The modern symbol for square root is no more—and no less—indicative of the operation to be performed than the earlier Rx (radix, root) generally written in the abbreviated form used in medical prescriptions ; this earlier form could equally carry small numbers to indicate higher roots, like the fourth or sixth.

who did not invent at least one symbol still in use : thus Robert Recorde, a teacher, not an original mathematician, was the first to use the modern sign for equality, though it had been used earlier as a non-mathematical commercial symbol ; in the *Whetstone of Wit* (1557) he explained that, in his view " nothing could be more equal " than two parallel equal lines. Nothing illustrates better the complexities of evaluating contributions to symbolism than the work of Simon Stevin on decimal fractions. His little work on the subject, originally published in Dutch in 1585 as *De Thiende*, then in French as *La Disme* (*The Tithe*), was influential in popularising the use of decimal fractions to simplify arithmetical computations ; but his notation was clumsy and was soon superseded. The first suggestion that there might be general rules established for algebraic notation came from François Viète (Vieta, 1540–1603) ; he advocated the use of vowels for unknown quantities and consonants for known or constant quantities. This principle was finally accepted (in a different form) when Descartes adopted the use of the letters at the end of the alphabet (especially x) for unknowns, and the initial letters of the alphabet for constants, a rule rapidly assimilated into seventeenth-century practice.

Of more importance in the long run than the development of the cossic art of symbolism was the discovery of general methods of handling algebraic powers and complex equations. The Greeks had solved quadratic equations geometrically ; the Islamic mathematicians followed suit, and achieved the solution of certain forms of cubic equations. But many of these latter were incapable of solution by the mathematical methods of the sixteenth century ; indeed, few quadratics could be solved by algebraic as distinct from geometrical methods. Thus Pacioli could give simple general rules for such equations as $x^2 + x = a$; but for more complex equations he had to provide a cumbersome geometrical solution. The aim was to find simple methods, such as anyone could learn to apply, but these simple methods were not readily found out. Few would now regard the following as a problem of higher mathematics : " Find me a number which, multiplied by its root plus three, will make twenty-one ? " (That is, find x^2 when $x^3 + 3x^2 = 21$.) Even if we cannot remember how to solve it, we know that the method is readily available. Yet Cardan, who prided himself on his algebraic skill, could not solve it when, among others of the same type, it was put by him to Tartaglia in 1539.[19] Indeed Tartaglia suspected as much, and suspected, too, that Cardan was trying to make him divulge his newly achieved power to solve most ordinary cubic equations.

Tartaglia's reputation as a professional teacher of mathematics (he lectured at Verona and Venice) as well as his livelihood depended on his being able to demonstrate his ability in a public challenge such as was common in the sixteenth century (and indeed continued common for another century and a half). It was almost necessary for such a man to keep a few techniques secret, so as to win renown and impress his colleagues. Tartaglia had been often challenged in the years before 1539 ; in each case, suspecting that cubic equations would be involved, he had worked out rules for solving one or more types ; and in each case he successfully responded to the challenge of solving such problems as " Find me four quantities in combined proportion, of which the second shall be two ; and the first and fourth added shall make ten," while at the same time baffling his competitors with his own problems. No wonder that he wrote on applied mathematics only, preferring to reap more honour and glory before telling the rest of the mathematical world how to solve such problems. The only wonder, in fact, is that, approached by Cardan in 1539 with problems that had formed part of a contest between Tartaglia and another mathematician two years before, Tartaglia should have yielded to importunity and given Cardan the answer which Cardan was incapable of working out for himself. To be sure, he made Cardan promise not to reveal the secret, a promise Cardan cheerfully broke when he published his algebraic treatise *Ars Magna* (*The Great Art*) six years later. Though Cardan gave full credit to Tartaglia, the latter was bitterly and publicly annoyed, and revenged himself by publishing the whole story in vivid detail. He had reason to feel bitter, and Cardan has emerged unfairly well in the eyes of historians of mathematics. To be sure, having been given the method of solution, Cardan showed ability in analysing the various kinds of cubic equations and in recognising negative roots for the first time as valid, but he did not originate the methods he describes.

Algebra continued to make progress in the later sixteenth century, if less boisterously, especially in the work of Viète and of Thomas Hariot (1560–1621). Both worked on cubic equations, devising new methods for their solution as well as for the solution of equations of higher degree.* Another considerable step forward was Viète's ordering of equations, that is, devising methods of reducing complex equations to their most workable form. Viète also spent much time on areas contained under complex curves, which he expressed by means of infinite series. Once again, national difference developed : Viète's work influenced primarily the French mathematicians, while the English mathematicians, especially John Wallis, preferred to draw their ideas and methods from Hariot's *The Art of Analytic Practice* (*Ars Analyticae Praxis*, 1631).

Arithmetic was useful in the affairs of the counting-house and market place ; algebra provided the solution to ingenious and interesting problems that might conceivably relate to com-

---

* They reduced cubic equations to the form $y^6 + y^3 = a$, which can be handled like a quadratic equation ; higher equations they handled by methods of approximation.

mercial enterprise ; but neither had much direct relevance for science at this stage. Arithmetic, of course, was used in astronomical calculation, but it was a cumbersome help at best. Astronomical calculation for long remained a drudgery which few can have undertaken cheerfully. Fortunately for astronomy there have always been scientists who enjoyed the sheer mechanics of wrestling with long and complex sums, a notable example in this period being Kepler. Even fairly simple astronomical calculations involved another branch of mathematics, of interest only to astronomers, the ancient art of trigonometry. This developed among Greek astronomers—notably Hipparchos and Ptolemy—out of the need to measure linear as well as angular velocity. Greek trigonometry was originally concerned with determining the length of an arc by measuring the length of the chord of the circle concerned. Thus, in figure 8, if a body moves from A to B along the arc of the circle, the distance traversed can be determined either by measuring the angle AOB, and knowing the length of the radius AO ; or from the length of the radius and that of the chord AB. The tables of chords drawn up by Ptolemy gave their lengths as parts of the diameter of the circle, and the related length of the arc corresponding to the chord. Various Hindu and Arabic developments led to the innovation of dividing the triangle in half, to give a right-angled triangle in which the important relationship was that of half the angle at the centre of the circle (the angle at O in the figure) with the radius. This is the familiar modern trigonometric sine,* though it only appears in its modern form of a decimal fraction in the eighteenth century. Tangents developed out of shadow measurements for time-reckoning. The fifteenth century saw the complete substitution of the right-angled triangle for the triangle inscribed in a circle ; this in turn suggested the introduction of

---

* The word itself is a Latin mistranslation of the Arabic transliteration of the Hindu word for the half-chord.

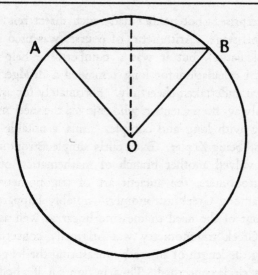

FIG. 8. THE GEOMETRICAL ORIGIN OF TRIGONOMETRIC SINES

the cosine (complement of the sine) as a useful trigonometric function. The secant and co-secant were also introduced in the fifteenth century, these as a by-product of navigational tables; like cosine and tangent they received their modern names in the course of the sixteenth century. Spherical trigonometry, which treated of triangles formed by the intersection of circles on a sphere, was widely used in astronomical calculation. Hence the "doctrine of the sphere," which began as the simplest branch of mathematical astronomy involving merely the naming and locating of the great circles of the universe, could also be a complex branch of mathematics.

Progress in trigonometry proceeded in orderly fashion, for the most part in conjunction with mathematical astronomy. Peurbach and Regiomontanus, studying Ptolemaic astronomy, studied trigonometry as well, and as companions to their astronomical treatises produced trigonometrical treatises. Peurbach was content with a new table of sines ; Regiomontanus produced *On Triangles* (1464, published in 1533), a complete survey of plane and spherical trigonometry. Copernicus annexed new trigonometric tables to the first book of *De Revolutionibus*, in exact imitation of Ptolemy, and his tables were, in turn, improved by Rheticus. In the later sixteenth century it was realised that trigonometric knowledge could be presented to non-mathematicians, and the more advanced and up-to-date sailing manuals taught the seaman how to use simple trigonometry. William Borough (1537–1598), an expert seaman who had learned the usefulness of mathematical knowledge by practice, exhorted the readers of his *Discourse on the Variation of the Cumpas* (1581) to compare his discussion with that of Regiomontanus, evidently his accepted authority. He was rather scornful of the table of sines prepared by Rheticus, which he thought inferior to those of Reinhold. He hoped to be able to publish even better ones " for the commoditie of all such as shall have occasion to use the same for Navigation and Cosmographie." [20]

The most important new development for both trigonometry and astronomical computation was the invention of logarithms by John Napier (1550–1614). In the sixteenth century sines were still expressed as lengths and, in order to avoid the fractions which made calculation so tedious, the radius of the circle in which the sine was inscribed was taken as a very large value ; this permitted the sine to be computed in units. Though this gave sufficient accuracy and avoided the use of fractions, it still made operations involving the multiplication and division of sines formidably complex and long-winded. Napier, searching for a means of devising tables which would permit rapid determination of the products of any two sines, began with a complex analysis of the relations between arithmetical and geometrical progressions of large numbers. He gradually found that the labour involved was so enormous that some other means must be invented ; analysing his results he discovered that he could achieve his desired end by the use of ratios, which he called logarithms. After twenty years of work he published in 1614 *A Description of the Marvellous Rule of Logarithms*: here he gave tables of the logarithms of sines and tangents, and explained how one could multiply sines by adding their logarithms and divide sines by subtracting their logarithms.[*][21] Napier's Latin version was translated by Edward Wright into English and published in 1616, to be followed three years later by Napier's description of the methods whereby he had calculated his tables. Meanwhile Henry Briggs (1561–1630), professor of geometry at Gresham College, had visited Napier and suggested the use of a decimal base together with the calculation of logarithms for ordinary numbers as well as trigonometric ones. Napier had previously intended to calculate tables to the base 10, and he gladly relegated the task to Briggs. In 1617 appeared the first of a series of tables prepared by Briggs, covering the first thousand numbers ; later tables were more extensive and included logarithms of trigonometric functions. In the form used by Briggs logarithms were seen to be immensely helpful, and they were soon widely used in long calculations. More tables followed, including a series by the Dutch printer Adam Vlacq in 1628, to fill up the gaps left by Briggs.

Logarithms were at once a triumph of pure mathematics and a gift to the practical mathematician, and both could appreciate them. What they could do was neatly summed up by the anonymous versifier who wrote in the preface to Wright's translation of Napier's *Description* that,

---

[*] Since a logarithm of a number to a given base is the power to which the base must be raised to give the number (so that the logarithm of 8 to the base 2 is 3, since $2^3 = 8$) logarithms obey the laws of exponents, whereby $a^n \times a^p = a^{n+p}$.

Their use is great in all true Measuring
Of Lands, Plots, Buildings & Fortification
So in Astronomie and Dialling
Geographie and Navigation.
In these and like, young students soon may gain
The skilful, too, may save cost, time and pain.[22]

CHAPTER VIII

THE ORGANISATION AND
REORGANISATION OF SCIENCE

There was but one course left, therefore—to try the whole thing anew upon a better plan, and to commence a total reconstruction of sciences, arts, and all human knowledge raised upon the proper foundations.[1]

Sixteenth-century scientists were filled with the twin spirits of novelty and rebellion. Consciously turning away from established views, eagerly making discoveries and discussing new ideas, they came to feel more and more surely that totally new methods were required for the effective investigation of nature. The mid-sixteenth century was too fully occupied—either in defending the thesis that all Aristotelian science was wrong (like the logician Petrus Ramus at Paris) or in developing new hypotheses in particular sciences—to consider the general problem of scientific method. But by the end of the century scientists had begun that intensive investigation of a possible reorganisation which was to lead to the production of full-fledged philosophies of science like those of Bacon and Descartes.

Organisation was of two possible varieties : organisation of method, and organisation of men ; and both were to be intensively discussed in this period. For both what was known and the numbers of those who knew it were increasing far too rapidly for the old ways to continue unchanged. In the fifteenth and early sixteenth centuries the scientists, like the humanists, could count themselves a small band of scholarly colleagues, and contact could be maintained by correspondence. Analogously, the content of science was still simple and restricted enough to make a conventional university education, supplemented by thorough independent reading of the classical authorities, adequate to equip a man to consider himself a serious scientist. The achievements of the mid-sixteenth century changed all that. Scientists began to demand, publicly and frequently, that there be far more teaching of science, and on two levels : on a high mathematical plane for the would-be scientists, and on a broad vernacular basis for the general public, now that science had shown that it could be useful. Scientists also began to wish for more opportunity for personal contact, now that every country could boast a significant number of trained scientists, not all of whom could expect to spend time in one of the major university centres of Europe. At first this discontent expressed itself in mere discussion ; by the end of the sixteenth century the first steps had been taken which were to yield the concrete results characteristic of the mid-seventeenth century.

The science of the early modern period begins in the university ; but, however much it remained true that most scientists were university-trained, and that many scientists found positions as university teachers, the science of the later sixteenth century was not indigenous to the university. In this respect the early modern period differs markedly from the Middle Ages when almost all scientific discussion took place within the walls of the university, and within the framework of the standard university curriculum. That by and large this was no longer true in the sixteenth century is only partly the result of university conservatism which saw no need to adapt its curriculum to the ferment of new ideas. It was equally the result of the changing nature and content of science itself. The universities, providing general education in the liberal arts for all, had always grounded their students in general elementary science only ; even the medical school gave theoretical rather than practical instruction, and generally of a fairly elementary nature. The complex anti-Aristotelian ideas on dynamics of the fourteenth century, the new astronomical speculations of the fifteenth century, or the new medical theories of the early sixteenth century could be presented to general students after their initial introduction to natural science because all the newer ideas were closely related in form and content to the Aristotelianism from which they were derived in revolt.

The new science of the later sixteenth century, though it was equally anti-Aristotelian in intent, derived from other sources than " the Master of Those That Know," and a knowledge of Aristotelian science did not necessarily make the new speculations intelligible. Even mathematics was more closely related to Archimedes than to Euclid, and only Euclidean geometry was taught in the university. Similarly, the texts of Sacrobosco or even Peurbach were not sufficient preparation for the astronomy of Tycho ; nor did knowledge of Mondino necessarily prepare one to tackle Vesalius. This situation grew more acute when, after the turn of the century, excitingly new developments were introduced by such men as Kepler, Harvey and Galileo, developments so radical that they bore little or no immediate resemblance to traditional natural philosophy. Until the new learning of the scientific revolution could produce its own textbooks in the elements of mathematics and natural philosophy, the academic education of any man, even the would-be scientist, was bound to be nearly devoid of useful scientific content. The most that university-trained men of the late sixteenth and early seventeenth centuries would admit as a (doubtful) benefit was an intense disgust for the Aristotelianism of traditional learning, generated by years of training in scholastic thought and method. The knowledge so acquired might prove useful in polemic, but hardly in any other way. It also left them with a working vocabulary poorly suited to new ideas, though having the advantage of universal intelligibility to all university graduates. All this being the case, it is no wonder that the best teachers preferred giving private lectures to holding university posts, or that one and all, whenever they thought or wrote on the problem of how best to advance scientific learning, always did so in terms of institutions quite unrelated to universities.

Scientific education proper in the sixteenth century took various forms, but the chief derived from the practice of the fifteenth-century humanist whose "family" included young scholars, half apprentices and as such members of his household, and half pupils. So John Dee at Mortlake taught others besides Thomas Digges, and directed and influenced their later work by the quasi-parental relationship established. The most striking example, on such a large scale that it seems almost an establishment apart, was the constitution of the household at Tycho Brahe's Uraniborg : surrounded in any case by the host of servants essential to feudal housekeeping, and needing numerous subordinates to assist him in his multifarious scientific activities, it was easy to include a few more young men in the life of the island. There was plenty to do : astronomical instruments to be set up and calibrated ; astronomical observations to be made ; alchemical experiments to be tried ; a printing-press to be supervised. Many aspirants to either astronomy or the scientific instrument trade did come to spend a few months or years—not only from Denmark, but from many foreign countries—at the only scientific centre of its kind. When Tycho finally left Uraniborg, many of his disciples went with him, so that he was never without scientific assistants in training.

A quite different form of scientific group—though still with an educational purpose—was that held together by a wealthy and noble patron, who hoped to learn from the scientists whom he wholly or partially supported. A famous example in England is the group gathered around Sir Walter Raleigh, usually identified with Shakespeare's School of Night (mysteriously mentioned in *Love's Labour's Lost*). Most of these men were poets and playwrights ; but one was the mathematician Thomas Hariot who taught Raleigh and the occult-loving Earls of Northumberland and Derby a good deal of astronomy and mathematics, and there were one or two minor mathematicians in the group. Their known subjects of discussion were philosophical, and their reputation a dark one, for they were said to dabble in both magic and atheism. In fact, except that they did interest themselves in chemistry—Raleigh had a famous secret cordial that he was allowed to try on the dying Prince Henry—their interests were probably fairly rational.

There were many other such groups, little known unless they happened to include some notable scientific figure. In effect the Accademia dei Lincei, to which Galileo was so proud to belong, was much the same sort of group : in origin and activity it was far closer to such gatherings than to the later scientific societies. The Accademia dei Lincei began in 1603 when Duke Federigo Cesi (1585–1630) began studying natural history at his home in Rome with three minor scientists ; the best known is Francesco Stelluti (1577–1653) who is associated with the publication of the first microscopical figures, a couple of plates of bees printed by the Academy in 1625, and later reissued in 1630. Like Raleigh's group, Cesi and his friends were suspected of occult studies (they were reputed to communicate in cipher) and they seem to have discontinued their meetings until 1610. They were then more formally organised with a membership of thirty-two which included both Galileo and della Porta. The newly invented optical instrument of Galileo was a source of great comment and interest, and it was at a meeting of the Academy that the name telescope was first applied to it. Many of these men were not resident in Rome, and their plans for a studious, quasi-monastic (but anti-clerical) communal life were soon abandoned. Perhaps they took warning from the evil reputation that they acquired, which must have seemed strengthened by the enrolment of della Porta within their ranks ; he had long since attempted to organise an Academy of the Secrets of Nature which, not surprisingly if its members pursued his favourite studies, soon foundered on the rock of suspected witchcraft. Nevertheless, the Lynxes continued to exist, to meet occasionally, and even to publish books— including some controversial writings of Galileo—until Cesi's death in 1630 ; and Galileo always proudly indicated his membership on the title-page of all his works.

The true form of academy, organised by its members and run entirely by them, did appear in the sixteenth century, but few were scientific. The humanists still led in scholarly organisation and most of these first societies were literary, like the Accademia della Crusca, formed to adjudicate upon the purity of the Italian language. A few had somewhat tenuous connections with science, like the academy (or rather series of academies) formed by the Pléiade after 1550. Although nominally under the patronage of King Henri III, its cessation with his death was the result of the increased political chaos in France, rather than an indication of his influence. The academicians were primarily interested in literature and the arts, but there was a good deal of scientific interest expressed by its changing membership, and its especial emphasis on music gave room for mathematics and physical acoustics.

In the generation after 1589 there were many demands in France for a revived royal sponsorship of an academy devoted like the old ones to learning and the arts. Some of the interest in both the aesthetic and scientific aspects of music characteristic of the older group is apparent in the work of Marin Mersenne (1588–1648), whose *Harmonie Universelle* (1627) is very like what the Pléiade sought to develop. Mersenne came to Paris in 1619, and deplored the lack of any sort of formal organisation to which men of learning might resort ; he wistfully remarked that there seemed in 1623 no chance for forming such centres as the earlier academies, though he hoped that some day something better might be organised. Meanwhile, he did his best to supply a substitute, partly by inviting men interested in science to visit him at his monastic cell, partly by becoming what the seventeenth century called an intelligencer, a man who made it his business to know and correspond with the leading scientists of the day, to whom he dispensed news in return for more news.

There were other cases of individual patronage by a scientist or semi-scientist : one of Mersenne's correspondents was Peiresc (1580–1637), a wealthy amateur in the south of France. Peiresc was a friend of della Porta and a member of the Lincei ; consequently he was an early telescopic astronomer, and others besides himself made use of his instruments. A curious case is that of John Dee : at his house at Mortlake he had assembled an enormous library and many scientific instruments, which were used by others besides himself. He hoped that the Queen (who often stopped at Mortlake to inquire the latest news from the stars) would grant him a comfortable living somewhere in the country where he could organise a scientific centre, to be shared by his friends. But nothing came of this except a long and informative appeal by Dee to the Queen.

In spite of abortive attempts at formal scientific organisation, it remained true that the best places to gather news and exchange ideas were the cities in which scientists, drawn by the prospect of employment, were most thickly gathered. These were not university towns, in general. Nuremberg, centre of the astrological, instrument-making and printing trades was more of a genuine scientific centre than any of the university cities. Other cities attracted scientists because they possessed mathematical lectureships, often designed for the benefit of the practical man. There was much interest in mathematical lectures in Paris, as Dee reported from his own experience, in spite of the existence of the university. He was invited to lecture there by " some English gentlemen " and his lectures on Euclid, so he remembered later, drew so many auditors " that the mathematicall schooles could not hold them ; for many were faine, without the schooles at the windowes, to be auditors and spectators, as they best could helpe themselves thereto." [2] The King offered him a good stipend if he would stay and become a regular lecturer in mathematics. This in spite of the fact that there was a mathematical professorship, whose incumbent at this period was Oronce Finé, associated with the humanist Collège Royale, founded by Francis I ; though the college was intended as a centre for study of Greek, Latin and Hebrew, it soon acquired professorships of both mathematics and medicine, and the lectures were given twice, once in Latin and once in French. But there was apparently still an unsatisfied demand.

In London there were no university or public lectureships, but private mathematical lectures abounded, similar to those given by Robert Recorde earlier in the century. The first public scientific lectures were those on anatomy given by the Reader appointed by the united Company of Barbers and Surgeons ; after 1583 there were also the Lumleian lectures in surgery read at the Royal College of Physicians. Various schemes for the establishment of professorships and academies by the Crown were proposed, and came to nothing. The first successful lectureship in the mathematical sciences was established by a public-minded group of London citizens in 1588 : the lectures were to be for the instruction of the captains of the trained bands, though they were open to the public. The first and only incumbent, Thomas Hood, faithfully lectured for four or five years, and translated and wrote several works in elementary mathematics as well, but when the emergency eased, the interest appeared to slacken.*

Perhaps this was because an endowment for a much more munificent scheme had already been announced, and was soon to take effect. In 1575 Sir Thomas Gresham had drawn up a will in which he bequeathed much of his property (chiefly the Royal Exchange and a great house in Bishopsgate Street) jointly to the City of London and Company of Mercers ; after his death and that of his wife, the heirs were to support seven professors—of Rhetoric, Divinity, Music, Physic, Geometry, Astronomy and Law—who were to live and lecture in his house.† The lectures began in 1598 in what was soon to be known as Gresham College ; the geometrical and astronomical professors were especially notable. The first professor of geometry was Henry Briggs, not yet associated with Napier ; a Cambridge graduate, he held the chair until 1619 when, on the establishment of the Savilian chairs of astronomy and geometry at Oxford, he became professor of geometry—a progression to be followed by other Gresham professors, including Wren. The first Gresham professor of astronomy was an undistinguished Oxford graduate named Edward Brerewood ; he was succeeded in 1619 by Edmund Gunter (1581–1626), already known for work in navigation and practical mathematics. Gunter's successor (in 1626) was Henry Gellibrand (1597–1636), who continued Gunter's interest in navigation, though in a different way : Gellibrand was responsible for the discovery of the change of magnetic variation from true North with time, a discovery which he made through experiments conducted over a period of years in the garden of

* Hood's first lecture was printed as *A Copie of the Speache : Made by the Mathematicall Lecturer, unto the Worshipfull Companye Present . . . the 4 of November, 1588.*

† There seems to be no indication of the origin of Gresham's interest in founding his college, unless it arose from the fact that, before being apprenticed to a merchant, he had been allowed a short period at Caius College.

the Keeper of Stores at the Navy Yard at Deptford. Gresham College was particularly important in supplying a meeting place for scientists. Many scientists and physicians with varied interests met in Gresham's great house before and after lectures on astronomy, mathematics and medicine, long before a group of young scientists who were later to form the nucleus of the Royal Society adopted the practice. Gresham College was in some senses the university of the new learning.

Part of the scientists' eager desire to increase the possibilities for acquiring a scientific education, as well as their enthusiasm for friendly meetings for the exchange of ideas, derived directly from a new self-confidence, a belief that theirs was truly the way to an understanding of nature. Not many as yet took them at face value ; one of the few men who did so, and who tried to preach their ideas more widely, was Francis Bacon, not a scientist, not even a recognised patron of science, who yet became the most vocal prophet of science possible. Bacon shared to the full the self-confidence of the scientist ; in his early youth (he was born in 1561) he once wrote to his uncle, Lord Burghley, that he had taken all learning to be his province. His first taste of formal education did not assist him ; he spent a short time at Cambridge University from which he derived no other benefit than the conviction that scholastic modes of thought were utterly sterile and useless. Like Petrus Ramus (1515–42) in France, he was inclined to think that if only one altered the method of reasoning by introducing a new logic all would be well. His later training, legal and courtly, tended to confirm this view ; yet gradually he extended his knowledge by reading and meditation, until he came to the conclusion that only science could provide the key to the truth, and only empiricism could provide the key to science. He had a Faustian belief that knowledge was power, but his exact legal mind prevented him from equating knowledge with magic ; the closest he came to magic was to adopt the work of the natural magician which he soon transmuted into genuine experimental science. He was fired with the conviction that he had found the best, shortest and safest road to scientific certainty ; and he passionately longed to persuade the world of the value of his knowledge, and the error of earlier methods.

Bacon's aim—to reform all knowledge and create a "new learning" in place of the old—was one which he shared with Galileo (his contemporary) and Descartes (a generation younger, 1596–1650) ; and like them, he believed that in the reform of scientific method lay the possibility of improving all learning. Unlike Galileo, he was no professional or even serious original scientist ; indeed his knowledge of contemporary science was curiously uneven, and his ideas of scientific experiment were naïve and over-simple. Unlike Descartes, he was no mathematician or profound abstract philosopher, nor was he a gentleman of leisure. He was a man of law by training and profession, a busy public figure, always seeking place and advancement until he achieved his goal as Lord Chancellor. Only his final disgrace and enforced retirement gave him time and leisure to complete (nearly) the series of works he had planned. The last five years of his life were filled with writing and experiment. It was love of experiment which occasioned his death in 1626 : he stopped one snowy day near Highgate to buy a hen from a housewife, which he had killed and then stuffed with snow, the object being to test the action of cold on the preservation of food ; inadvertently, he tested the action of cold on the human body, and died of pneumonia. The experiment is curiously characteristic : a good idea in advance of its time, investigated spontaneously, unsystematically, inconclusively, but ardently.

Bacon's first attempt at describing his ideas about the deficiencies of current science, and the need for a new approach, was addressed to James I, and published in 1605 under the title *The Two Books of Francis Bacon, of the Proficiencie and Advancement of*

Learning, Divine and Human. After a preliminary bid for royal patronage, Bacon turned to the subject of his real interest, the appraisal of current knowledge in all fields, and the proposing of steps to be taken to improve it. Eager to criticise the school-men, Bacon was yet anxious to avoid any tendency towards anti-intellectualism, and was therefore careful to balance praise and blame, praise for learning rightly pursued, blame for the methods of the contemporary world. It was easy to show the benefits of learning : rightly pursued, it improves the mind, strengthens the character, ennobles the citizen and the state, and is a source of power, delight and utility to man. Learning as practised may appear none of these things, but that is because it is subject to abuse, pedantry, excessive reliance on authority, ignorance, the self-esteem of its proponents, the pitfalls of the human mind, mysticism and limitation of range. The pitfalls of the human mind, which Bacon was later to dramatise as Idols, particularly interested him, for they were the result of inherent human tendencies—like love of system building, the influence of custom, and the snare of words improperly used—against which the only weapons were recognition and vigilance.

The worst of all defects was that men generally have sought knowledge for the wrong reasons :

Men have entered into a desire of learning and knowledge, sometimes upon a natural curiosity and inquisitive appetite ; sometimes to entertain their minds with variety and delight ; sometimes for ornament and reputation ; and sometimes to enable them to victory of wit and contradiction ; and most times for lucre and profession ; and seldom sincerely to give a true account of their gift of reason, to the benefit and use of men.[3]

The " benefit and use of men " meant to Bacon many things : power, because it was synonymous with understanding ; truth ; control of nature ; and the " relief of man's estate," the application of science to the useful arts that could improve the material well-being of mankind. Because this was a novel idea, it has perhaps been too much stressed as Bacon's chief aim in the advancement of science, which it emphatically was not. No one ever inveighed more firmly than Bacon against the evils of purely " lucriferous " (money-grubbing) knowledge ; what he sought was " luciferous " (enlightening) knowledge. But he did believe that knowledge gave the power to improve the lot of mankind, and to increase the sum total of human happiness : in many ways, Bacon was the real progenitor of the eighteenth-century enlightenment. His greatest criticism was that men had sought learning for private and trivial reasons ; he was too impressed with the potentialities of true knowledge not to feel that its study should be undertaken only in a serious—even a solemn—spirit.

Of all the varieties of human knowledge, science had advanced the least, because it lacked any coherent method of procedure. It did not even, like the mechanic arts, build on past experience ; one age, Bacon complained, did not learn from another ; Aristotle was as good a scientist—perhaps a better one—than any since, and in the early seventeenth century one knew for certain little more than Aristotle had known. Or, as he put it,

The sciences stand where they did and remain almost in the same condition, receiving no noticeable increase, but on the contrary, thriving most under their first founder, and then declining. Whereas in the mechanical arts, which are founded on nature and the light of experience, we see the contrary happen, for these (as long as they are popular) are continually thriving and growing, as having in them a breath of life ; at first rude, then convenient, afterwards adorned, and at all times advancing.[4]

So the sciences should copy the mechanical arts in two respects : they should be " founded on nature " and they should learn to be cumulative. The greatest need was for the organisation of scientific method. Until the structure of scientific inquiry was understood, how could men know how rightly to proceed? And it was precisely for want of co-ordination of the various lines of scientific inquiry that the sciences had hitherto languished so miserably; or, as Bacon admonished,

Let no man look for much progress in the sciences—especially in the practical part of them—unless natural philosophy be carried on and applied to particular sciences, and particular sciences be carried back again to natural philosophy.[5]

To Bacon it seemed natural—since he was imbued with the notions and methods of the law *—that the first step was to classify the major divisions of learning, including natural philosophy, in order to see what needed to be done. He divided learning into three main headings: History, Poesy, and Philosophy, each of which required the exclusive use of one of the three Faculties of Memory, Imagination, and Reason. Imagination was all too much stimulated in Bacon's view—it must have been hard to be born without appreciation of poetry in the Elizabethan world—and hence had no role in the advancement of learning, however much it might delight and amuse. History and philosophy had been cultivated, but not properly, and not enough; and besides, natural history and natural philosophy must be separated from the other aspects of these two heads. This was easy to do with natural history, which had little tendency to become entangled with civil or ecclesiastical history, but natural philosophy presented a more difficult problem. Natural philosophy must be separated on the one hand from divine theology—for though the study of nature might lead to an enhanced view of the wonder and majesty of God, faith and natural philosophy had no connection and must be kept firmly

* Harvey was later to say, cogently enough, that Bacon wrote philosophy like a Lord Chancellor.

separate—and on the other hand from metaphysics, which concerned itself exclusively with final causes, those " barren virgins " who had no place in natural philosophy. Analysing and cataloguing further, Bacon divided natural philosophy into " the Inquisition of Causes, and the Production of Effects ; Speculative, and Operative ; Natural Science, and Natural Prudence," [6] each of which he was quite ready to subdivide further.

In spite of his legal mind, Bacon did not classify as a means of description, but as a means of demonstrating that " knowledges are as pyramids, whereof history is the basis : so of Natural Philosophy the basis is Natural History ; the stage next the basis is Physic ; the stage next the vertical point is Metaphysic," [7] Physic being the inquiry into material and efficient causes, Metaphysic the inquiry into formal causes. It was this belief in the pyramidal structure of learning that led Bacon to lay such great stress on natural history—the collection of facts about

> nature in course, . . . nature erring or varying, and . . . nature
> altered or wrought ; that is, history of creatures, history of
> marvels, history of arts.[8]

By " nature in course " Bacon meant nature as she appears to the observant eye and mind ; " nature erring or varying " was aberrant, marvellous nature, the study of wonders and monsters, a subject which the Royal Society was to find almost as interesting as Bacon did. By " nature altered or wrought " he meant on the one hand the curious discoveries of the trades and mechanic arts, and on the other hand what could be created in the form of an experiment by the curious investigator.

That the arts and crafts had much to teach the scholar was fairly widely recognised in the later sixteenth century, though no one had previously made it a principle ; that men should deliberately experiment—as distinct from merely observing nature—had seldom been advocated, certainly not with the systematic thoroughness which Bacon proposed. To Bacon, experiment was the one truly necessary ingredient of scientific endeavour : without it, he thought, natural philosophy was no better than metaphysical speculation, and the scientist no better than the metaphysician who spun webs of *a priori* hypotheses out of his own inside. With experiment, the scientist possessed the key he needed to unlock the secrets of nature ; the use of experiment

is of all others the most radical and fundamental towards natural philosophy ; such natural philosophy as shall not vanish in the fume of subtle, sublime or delectable speculation, but such as shall be operative to the endowment and benefit of man's life ; for . . . it will give a more true and real illumination concerning causes and axioms than is hitherto attained. For like as a man's disposition is never well known till he be crossed, nor Proteus ever changed shapes till he was straitened and held fast ; so the passages and variation of nature cannot appear so fully in the liberty of nature, as in the trials and vexations of art.[9]

In spite of the baroque style, Bacon's meaning is clear ; he was firmly convinced that the experimental method, properly developed, was the only true way.

There was another advantage to the experimental method, of peculiar importance for the organisation of the scientist as distinct from the organisation of science : it permitted co-operative endeavour, and it permitted various kinds of minds to contribute equally to the progress of science. Facts, whether derived from observations or experiments, were useful to the scientist who was investigating nature ; and it did not take the same kind of mind to collect facts as it did to make use of them. The encyclopedia was a preliminary to the scientific theory, and the man who collected its facts could feel that he had made a genuine contribution to the advancement of knowledge. It was a kind of demo-cratisation of knowledge, because it lessened the need for high intellectual powers such as were required in reasoning ; as Bacon once put it,

> the course I propose for the discovery of sciences is such as
> leaves but little to the acuteness and strength of wits, but
> places all wits and understandings nearly on a level.[10]

That this was exaggeration he knew ; and when he wrote the fragmentary account of a scientific Utopia in *The New Atlantis*, his Salomon's House—the island's scientific research centre—contained men who merely thought, devising experiments, analysing results, and drawing conclusions, as well as men who merely observed facts and performed experiments at the direction of others. Though he was unduly optimistic about the possibilities of co-operative science, Bacon ingeniously saw some of the possibilities for utilising lesser minds to help in the task of understanding nature which were, in fact, to emerge as modern science progressed.

Bacon by no means understood his experimental way to be pure empiricism. He had no use for random experimentation, undertaken without aim or guiding principle, however much he might at various times fall into the error of collecting diverse experiments, such as those posthumously published in his *Sylva Sylvarum*. Nor did he think well of investigating some one aspect of nature by performing all possible experiments relating to it, " as Gilbert with the magnet, and the chemist with gold " ; this he thought " blind and stupid. For no one successfully investigates the nature of a thing in the thing itself ; the enquiry must be enlarged, so as to become more general." [11] Experimenters should begin by trying to devise " experiments of light " from which general axioms or principles could be drawn by the method of induction, and this meant that the experiments must be planned beforehand.

The method of induction was Bacon's bid for a new logic to replace Aristotle's, and for that reason he called the work in which he discussed this logical method *Novum Organum* (1620). This book is Bacon's boldest claim for serious consideration as a philosopher ; the fact that generations of philosophers have criticised it as naïve, inconsistent and a failure shows that he did at least succeed enough to be taken seriously. *Novum Organum* was intended to be a book of restricted scope, for it was written as the second part of his " Great Instauration," the plan for the restoration of the sciences to their proper dignity and usefulness. The first part was a revised and extended Latin version of the *Advancement of Learning*, which explained the need for the new sciences ; *Novum Organum* explained the method ; and there were to be other works on natural history and natural philosophy illustrating the possibilities of the plan, by demonstrating how to proceed from fact to theory. The method was simple ; induction is that method of reasoning which " derives axioms from the senses and particulars, rising by a gradual and unbroken ascent, so that it arrives at the most general axioms last of all. This is the true way but as yet untried." [12] The method of induction is based upon the method of experiment, and is a method whereby reason and sense experience can learn to support one another.

The weakness of Bacon's proposals is that, however admirable in themselves, they do not carry conviction, and this chiefly because one cannot read long without realising that Bacon could not judge the worth of any individual experiment or experimental discovery, even while proclaiming the value of experiment. Partly this was because he expected too much ; convinced of the value of the experimental way, he could not believe that it had been rightly applied if it failed to give immediately satisfying answers. So (in 1620, when nothing had as yet been published on the subject) he doubted whether the microscope would prove a permanently useful scientific instrument, because it had been used only on trivial subjects ; yet, he added, " great advantages might doubtless be derived from the discovery " if only " it could be extended to larger bodies, or to the minutiae of larger bodies, so that the texture of a linen cloth could be seen like network, and thus the latent minutiae and inequalities of gems, liquors, urine, blood, wounds &c. could be distinguished," [13] a very reasonable expectation. He was not even certain of the permanent value of the telescope ; Galileo's discoveries were admirable, but the telescope had made no more discoveries ; and he could not help regarding even the first ones " with suspicion chiefly because the experiment stops with these few discoveries, and many other things equally worthy of investigation are not discovered by the same means." [14] Bacon clearly had no feeling for patience in scientific inquiry, and no notion of how long any actual " discovery of nature " might be expected to take.

In a different way, he deprecated the work of Gilbert because, after a wealth of experiment, " he made a philosophy out of the loadstone " and indulged in extravagant speculations ; if he could speculate so wildly, how could one be quite sure that his experiments were truly performed and reported ? (Had Bacon lived to read Galileo's most important works, he would have deplored Galileo's addiction to mental in place of physical experiment.) Though he drew information from Gilbert, he could not help regarding him as a perpetual awful warning. Because he so feared abstract reasoning not based upon experience, Bacon was particularly mistrustful of such purely theoretical science as Copernican astronomy. From his reading (which probably included some of Tycho Brahe's works) Bacon had learned that Copernicus had wrongly attributed an extra and unnecessary motion to the Earth, and he had also learned that astronomers had begun to believe that it might be possible to construct a system of the world devoid of mathematical epicycles, strictly physical in construction.

But the strongest argument against Copernicus was that he had no observational evidence for his system :

It is easy to see, that both they who think the Earth revolves, and they who hold the *primum mobile* [to do so] . . . are about equally and indifferently supported by the phenomena.[15]

In fact, he preferred Tycho's system, very probably because of Tycho's greater insistence upon astronomical observation than upon astronomical calculation. He joined Tycho and other radical contemporary astronomers in denying the special character of the heavens, and he went beyond most in asserting that the physics of the celestial and the terrestrial spheres is identical. Or as he said picturesquely :

For those supposed divorces between ethereal and sublunary things seem to me but figments, superstitions mixed with rashness ; seeing it is most certain that very many effects, as of expansion, contraction, impression, cession, collection into masses, attraction, repulsion, assimilation, union and the like, have place not only here with us, but also in the heights of the heaven and the depths of the earth.[16]

It was observation alone that would demonstrate this, as it could settle other questions : whether the stars are scattered at different distances throughout space ; whether (as Gilbert thought) the stars revolve ; whether there is a true system of the world, or whether there are only stars and planets moving and existing independently in space. The astronomers might protest that Bacon ignored what they had done, and asked them to settle points on which they had no evidence ; but Bacon's notion of astronomical problems is both varied and complex, and a useful antidote to the new self-confidence of the astronomers.

Bacon's estimation of the problems that the scientists should immediately try to solve was often wrong, and he was unduly optimistic in thinking that it was only misunderstanding of scientific method which obstructed an immediate answer. But he was by no means always wrong about what were the most interesting problems, nor what the most correct approach to their solution might be. Of all parts of natural philosophy the most important—and the least studied up to his day—was, Bacon held, " the discovery of forms " ; and yet it was " of all other parts of knowledge the worthiest to be sought." [17] Here, as elsewhere, Bacon adopted the terminology of the familiar Aristotelian theory of causation—it was the only one available to him—but he modified it as he used it. Since all bodies are composed of both *form* and *matter*, there must be both formal and material causes of all things. Bacon believed that science should concern itself most with cause and effect ; but the formal cause, though it was not sufficient in itself, needed elucidation. Aristotle had defined the formal cause as what makes a body's essential nature, the sum of the attributes which make an object belong to a particular category or class, so that perceiving it we instantly give it a name. Originally Aristotle had in mind such things as the attributes of a piece of furniture, which make it recognisable as a chair, or of a piece of bronze which we recognise as a statue. In natural philosophy " form " came to have a more extended meaning : the formal cause for the kettle's growing warm on the fire was the heat which the fire contained, and similarly gold was characterised by such forms as density, malleability, resistance to corrosion, yellowness and the like.

The sixteenth century had introduced a vast new number of forms to account for chemical changes, and the new physics was to introduce even more. Bacon took a different view of what a form might be. As he explained :

For though in nature nothing really exists besides individual bodies, performing pure individual acts according to a fixed law, yet in philosophy this very law, and the investigation, discovery, and explanation of it, is the foundation as well of knowledge as of operation. And it is this law, with its clauses, that I mean when I speak of *Forms*.[18]

Lacking an understanding of mathematical law, Bacon was forced to introduce the somewhat cumbersome device of the law of forms. Yet what he meant is clear enough, and refutes the notion that he did not seek general principles of nature. Indeed, Bacon laid almost too much stress on the importance of discovering the natural laws of forms : from that alone, he thought " results truth in speculation and freedom in operation." [19]

The discovery of forms meant, generally, the study of the physical properties of matter. Such things as heat, colour, whiteness and blackness, rarity and density, attraction and repulsion, must be not just accidental attributes of matter, but the result of the obedience of matter to certain laws. Much of Book II of *Novum Organum* is devoted, in almost painful detail, to Bacon's views as to how one investigates and discovers these laws by induction. First comes the elaborate compilation and classification of the natural history of the particular form in question, the collection of a body of fact ; next comes the even more elaborate comparison, check and elimination which should, almost mechanically, leave one with the correct answer. He showed in detail how this should be done by performing it for the form of heat, giving four separate tables of " instances " of the production of heat all laboriously compared and cross-indexed. It is difficult to believe that anyone could imagine that this was the proper way to proceed in scientific inquiry ; yet Bacon's results were exactly those of the best seventeenth-century scientists, who derived their conclusion in quite a different way, namely that, " Heat is a motion, expansive, restrained, and acting in its strife upon the smaller particles of bodies." [20] And he ventured to predict that a number of other properties—colour, whiteness, chemical action—would similarly be found to result from the motion of the small parts of bodies. Obscure, almost mystic, as his language might be, Bacon was clearly formulating the premisses of what later came to be an almost universally accepted principle of science under the general name of the mechanical philosophy.

The nature of Bacon's achievement in this regard is more than a little obscured by the tediousness of his presentation. He discussed the investigation of forms at length only in a work primarily intended to explain the workings of the inductive method, and the sections on forms were introduced primarily as examples of that method. As he warned the reader, " It must be remembered however that in this Organum of mine I am handling logic, not philosophy." [21] Endeavouring to keep to the point, he resisted the temptation to desert induction for forms, and never explicitly discussed his tacit belief that the general method of explanation of forms lay in the study of matter and motion—though this was a view which later scientists like Robert Boyle derived from him.

Though philosophers—and his nineteenth-century editors—have often regarded Bacon's doctrine of forms as extraneous to his philosophy, this is because what they call philosophy he called logic. In his view, a rational experimental investigation of the properties of bodies was essential to the development of the new natural philosophy. In this he was prescient : for the mechanical philosophy, the derivation of physical properties from the mere structure and motion of matter—from the size, shape and motion of the invisible particles which must compose visible bodies—was to become one of the great organizing principles of seventeenth-century science. Bacon was one of the first to adopt and proclaim the view that one of the fundamental problems of natural philosophy was to find a method of explaining " occult properties " in rational terms.

The same point was made, more explicitly though in less detail, by Bacon's almost exact contemporary Galileo. In his famous polemical work The Assayer (Il Saggiatore, 1623) Galileo simultaneously developed an erroneous theory of the nature of comets and a brilliant analysis of his own, most fruitful, scientific method. As a consequence of the necessity to attack his opponent's views on the nature of heat, he had occasion to distinguish between the objective and the subjective attributes of physical bodies. He wrote :

But first I must consider what it is that we call heat, as I suspect that people in general have a concept of this which is very remote from the truth. For they believe that heat is a real phenomenon, or property, or quality, which actually resides in the material by which we feel ourselves warmed. Now I say that whenever I conceive any material or corporeal substance, I immediately feel the need to think of it as bounded, and as having this or that shape ; as being large or small in relation to other things, and in some specific place at any given time ; as being in motion or at rest ; as touching or not touching some other body ; and as being one in number, or few, or many. From these conditions I cannot separate such a substance by any stretch of my imagination. But that it must be white or red, bitter or sweet, noisy or silent, and of sweet or foul odour, my mind does not feel compelled to bring in as necessary accompaniments. Without the senses as our guides, reason or imagination unaided would probably never arrive at qualities like these. Hence I think that tastes, odours, colours, and so on are no more than mere names so far as the object in which we place them is concerned, and that they reside only in the consciousness. Hence if the living creature were removed, all these qualities would be wiped away and annihilated.[22]

This is a remarkably clear statement of what Locke was later to make famous as the distinction between primary and secondary qualities, primary qualities being those attributes of bodies which produce in us the sensations which we commonly ascribe to the bodies. Everyone has, usually as a child, encountered the old logical problem of whether when a tree falls in the centre of an uninhabited forest its fall produces a noise, when no one is there to hear ; Galileo offered the first explicitly satisfactory answer. Having raised the question, he concluded :

> To excite in us tastes, odours, and sounds, I believe that nothing is required in external bodies except shapes, numbers, and slow or rapid movements. I think that if ears, tongues and noses were removed, shapes and numbers and motions would remain, but not odours nor tastes nor sounds. The latter, I believe, are nothing more than names when separated from living beings.[23]

Having established the general principle, Galileo, like Bacon, turned to heat as a specific example. His choice was dictated by the considerations of polemic : he wished to demonstrate that his opponent was quite wrong to believe that heat could be generated by friction alone, and consequently to show that comets could not shine merely because they passed very rapidly through the atmosphere. He did not mean to argue that friction did not make a body hot, but merely to demonstrate that friction alone was not enough. In particular, Galileo wished to stress that something moving rapidly through the air did not necessarily grow hot, the old stories that the Babylonians cooked their eggs by whirling them in slings notwithstanding. At least, one could not achieve that result nowadays ; and consequently it must have been done by some other method than merely whirling them about. As Galileo argued :

> To discover the true cause I reason as follows : " If we do not achieve an effect which others formerly achieved, then it must be that in our operations we lack something that produced their success. And if there is just one single thing we lack, then that alone can be the true cause. Now we do not lack eggs, nor slings, nor sturdy fellows to whirl them ;

yet our eggs do not cook, but merely cool down faster if they happen to be hot. And since nothing is lacking to us except being Babylonians, then being Babylonians is the cause of the hardening of eggs, and not friction of the air." [24]

The purposes of polemic aside, Galileo had concluded that heat was the result of the impinging of the moving particles of fire, contained in matter, upon our organs of sense ; or as he described it :

Those materials which produce heat in us and make us feel warmth, which are known by the general name of " fire," would then be a multitude of minute particles having certain shapes and moving with certain velocities. Meeting with our bodies, they penetrate by means of their extreme subtlety, and their touch as felt by us when they pass through our substance is the sensation we call " heat." [25]

And he added, " Since the presence of fire-corpuscles alone does not suffice to excite heat, but their motion is needed also, it seems to me that one may very reasonably say that motion is the cause of heat."

Only one other thinker in this period dealt with the explanation of the properties of bodies in terms of the structure and motion of their ultimate particles : the obscure Dutch schoolmaster Isaac Beeckman (1588–1637).* He was not interested in publication, being satisfied to discuss his scientific speculations with friends, but he kept an elaborately detailed diary in which he noted a multitude of scientific observations, experiments and conjectures. He had developed a complex theory of matter before 1618, partly derived from atomic theories of Greek

* There were, of course, attempts to revive the atomic theories of the Greeks in this period ; examples are Nicholas Hill's *Epicurean Philosophy* (1601) or Sebastian Basso's *Anti-Aristotelian Natural Philosophy* (1621, based on Plato and Democritus) or Daniel Sennert's Democritean attempts to reconcile atomism and Aristotle (1619) ; but they were neither original, successful, nor, ultimately, influential.

antiquity, but directed to the same purpose as Bacon's discovery of forms and Galileo's speculations on the causes of sensations. Like Galileo, Beeckman believed that heat was caused by the motion of fire particles in a body ; and like Bacon and Galileo both, he believed that particles in motion were responsible for the physical properties of bodies. Or as he said, " all properties arise from motion, shape and size of the atoms, so that each of these three things must be considered." [26] Although he never worked out these speculations in detail, he did now and again revert to them, and he continually tried to explain seemingly mysterious natural events in terms of matter and motion. (So he attributed the action of suction pumps to the pressure of the incumbent atmosphere.) Beeckman's statements would be of little more than antiquarian interest were it not for the fact that he was a friend and associate of Descartes, whose views on science he greatly influenced ; though in this case Descartes, accepting the general principle, that the properties of bodies lie in the structure and motion of small particles, rejected the particular conclusion that matter was composed of atoms and vacuum.

By 1630, two generations of effort had hardly seemed to be rewarded, and science appeared as much unorganised as ever. Yet this was a false appearance, and the next generation was to demonstrate the fact. For the attempts at organisation of science, by the formation of societies, by the development of workable forms of scientific method, by the acceptance of the mechanical philosophy all reached fruition in the next thirty years.

CHAPTER IX

CIRCLES APPEAR IN PHYSIOLOGY

The animal's heart is the basis of its life, its chief member, the sun of its microcosm ; on the heart all its activity depends, from the heart all its liveliness and strength arise.[1]

The anatomical work of the first three-quarters of the sixteenth century satisfactorily established the gross structure of the heart, as it did the structure of most of the organs ; it could not so satisfactorily establish its function. Physiology, as the attempts of Fernel showed, could not advance much beyond Galen without further anatomical study ; but equally, as the elaborate attempts of Vesalius showed, it was not easy on the basis of exact anatomy alone to advance much beyond Galen. *On the Use of the Parts* remained the great standard work of reference and instruction, though now brought up to date and corrected in many special treatises : there is something curiously old-fashioned about the continual references to Galen as authority from men whose anatomical knowledge, as they were boastfully aware, now far exceeded Galen's. But until they could develop a new approach to physiological problems, Galen's system was not obsolete.

Temporarily, the anatomist was fairly well satisfied with his understanding of the structure and function of the venous and digestive systems ; Vesalius, finding Galen full of errors, was quite certain that he had been able to eradicate them. Vesalius had been content to leave Galen's picture of the structure of heart, arteries and lungs virtually unchanged; perhaps partly for this reason his contemporaries turned their attention particularly to the physiology of the heart and lungs, the physiology of respiration. They could not know that this was to remain an obscure problem until the pneumatic investigations of the mid-seventeenth century offered a new method of inquiry, and a supply of new facts. Even without new facts, sixteenth-century anatomists found Galen's account of the way by which the vital spirits got from the air to the lungs to the arteries to the tissues neither clear nor convincing. And what made Vesalius belligerently reluctant to enter too far into a subject rendered delicate by its theological implications—for "spirit" and "soul" were closely associated, since breath, vital to life, was the vehicle by which the soul left the body at death—made others, less place-seeking and less philosophically orthodox, peculiarly attracted to the problem.

It is more than probable that only theological interest ever led Michael Servetus (c. 1511–53) to wander into the complexities of the anatomy and physiology of the arterial system; for until he had occasion to discuss the mechanism whereby the vital spirit got to the heart, he had shown little interest in this branch of medicine. It is true that he had studied anatomy with Vesalius' teacher, Guinther of Andernach, and had even, like Vesalius earlier, served as his assistant; but what Servetus learned from his teacher was a profound interest in the philological aspects of medical humanism and a pronounced Galenism, rather than any bent for anatomical or physiological research. Indeed, Servetus appears to have turned to medicine as a profession only when literary and philological scholarship proved insufficiently rewarding.

The details of Servetus' life are as intricate as his theology; but the main outlines indicate a passionate addiction to radical unitarianism, coupled with a humanist devotion to the works of the ancients of a rather arid and rigid sort. He was born in Navarre and brought up in Catalonia ; sent by his father to study law at the University of Toulouse, he acquired a profound distaste for Catholic doctrine, a passionate anti-clericalism only nourished by a journey to Rome, a preference for France as a country of domicile, and no obvious taste for the law. His violent unitarianism soon found expression in two books on the errors of the Trinity, published before he was twenty-one, books so radical that they gave him a dangerously heretical reputation even in Protestant centres like Strasbourg and Basle. With a discretion he never showed again, Servetus changed his name to Villanovus (after the town in which his childhood had been passed), went to Lyons and there worked in the publishing trade for some years, interrupted by intermittent periods of study in Paris. He seems to have begun with some interest in applied mathematics if his first published work, a new edition of Ptolemy's *Geography* (1535) is any clue ; he also achieved a competence in, and profound acceptance of, judicial astrology. He indulged his addiction to controversy by writing a violent pamphlet attacking the botanist Fuchs, who was himself engaged in an equally violent controversy with a physician of Lyons named Champier. What the grounds of the dispute were is by no means clear, for both men were Galenists, and both opposed the Arabists ; but it provided much material for the printers.

Perhaps it was this which drew Servetus' attention to medicine ; at any rate, in 1536 he was attending Guinther's lectures at Paris, and the next year wrote *A Complete Account of Syrups carefully Refined According to the Judgement of Galen*, a prolonged discussion of the role of " syrups " (" sweet, prepared potions ") as aids to digestion. It was after this that he matriculated formally as a medical student, only to be in danger of censure for his renewed practise of judicial astrology. Although his defence was fairly successful, he nevertheless—for the second time in his life—gave up the unequal battle and retired into obscurity as a practising physician in a small town at some distance from Lyons. After a few years he moved to Vienne, resumed his connection with the Lyons publishers, and continued medical practice. His editorial work involved assistance on a new edition of the Bible, which perhaps revived his interest in theological controversy. At any rate, by 1546 he had written *On the Restitution of Christianity* whose publication, under the name Servetus in 1553, led to an immediate uproar in both Catholic and Protestant circles. Forced to flee from Catholic France, he travelled by way of Geneva where he was recognised as the Servetus who had attacked Calvin twenty years before in his first theological treatises ; he was promptly arrested, tried and executed as a heretic.

The seven books of the *Restitution of Christianity* were indubitably radically unitarian and wildly heretical. It is difficult to see how a physiological discussion belongs in these impassioned pages. But there is a connection : Book v deals with the Trinity, and here, discussing the nature of the Holy Spirit and the way in which God imparts the divine spirit to man, Servetus suddenly turned to physiology. The connection is primarily a philological one : because the Hebrew word for " breath " is the same as that for " spirit," Servetus argued that the two were one. For this reason, he says,

> So that you, reader, may have the whole doctrine of the divine spirit and the spirit, I shall add here the divine philosophy which you will easily understand if you have been trained in anatomy.[2]

The "spirit" as distinct from the "divine spirit" is a combination of the three Galenical spirits, which Servetus regarded as originally one. Vital spirit, he thought, when " communicated through anastomoses from the arteries to the veins " became natural spirit, and then belonged in the liver and veins, differing from vital or animal spirits only in location. This amalgamation of the spirits enabled Servetus to reconcile the primacy of the heart, the necessity of air for life, and the Biblical identification of blood with life, with orthodox medical physiology.

Before he could proceed further, Servetus felt it essential to explain how and where the vital spirit " which is composed of a very subtle blood nourished by the inspired air " came into existence. It was usually considered that the vital spirit was produced in the left ventricle of the heart ; Servetus at first accepted this view, only to contradict himself later. He defined the vital spirit as " a rarefied spirit, elaborated by the force of heat, reddish-yellow and of fiery potency " ; and declared that the " reddish-yellow colour is given to the spirituous blood by the lungs ; it is not from the heart " because the heart is not endowed with organs suitable for mixing air with blood to create this reddish-yellow fiery spirit.★ This fiery spirit must, he thought, be

generated in the lungs from a mixture of inspired air with elaborated, subtle blood which the right ventricle of the heart communicates to the left. However, this communication is not made through the middle wall of the heart, as is commonly believed, but by a very ingenious arrangement the subtle blood is urged forward by a long course through the lungs ; it is elaborated by the lungs, becomes reddish-yellow and is poured from the pulmonary artery into the pulmonary vein.† Then in the pulmonary vein it is mixed with inspired air and through expiration it is cleansed of its sooty vapours. Then finally the whole mixture, suitably prepared for the production of the vital spirit,

★ Servetus, with a large vocabulary of Latin adjectives to choose from, selected *flavus* to describe the colour of arterial blood, perhaps to emphasise its similarity to fire and flame.

† Servetus, like all anatomists of the period, writes " artery-like vein " (*vena arteriosa*) and " vein-like artery " (*arteria venosa*).

is drawn onward from the left ventricle of the heart by diastole.

This is as clear a statement as one could expect of the pulmonary or lesser circulation—the route by which, in fact, the blood goes from the right side of the heart to the left—and the first to be published in modern times.* How Servetus arrived at his correct conclusion it is impossible to conjecture. It is true that many anatomists besides Vesalius had doubted that blood traverses the septum or middle wall of the heart through visible pits ; but they were not prepared to offer an alternative path. (Ironically Servetus, like other proponents of the pulmonary circulation, was inclined to think that " something may possibly sweat through.") Servetus was thoroughly alive to the novelty of his conclusion ; anyone who carefully compared his statements with those of Galen in *On the Use of the Parts* would, he boasted, " thoroughly understand a truth which was unknown to Galen."

It is most probable that Servetus arrived at his conclusion by pure ratiocination. He was, after all, primarily concerned with where and how air (containing the all-important spirit) became mixed with arterial blood ; it was not unnatural to suppose that the lungs, the organ with which we breathe, should play an important role. And as Servetus shrewdly noted, there was no reason to assume that the left ventricle of the heart, which resembles the right ventricle in appearance, had a more complex function than the right ventricle. It was easier and simpler (as well as more consonant with his theology) for Servetus to assume a pulmonary route. All the vessels under consideration were perfectly well known, even to a man who had more Galenical than empirical anatomical knowledge ; no other knowledge was

* It has recently been found that a thirteenth-century Persian commentator on Avicenna, Ibn al-Nafis, had suggested this route for the blood ; but his commentary was unknown in Europe until the twentieth century.

required for the formulation of the pulmonary circulation. And having made his point, Servetus continued with the physiology of the soul.

The circumstances surrounding Servetus' enunciation of the theory of pulmonary respiration, and his subsequent martyrdom, have won him a highly sympathetic response from historians of science ever since 1694, when William Wotton stated in *Reflections on Ancient and Modern Learning* that he had been told by a friend, who had a transcript of the relevant passage made by someone else, that the first discussion of the lesser circulation occurred in *On the Restitution of Christianity*. Certainly Servetus was the first to write of it ; whether later anatomists who also did so should rightly be accused of plagiarism and lack of originality is questionable. How any anatomist could have known of what Servetus wrote is not clear ; for it is doubtful if many of them were sufficiently interested in theological controversy to risk reading a wildly heretical, unitarian book by an unknown author (for it was Villanovus who was the medical man and scientific author, not Servetus, who only wrote on theology). Since many copies were destroyed unbound, and more burned with the author (so that a mere handful are known to have survived), it is doubtful whether any anatomist could have read the book had he wished to do so.

It is probable, therefore, that one ought not to count Servetus as the propagator of the doctrine of the pulmonary circulation, though he may have been its progenitor. It is most likely that acceptance of Vesalius' conclusion—that there was no passage for blood through the septum of the heart—naturally suggested the need to find an alternative passage. Once an anatomist tried to postulate such an alternative route, he would naturally consider that through the lungs, since *part* of the blood had always been supposed to follow this route. This was particularly likely to be the case as anatomists turned more and more to the problem of respiration. There is even the faint possibility that each anatomist who said—as they constantly did—"nobody has ever observed this before me" meant to speak truthfully, even though standards of truth were low when claims for scientific discovery were made.

The most important sixteenth-century statement of the pulmonary route of the blood from the right to the left ventricle of the heart—because it was the most widely read—is in *Fifteen Books on Anatomical Matters* (*De Re Anatomica Libri* xv, 1559) by Realdus Columbus (1516–59).[3] Published posthumously, it is apparently the text of the anatomy lectures he gave for many years, first at Padua (he was the alternate for the post in Surgery in 1541; lectured when Vesalius, the incumbent, was absent; and when Vesalius finally gave up the job in 1544, held it for a year), then at Pisa (1545–8) and finally at Rome (1548–59) where his audience was composed of eminent laymen as well as medical students. Columbus, as a competitor for the position which Vesalius won and then soon abandoned, was not above pointing out with great enjoyment that his successful rival had committed many errors, including some associated with the function of respiration. Like Vesalius, Columbus modelled his book on Galen's *Anatomical Procedures*, while at the same time fulminating against Galenists more devout than himself. As he put it:

> To think that some folk in our time swear to the dogmas of Galen about anatomy so that they dare to assert that Galen ought to be taken as gospel, and that there is nothing in his writings which is not true! It is wonderful how men are carried away by this doctrine; and the princes of anatomy offer it to the rabble.[4]

It was in considering the structure of the heart that Columbus differed most easily and convincingly from the beliefs of the "Galenists." After describing the anatomy of the right and left ventricles he continued:

Between these ventricles there is placed the septum through which almost all authors think there is a way open from the right to the left ventricle ; and according to them the blood is in transit rendered thin by the generation of the vital spirits in order that the passage may take place more easily. But these make a great mistake ; for the blood is carried by the artery-like vein to the lungs and being there made thin is brought back thence together with air by the vein-like artery to the left ventricle of the heart. This fact no one has hitherto observed or recorded in writing ; yet it may be most readily observed by anyone.

That his statement is similar to that of Servetus need occasion no surprise ; it is difficult to see how else it could be put, since both are describing a flow of blood from right ventricle to lungs to left ventricle, and its assimilation of air in the lungs. Columbus was not interested in the theological aspects ; for him, the soul was quite distinct from the air, and he was only concerned with the physiological role of the vein-like artery (pulmonary vein), because, he thought, its function had been misunderstood by Galen, and by those who followed Galen too closely. As he put it, in the customary polemical tone of the day,

Anatomists, not very wise, begging their pardon, . . . think that the use of this [the vein-like artery] is to carry the changed air to the lungs which, like a fan, ventilate the heart, cooling this organ—and not, as Aristotle thought, the brain. The same writers think that the lungs receive I know not what smoky fumes . . . discharged from the left ventricle. About this, all one can say is that it pleases them, for they certainly seem to think that the same state of things exists in the heart as in a chimney, as if there were green logs in the heart which give out smoke when burnt. . . . I for my part hold quite a different view, namely that this vein-like artery was made to carry blood mixed with air from the lungs to the left ventricle of the heart. And this is not only most probable, but is actually the case.

The function of the lungs was not merely to supply air for temperature regulation, but to elaborate and retain the vital spirits drawn from the air,[5] the left ventricle of the heart then being merely a receptacle, not a chamber for converting air into vital spirits.

Columbus might have modified his poor opinion of other anatomists had he known that they would read his book in considerable numbers, and soon generally adopt the mechanism of the pulmonary circulation. It had the further advantage, as the devout Aristotelian botanist Cesalpino (1519–1603) recognised, of suiting Aristotle's theory of the primacy of the heart better than the traditional view, and of offering an anti-Galenist argument which, to a real Peripatetic, meant another blow for Aristotle. Cesalpino was still a student at Pisa when Columbus began lecturing there (the precocity of Columbus was notable, in an age which still expected doctors to be greybeards), and no doubt heard of the pulmonary circulation from him. Italian nationalism has decreed that Cesalpino had a clear notion of the systemic circulation a generation before Harvey ; but Cesalpino's statements are far from clear, and are often contradictory. Certainly he developed arguments which might have led him to postulate the " motion of the blood in a circle " : he noted correctly that when a vein is ligatured, it swells on what he called " the far side," from which he argued that the movement of the blood is not all outwards from the viscera to the various organs. Confusingly, he concluded from this only that it explained Aristotle's view of sleep. As Cesalpino commented,

Here is the solution of the doubt arising from what Aristotle writes concerning sleep when he says : " It is necessary that what is evaporated should be driven to some place and then turned back and changed like Euripus. For the heat of every living thing ascends by nature to a higher place, but when it has reached the higher place, it in many cases turns back again and is carried downwards." This is what Aristotle says.

Cesalpino's explanation of this rather confused doctrine is as follows :

Now when we are awake the movement of the native heat takes place in a direction outwards, namely, to the sensory regions of the brain. When we are asleep, however, it takes place in the contrary direction towards the heart. We must therefore conclude that when we are awake a large supply of blood and of spirits is conveyed to the arteries and thence to the nerves. When we are asleep however the same heat is carried back to the heart not only by the arteries but by the veins. For the natural entrance into the heart is furnished by the vena cava, not by the arteries. A proof of this may be seen in the pulses, which when we are wide awake are full, vehement, quick, with a certain rapidly repeated vibration, but when we are asleep are small, languid, slow and infrequent. For in sleep the supply of native heat to the arteries is diminished, but it bursts into them with vehemence when we wake. The veins however behave in an opposite manner ; for when we are asleep they are more swollen, when we are awake they are shrunken, as anyone may see who watches the veins in the hands.[6]

It seems hard to conclude from this that Cesalpino *clearly* understood anything about the physiology of the venous and arterial system. Setting aside the question whether " native heat " flow is necessarily the same as flow of blood, it appears that he thought the blood and heat behaved quite differently in sleep and in waking. Nor did he repudiate the importance of the venous system to digestion, remarking firmly in his last book, published posthumously in 1606, that the vena cava was as important in its physiological role as the aorta.[7] Cesalpino's discussion should really be read as one of a number of discussions (by Ruini, among others) on how the blood *might* get from the right side of the heart to the left if the anatomists were correct in thinking that none got through the septum, rather than a firm statement of how it did so. Neither the suggested solution of Cesalpino, nor that of any other sixteenth-century anatomist, aroused the least interest among contemporaries. These explanations cannot have seemed especially cogent, and certainly they all lacked the relentless pressure of argument that was to drive Harvey's conclusion home.

When William Harvey (1578–1657) was a very old man, the then very young scientist Robert Boyle consulted him professionally. The consultation did not produce any useful medical results, but both men enjoyed the scientific contact. Among other topics, they talked of the circulation of the blood, and Harvey was asked how he had come to think of such a thing : he replied that it was from a consideration of the action of the valves in the veins in sending all the venous blood to the heart. In his published work, Harvey ascribed the first " depiction " of the valves to either Fabricius of Padua, or Sylvius (1478–1553) of Paris. In fact, the venous valves were discovered by a number of sixteenth-century anatomists ; * that is to say, they described little membranes found in a number of veins, although they did not speculate successfully on their function.

The most complete account of their structure and possible function is that contained in a little pamphlet of twenty-four pages *On the Valves in the Veins* (*De Venarum Ostiolis*) published in 1603 by Fabricius of Aquapendente (*c.* 1533–1619). Fabricius, who had taken his M.D. from Padua in 1559, remained in and

---

* Notably by Giambattista Canano (1515–1579), who mentioned them in lectures given in the 1540's and told Vesalius of his discovery ; by Charles Estienne, who mentioned them in his anatomical treatise of 1545 ; by Amatus Lusitanus (1511–1568), who described Canano's work in 1551 ; and probably by others. Sylvius described some valves in his *Isagoge* of 1555.

about the University, giving private lessons in anatomy for many years. In 1565 he became professor of surgery, a post which he held until 1613. When Harvey first attended his lectures in 1600, Fabricius (according to his own later account) had been describing the valves for sixteen years, and there is independent evidence that he had mentioned them in the anatomy lectures for the year 1578–9. Unlike his predecessors, he did more than merely describe their existence ; he investigated all the veins, to find out which possessed these membranes ; he provided anatomical illustration of their structure and action ; and he attempted to explain their function. This he described as follows :

The mechanism which Nature has here devised is strangely like that which artificial means have produced in the machinery of mills. Here millwrights put certain hindrances in the water's way so that a large quantity of it may be kept back and accumulated for the use of the milling machinery. These hindrances are called . . . sluices and dams. . . . Behind them collects in a suitable hollow a large head of water and finally all that is required. So nature works in just the same way in the veins (which are just like the channels of rivers) by means of floodgates, either singly or in pairs.[8]

Perhaps the most important aspect of Fabricius' description was his conception of the venous and arterial systems in hydraulic terms, so that he was able to perceive that these membranes must act to control the blood supply, which was analogous to a water supply. (This analogy was to be happily applied again by Harvey.) But though it is quite clear that Fabricius was trying to deal with the problem of the blood as a problem of simple hydraulics, it is equally clear that he did not really understand that the blood supply required valves to regulate the direction of flow. He did not think of valves at all ; the word he used (*ostiola*) obviously, from his analogy, meant to him "floodgate" rather than "valve," and he thought these served to regulate the blood supply in volume rather than direction. Indeed, he was quite explicit about this, stating that,

> Nature has formed them to delay the blood to some extent, and to prevent the whole mass of it flooding into the feet, or hands and fingers, and collecting there.[9]

Which is to say that the membranes were placed in the veins to ensure a fair and even distribution of blood for the nutrition of the various parts of the body. Fabricius did not see that he should have chosen a pump, rather than a mill, as his model. He was not really much closer than his predecessors to a true view of the function of these membranes, though he did see that they played a mechanical role, and were connected with the problem of the motion of the blood. Still less did he detect any connection between the existence of the valves and the action of the heart. This was partly (as Harvey was to note later) because Fabricius approached the relation of the heart and lungs purely from the point of view of respiration (on which he published a book in 1615) ; and because he never dealt with the mechanism of the heart itself.[10] It was also because these membranes were found in the veins ; had they existed in the arteries Fabricius might have seen a connection with the heart, but the venous system was centred on the liver, until the work of Harvey centred it at last on the heart.

There was nothing in Harvey's training to make him approach the problem differently : thoroughly indoctrinated with Galenic medicine at Caius College, he studied under Fabricius at Padua for two years before taking his M.D. and becoming a successful London physician. From Fabricius Harvey certainly learned something of the advantages of a mechanical approach to physiology, and something as well of the currently fashionable tendency towards replacing Galen's primacy of the liver by Aristotle's primacy of the heart. Quite original was Harvey's profound interest in the structure of the heart and in such problems as that of the different functions of the two structurally identical ventricles (one of which controlled the flow of spirits, the other the flow of blood). He wondered why the artery-like vein served to nourish only the lungs, while the similar vein-like artery nourished the whole body ; why (as Columbus had asked) the lungs should need so much more nourishment than other organs ; why, since the lungs moved, the right ventricle also moved ; and why the right ventricle should exist exclusively for the use of the lungs.[11] This was hardly what later scientists were to call the mechanical approach, and it plunged deep into the search for final causes ; yet here, in spite of Bacon, final causes proved fruitful of experimental results.

When, in 1616, Harvey began his lectures as Lumleian professor to the Royal College of Physicians, he had thought long and carefully about the structure and function of the heart and lungs. (He had already also formulated for himself a new and surprising code of behaviour ; for in his lecture notes he carefully reminded himself to avoid contentiousness and the common practice of attacking all other anatomists, " for all did well, and there was some excuse even for those who are in error." [12]) Here in 1618 Harvey first set out his new theory of the motion of the blood in compact and certain terms :

It is plain from the structure of the heart that the blood is passed continuously through the lungs to the aorta as by the two clacks of a water bellows to raise water. It is shown by the application of a ligature that the passage of the blood is from the arteries into the veins. Whence it follows that the movement of the blood is constantly in a circle, and is brought about by the beat of the heart.[13]

Here Harvey obviously had a firm grasp of the actual motion of the blood, and that with a clarity of thought which makes the speculations of his predecessors seem murky indeed. In the first place, he understood that the membranes of the veins and heart acted like clack-valves, which opened to allow the blood to go from the lungs to the left side of the heart, but which would not open in a reverse direction. In the second place, he understood that this occurred throughout the body, so that the blood always travelled from the arteries to the veins " in a circle " and back to the heart and lungs. There is a real sense in which Harvey, rather than any earlier anatomist, is the true "discoverer" of the valves in the veins, since he was the first who truly understood that they were valves.

For the next ten years Harvey continued to pursue his study of the motions of the heart and blood in animals ; the results were published in 1628 in *De Motu Cordis* (*Anatomical Exercises on the Motion of the Heart and Blood in Animals*). As its title implies, Harvey here treated the matter primarily from the anatomical point of view ; like any sixteenth-century anatomist he insisted that " I profess to learn and teach anatomy not from books but from dissections, not from the tenets of philosophers but from the fabric of nature." [14] Like any sixteenth-century anatomist too he began with what Galen had taught, and managed to interpret Galen's words to win support for his own new doctrine. He wrote,

The proof which Galen adduces for the passage of blood from the vena cava through the right ventricle and into the lungs can more rightly be used, if only the names are changed [!], for the passage of blood from the veins through the heart into the arteries, and I should like to use it in that way.[15]

Like many truly original thinkers, Harvey had no desire to be outrageously revolutionary ; he was fully aware of his originality, but did not wish to alienate his adversaries more than need be. Secure in his knowledge that he really had done and seen what no one had done or seen before, he was willing to present his new discoveries in conventional language.

Yet the whole of *De Motu Cordis* is a singularly tough and tight piece of scientific reasoning, firmly based upon experimental evidence. Harvey was not content to state the fact of the circulation of the blood, even with the support of plausible arguments; he was determined to demonstrate it convincingly and unarguably, which to him meant the total explanation of the function of the heart, its purpose, and the means by which it achieved its purpose. The result is a short but remarkably trenchant book, in which the reader is bludgeoned with illustrative and demonstrative argument until he can hardly help agreeing with Harvey's concluding claim:

All these phenomena to be seen during dissection, and very many others, appear if rightly assessed to elucidate well and to confirm fully the truth which I stated earlier in this book, and at the same time to oppose the commonly accepted views. For it is very difficult for any one to explain in any other way than I have done the reason why all these things have been arranged and carried into effect in the manner that I have described.[16]

Because Harvey was interested in the function of the heart, rather than in that of the lungs, he was able to make use of cold-blooded animals which were well suited to vivisection experiments; consequently he was able to study the action of the heart in a way that had not been attempted before. As he commented,

Since it is probable that the connection of the heart with the lungs in man provided, as I have said, the opportunity for going astray, those persons do wrong who while wishing, as all anatomists commonly do, to describe, demonstrate and study the parts of animals, content themselves with looking inside one animal only, namely man—and that one dead.[17]

Cold-blooded animals provided ideal subjects, and the conclusions drawn from them about the action of the heart in animals could be confirmed on warm-blooded animals whose heart action slowed down with approaching death. Harvey began his investigations by analysing the motions and characteristics of the heart. He established that the heart was a muscle (a much debated point previously) ; and that it was active in systole—when it contracted, the moment when the blood was expelled—rather than, as had been held up to then, in diastole. That is to say, the action of the heart consisted in the expulsion of blood, rather than in the sucking up of blood. Harvey further found that he could correlate the dilation of the arteries, the systole of the heart and the beat of the pulse. He did not compare the heart to a pump, as he might have done, but to a machine : the function of the heart was " the transmission of the blood and its propulsion, by means of the arteries, to the extremities everywhere." [18]

Having established the mechanics of the heart action, Harvey was next able to consider the question of the pulmonary circulation, " the ways by which blood is carried from the vena cava into the arteries, or from the right ventricle of the heart into the left one," [19] still by means of anatomical experiment. He demonstrated the case by dissection of a living fish, a toad, frog, snake, lizard and a mammalian foetus before considering the more complex case of mammals. The only problem he found difficult of resolution was the means whereby the blood was transferred through the tissues of the lungs, and here he was forced to argue merely by analogy from other secretions. Now he was ready to prove the existence of the systemic circulation, the true circular motion of the blood from the left side of the heart, through the arteries, into the veins, to the right side of the heart, through the lungs back to the left side of the heart again. This he did once more by a wealth of anatomical evidence, supported by teleological arguments of the fitness of various structures to the functions he ascribed to them.

But Harvey also introduced more novel kinds of reasoning. The most interesting is his use of a quantitative argument, based on the analogy between blood supply and water supply : he considered the size of the ventricle, the amount of its contents that it ejects with each contraction, and the rate of its beat. From this he concluded that, inevitably, the amount of blood sent by the heart into the arteries in half an hour is greater than the amount of blood in the whole body ; and that in a day the heart would eject a greater weight of blood than the weight of the whole body. This being impossible, it must be that the same blood is continually passing through the heart ; that is, there is a circulation. Quantitative arguments are rare in the seventeenth century, even in physical science ; it is even rarer to find them used in so appropriate a connection.

The fact of circulation at last made plain the purpose of the valves in the veins : " they completely prevent any backflow from the root of the veins into the branches, or from the larger into the smaller vessels " ; [20] they serve to ensure that all the blood flows towards the heart, from the small veins to the larger ones. Indeed, the valves were, Harvey found, so strong that he could not pass a probe through them in the wrong direction. Only one problem remained unsolved : the means by which the blood passed from the very small arteries to the very small veins. Here, and here alone, Harvey was reduced to pure speculation : he was forced to conclude that there must, as in the lungs, be an area of spongy tissue between the arteries and veins through which the blood seeped until it found the entrance to a vein—a poor mechanism, on the whole, since it did not explain the continued and necessary unidirectional flow.

Nevertheless, Harvey had every reason to feel satisfied that " calculations and visual demonstrations [had] confirmed all [his] suppositions " and that one could not by any means fail

to conclude that in animals the blood is driven round a circuit with an unceasing, circular sort of movement, that this is an activity or function of the heart which it carries out by virtue of its pulsation, and that in sum it constitutes the sole reason for the heart's pulsatile movement.[21]

But in fact he could never be quite satisfied, nor cease from searching for more arguments to support his view. Having reviewed all other kinds of arguments, those drawn from anatomy, from experience in phlebotomy, from reason, he fell back on the purpose of the circulation, and the relation between structure and function of the various organs. It appeared obvious, he said, that " It will not . . . be irrelevant to add that, according to certain common reasonings, it should both fittingly and necessarily be thus." [22] And so he shows that he can provide arguments on this level too.

Most of his arguments were drawn from Aristotle ; it is curious that as anatomists learned more, and rejected Galen, they turned to Aristotle for philosophical support of their new-found knowledge. Aristotle seemed particularly helpful to Harvey, because of his doctrine of the supremacy of the heart. This was a view which Harvey accepted wholeheartedly : the heart, he thought, " deserves to be styled the starting point of life, the sun of our microcosm, just as much as the sun deserves to be styled the heart of the world." Indeed, Aristotle's scientific discussions in other areas offered support for circular motion everywhere :

We have as much right to call this movement of the blood circular as Aristotle had to say that the air and rain emulate the circular movement of the heavenly bodies. The moist earth, he wrote, is warmed by the sun and gives off vapours which condense as they are carried up aloft and in their condensed form fall again as rain and remoisten the earth, so producing successions of fresh life from it. In similar fashion the circular movement of the sun, that is to say, its approach and recession, give rise to storms and atmospheric phenomena.

It may very well happen thus in the body with the movement of the blood.[23]

De Motu Cordis is, in fact, filled with as many dithyrambs in praise of the heart and the circular motion connected with it as an astronomical treatise by a Copernican might be filled with praise of the Sun. The heart, like the Sun, provides living creatures with warmth, essential to life and digestion, it is the king and ruler of the microcosm. It is all very mystic ; but the mysticism is an enthusiasm which led Harvey to investigate the function of the heart in an eminently non-mystic fashion. Harvey must have had some sympathy with the work of Gilbert, and even of Kepler ; for he was of their philosophical persuasion.

One aspect of Harvey's work remains a puzzle : the small effect his discovery had on medical practice. No one—not even Harvey—suggested that the discovery of circulation revealed a fallacy in the age-old practice of phlebotomy. Though Harvey had noted the fact that an animal could bleed to death through a cut artery as evidence in favour of the existence of the circulation, he emphatically never took the next step of recognising that excessive bleeding might be positively harmful. He did, nevertheless, discuss how the existence of the circulation could be used to explain certain peculiar medical facts : why a poisoned or infected organ can cause illness to the whole system ; why some fevers, attacking the heart, affect respiration ; why medicines applied externally can have an influence upon the internal organs —as for example, when " garlic bound to the soles of the feet helps expectoration." [24] All these were attempts to rationalise " facts " previously believed to require a mystic or occult explanation.

Opposition to Harvey's views was often violent and partisan, but it was by no means universal. Even in France, where the conservative medical faculty totally rejected the new doctrine as late as 1650, there were still many who accepted it, and wrote in its favour. In fact, it became one of the basic tenets of the " new science," and was accepted as such by figures so diverse as Descartes and the clever young physicians who, about 1645, began those meetings at Gresham College which were the seed from which the Royal Society was to grow. The circulation of the blood ranked as an important example of the new experimental natural philosophy. Bacon did not live to read of it, but later scientists, who could give no higher praise, thought that Harvey admirably exemplified what they took to be the Baconian method.

CHAPTER X

CIRCLES VANISH FROM
ASTRONOMY

> Tycho did what Hipparchus did : it serves as the founda-
> tions of the building. Tycho endured the greatest labour.
> We cannot all do everything. A Hipparchus needs a
> Ptolemy who builds up the theory of the other five planets.
> While Tycho was alive I achieved this : I built up a theory
> of Mars, so subtle that the calculations completely accord
> with the observations.[1]

Of all the astronomers of the post-Copernican period, the most
difficult to appraise and appreciate is Johannes Kepler. Not a
great observational astronomer—poor eyesight would have
hindered him had he tried to be one—he yet insisted upon
closer agreement between theory and observation than any
astronomer before his time. A passionately devoted mathematical computer, and an extreme neo-Platonist mathematical
mystic, he cared only for those mathematical representations of
the heavens which offered the possibility of interpretation in
physical terms. Mystic and rational, mathematical and quasi-
empirical, he constantly transformed apparently metaphysical
nonsense into astronomical relationships of the utmost importance and originality. Immensely arrogant in his conviction that
he held a sure key to the mysteries of the universe, and even to
the structure planned by God at the Creation, he always acknowledged his debt to his predecessors. He took his achievements
with the utmost seriousness, and left an elaborate trail of the

procedure whereby he arrived at the eponymous laws of planetary behaviour by which he is remembered ; yet he never called them laws, nor did he distinguish them from others, to him equally precious, most of which are now rightly forgotten. Totally dependent for his best work on the observations of Tycho Brahe, he was a firm and unwavering Copernican. He was a prodigious worker, author of a couple of dozen books on astronomy, optics, mathematics and religion, and at the same time conducted a voluminous correspondence. Yet his theories had little influence on his contemporaries, for the works in which they were embedded were in a style alien to the ablest astronomers of his day, men too clear-headed and too scornful of occult and mystic notions to trouble to analyse them properly ; and the generation of mystic Copernicans like Digges and Gilbert was nearly all dead by 1600—before, that is to say, Kepler had made any real contributions to theoretical astronomy. Paradoxically, Kepler's ideas were first really appreciated by the intensely rational generation of scientists working after 1660, who saw that they could be applied to the mechanical systems of the universe and removed from the mystical context in which Kepler had set them.

Kepler belonged to the first truly Copernican generation, for he learned the elements of Copernicanism as a student under Maestlin. Though Maestlin had for long lectured only on the Ptolemaic system, his lectures at Tübingen in Kepler's day included a thorough presentation of the new as well as the old astronomy. And this in the introductory course of lectures ; for Kepler was not at first especially drawn to astronomy. Born in 1571 in a small town in Württemberg, of a respectable but decaying family, Kepler benefited from the Lutheran belief in the value of education. The family was devout if not hard-working, and though not very admirable parents, his mother and father saw the advantages of training their eldest son for the ministry.

This led to education at a seminary in preparation for entry into the Protestant University of Tübingen. Kepler seems to have begun as a diligent and orthodox student, but apparently could not maintain the required rigidity of Lutheran doctrine ; indeed his religious views were never subsequently wholly acceptable to the Lutheran congregations of the various towns in which he lived. Nor was his temperament such as to suggest the successful pastor. At the same time, it was noticeable that he had a marked bent for mathematics. Wisely, the faculty at Tübingen urged him to accept the offer of a post as district mathematician and teacher of mathematics at the Protestant seminary in the Austrian city of Graz ; in default of any clerical post, Kepler reluctantly accepted the position in 1594.

However he may have felt at the time, Kepler was eminently well prepared for his new job. Maestlin was one of the most esteemed astronomical teachers in Germany, and one of the few who publicly taught both the Ptolemaic and Copernican astronomy. Maestlin obviously encouraged his pupils to weigh the pros and cons of the two systems seriously ; as Kepler wrote in the preface to his first book—ever anxious to expose the history of his ideas, a practice he continued throughout his life—

Ever since the time at Tübingen, six years ago, when I enjoyed association with the most celebrated master, Michael Maestlin, I felt how little satisfactory was the usual concept of the many motions of the world. At the same time I conceived such an enthusiasm for Copernicus, whom my master often mentioned in his lectures, that not only did I frequently defend his views in the disputations on physics with other students, but even composed an entire disputation in defense of the thesis that the " first movement " came from the rotation of the Earth.[2]

It is clear that Kepler had received a far better and more

up-to-date training in theoretical astronomy than any astronomer of his age.

The professional duties at Graz were not onerous. Kepler was expected to prepare yearly calendars, containing full astronomical information liberally spiced with astrological prediction (to which he was not at all averse), and to teach such students as presented themselves. These were not many ; astronomy was not a required subject, as it would have been at a university, and Kepler was not a good enough lecturer to arouse any interest in his subject. No one minded whether Kepler had students or not ; it was enough that his services were available if required, and that he provide the yearly calendar. Hence he was free to devote himself to his own interests ; to his private life (he indulged in a long and slightly farcical courtship which culminated in his marriage to a young widow a few months after the publication of his first book in 1596) and to research in theoretical astronomy. All during 1595 Kepler was busy elaborating a new theory of the mathematical relationships involved in the sizes of the planetary orbits ; the results were proudly sent to Maestlin, who saw *The Cosmographical Mystery* (*Mysterium Cosmographicum*) through the press late in 1596.

Kepler was naturally immensely proud of his work, and took care to send copies to princes—who might offer him a better and more secure job—and to distinguished scientists, among them Tycho Brahe. At least one copy reached Italy, where it was seen by Galileo, still professor of mathematics at Padua and not yet known as an astronomer ; the professor wrote kindly to the younger man, explaining that the promptitude of his writing prevented him from having read the work properly (there is no evidence that he ever did so), but praising Kepler for his faith in the Copernican theory, to which he admitted himself to be an adherent. (It was on the strength of this letter that Kepler later attempted to establish a regular correspondence between himself and Galileo.) Tycho also was encouraging ; the mysticism of the book could not repel anyone as committed to the belief that both alchemy and astrology were sound roads to truth as Tycho was, and he saw that Kepler was already an exceptionally able and industrious astronomical computer. He would make an ideal assistant at Uraniborg ; Tycho therefore urged Kepler to join him, promising him access to his vast collection of observations. But Kepler refused, then as later reluctant to leave German-speaking lands, and seeing no reason why the Graz post should not continue.

But the situation in Austria, and even in South Germany, was to grow rapidly uncomfortable as the forces of the Catholic Counter-Reformation slowly pushed back the Protestant frontiers. In the autumn of 1597, all Protestant clergy and teachers were ordered to leave Graz. An exception was specifically made for Kepler, as it was to be on subsequent similar occasions ; he was known to be regarded as unsound among orthodox Lutherans, who thought him too liberal (and too inclined to Calvinist doctrine, but perhaps the Catholic authorities did not know that). He was also on good terms with many Catholics, including a number of Jesuits, the vanguard of the Catholic force. Kepler was always ready for religious discussion ; true, he always in the end firmly rejected Jesuit pressure for conversion, blandly claiming to be already a Catholic since he was a Christian, but he was obviously by no means so intransigently anti-Catholic as most of his co-religionists. No doubt also the civil authorities felt that it was more important to have a district mathematician who was an able astronomer and caster of horoscopes than one who was better at attending mass than waiting on the stars. So Kepler stayed on in Graz for another three years, speculating on the mysteries of the solar system.

But by 1600 the demands for religious uniformity were greater. Besides, Tycho had now left Denmark to settle in

Bohemia as Imperial Mathematician, a post which would enable him to employ assistants. Kepler decided to see if Tycho's original offer still stood, and whether things could be arranged on a basis satisfactory both to his pride and his pocket ; he went to Prague, and after much negotiation the situation was resolved in a manner acceptable to both astronomers. Kepler was soon settled, and busily investigating the orbit of Mars on the basis of data collected over many years, an investigation destined to lead him to such remarkable discoveries that he was quite justified in considering that he had created a new astronomy.

Tycho's death in 1601 produced little disturbance, except for time lost in negotiation with the Emperor and with Tycho's heirs. In the end, Kepler retained access to Tycho's papers, and continued his work, at the same time succeeding to Tycho's post as Imperial Mathematician. This was not an onerous position except for the amount of energy required to secure any substantial part of the promised salary ; Kepler cast a few horoscopes for the Emperor, but he was chiefly intended to complete the planetary tables based on Tycho's work which had been promised long before Tycho's death. Kepler worked on these fitfully through the years, never wholly neglecting them, but always turning aside to pursue other interests ; they were finally completed in 1623, only to be delayed for five more years as a result of the chaos of war. When they appeared in 1627, the *Rudolphine Tables* were a fitting monument to Tycho from his worthiest disciple.

Meanwhile, without totally neglecting his official work, Kepler busied himself with a multitude of problems : with work on the orbit of Mars, with the optics of refraction, with the new star (nova) of 1604, with mathematics, with consideration of Galileo's new astronomical observations, with scientific correspondence. But Kepler was not entirely happy at Prague ; the Emperor was growing old (he died in 1612), and the political disturbances which preceded the outbreak of the Thirty Years' War indicated that Prague was no longer a suitable environment for scholarly pursuits. Alarmed, he sought a more secure post, which he found as district mathematician in the Austrian city of Linz, a better paid job than his earlier position at Graz, and near South Germany, to which he always hoped—in vain—to be able to return. Kepler remained at Linz for fourteen years, until the actual presence of war drove him to search for a suitable place of publication for the *Rudolphine Tables*, now complete, and of residence for himself and his family. The *Rudolphine Tables* were published at Ulm in 1627, after which Kepler was free to search for new patrons—Wallenstein was briefly one of them—for the restoration of property and his long-overdue salary from various sources ; still searching, he died at Regensburg in 1630.

When he died, Kepler was widely known as a scientific heir to Tycho Brahe—as much on the strength of his writing on the nova of 1604 as on account of his profound calculations based on Tycho's observations. Indeed, in 1610 the poet John Donne could write in the satirical *Ignatius his Conclave* of "Keppler, who (as himselfe testifies of himselfe) ever since Tycho Brahe's death hath received it into his care, that no new thing should be done in heaven without his knowledge." This was the view of the non-scientist ; astronomers knew that Kepler had written four works on theoretical astronomy—the *Cosmographical Mystery* (1596), the *New Astronomy* (1609, *Astronomia nova* or commentaries on the motion of Mars), the *Epitome of Copernican Astronomy* (1617–21), and the *Harmony of the World* (1619 *Harmonices Mundi*)—all intensely Copernican, and all concerned with a new, daring and decidedly obscure form of mathematically based astronomical theory. It was on these that Kepler's reputation was ultimately to rest ; but they were probably the least appreciated of his works when he died.

What fascinated Kepler about the Copernican system was the superior order and harmony which it appeared to him to display. From his earliest years as an astronomical speculator, Kepler had been convinced that there was more order and harmony in the universe than customary astronomical methods could display. By order and harmony Kepler meant two slightly different aspects of the cosmos : one, a reflection of the properties of the divine creator, the other, a set of mathematico-physical relationships : a mystic harmony and a mathematical one. It was this concept and vision that led Kepler to pursue the attributes of God through the observations of Tycho Brahe, never relinquishing either the mystic vision or the observational fact, but labouring unceasingly until the two were one. As he informed the reader of the *Cosmographical Mystery*,

> there were three things of which I pertinaciously inquired the causes of their being as they are, and not otherwise : the Number, Size, and Motions of the Orbs. This I ventured to do because of the wonderful correspondence of things at rest—the Sun, the fixed stars, and the intermediate space—with God the Father and the Son and the Holy Ghost : an analogy which I shall pursue further in my *Cosmography*.[3]

This was a favourite analogy, which to Kepler provided at once a reason for research and a true mystic vision ; over twenty years later in the *Epitome of the Copernican Astronomy* he again compared the centre of the world to the Father, the sphere of the fixed stars to the Son and the planetary system to the Holy Ghost. This was not a pantheistic mysticism alone, as for Giordano Bruno ; it was at once religious comfort and a powerful stimulus to investigation—*physical* investigation ; for the harmonies of the world were meaningless to Kepler unless they conformed to accurate observation. Even in his first attempt, before he had worked with Tycho, Kepler repeated his computations over and over, all during a whole summer, until at last he found a mathematical representation sufficiently close to the best data he could obtain. It was because he hoped that even more accurate data would permit him to discover even more wonderful relationships that he wanted to work with Tycho ; as he was to remark some years later, " since, with divine goodness, God has granted a supremely careful observer like Tycho Brahe . . . it is only proper for us gratefully to recognise and benefit from this gift of God." [4] Kepler recognised that it was Tycho's observations which made the errors of earlier theories perceptible. And whether it helped or hindered him, Kepler loyally accepted Tycho's work as the exact basis on which he must build, cheerfully throwing away the labour of months if the calculations showed that his theory was not as exact as the observations required. Never before had an astronomer taken such a rigid view of the concord necessary between observation and theory ; equally, never before had any calculator had such accurate and consistently reliable observations from which to work. It was Kepler's good fortune to have access to such an extensive body of observations ; it was a part of his genius that he took advantage of their precision. No one but Kepler could have drawn the full advantage from Tycho's work, as he gratefully acknowledged ; but not even loyalty could persuade him to accept the Tychonic system rather than the Copernican.

Devout Copernican that he was, Kepler obeyed Tycho's injunctions and justified his use of Tycho's work by increasing his computational labours : in the *New Astronomy* (his first publication using Tycho's data) he calculated all the elements of the orbit of Mars for three systems, the Ptolemaic, the Tychonic and the Copernican. As late as 1619 he noted that he derived his study of " the most perfect harmony of the celestial motions and their origin from the same cause of the Eccentricities, Semi-diameters and periodic times " according to both the Tychonic and the Copernican systems. (By the time he came to draw up the title page for the fifth book of the *Harmony of the World* he no longer felt it worth while to bother with the Ptolemaic system.) He thus salved his conscience while remaining a Copernican; but in other ways he was more nearly a follower of Tycho than of Copernicus. Like Tycho, he rejected the notion of crystalline spheres (though, again like Tycho, he retained the enclosing sphere beyond the fixed stars, which gave the world boundary and limit); this permitted him to accept Tycho's theory of comets as being bodies situated in the heavens, travelling in a possibly closed orbit between the orbits of Venus and Mars. Like Tycho, he transformed the planetary orbs from spheres into orbits; but, unlike Tycho, he saw that this led to the necessity of finding a physical cause for their persisting in these orbits. For it was not enough to find that the planets did revolve in fixed and continuous orbits, of constant size and determinate shape, located in space at fixed distances from the centre of the universe: Kepler thought it essential to search for the reason why they did this, and he was prepared to search until he found the reason out.

Indeed Kepler, like a small boy, was obsessed with the possibilities inherent in the word *why*. Why was half the universe (the centre and circumference) at rest and the other half in motion? Why did the outer planets move more slowly than the inner ones? Why did the planets have orbits of certain sizes and not others? Why were there just six planets, no more and no less? To all these questions there must be an answer; and for Kepler this answer must be expressed in physical as well as in mathematical terms. To say that six is a "perfect" number (as the Greeks had called numbers which were the sum of their factors) was not enough; once again, Kepler asked why? What was the *physical* significance of this mathematical fact? It was this obsessive curiosity which was the basis of his first book, the *Cosmographical Mystery*. Could geometrical figures be employed to give con-

crete reality to a numerical relationship ? About the apexes of
the triangles designating various positions of the conjunctions *
of Jupiter and Saturn one could construct a circle ; but unfortunately its dimensions did not appear to accord with the true
size of the planetary orbits. Hence, although the geometrical
" harmony " was attractive, Kepler reluctantly abandoned this
calculation. In any case, it ought to be possible to find a
geometrical relationship involving all the planets ; or, as Kepler
put it,

For if (I thought) for the size and proportion of the six
celestial bodies which Copernicus established, there could be
found five figures among the infinite number of others which
would have just these particular properties, this would give
what I wished. And so I pressed on. What had plane
figures to do with solid orbs ? They rather resemble solid
bodies. Behold, Reader, the discovery and material of all
this little work.[5]

Rather naïvely, perhaps, Kepler regarded solid bodies as being
more physical, less purely geometrical, than plane figures, and
hence as having greater significance. He was so delighted with
his final discovery, to which he was led by consideration of the
regular Platonic solids, that he could not forbear to present it
to the reader in the preface, although it was to be discussed
in immense detail in the body of the book. This was his discovery :

The sphere of the Earth is the measure of all. Circumscribe
about it a Dodecahedron : its circumscribed sphere will be
[that of] Mars. Circumscribe a Tetrahedron about [the
sphere of] Mars : its circumscribed sphere will be [that of]
Jupiter. Circumscribe a Cube about [the sphere of] Jupiter :

* A planet was " in conjunction " when a line drawn from the Earth to
the Sun could be extended to include the planet ; that is, when Earth, Sun
and planet all lay on the same straight line, in that order.

its circumscribed sphere will be [that of] Saturn. Now inscribe an Icosahedron within [the sphere of] Earth : its inscribed sphere will be [that of] Venus. Inscribe an Octahedron in [the sphere of] Venus : its inscribed sphere will be [that of] Mercury. Here you have the reason of the numbers of the planets.

And here, to prove it, was an enormous double folio illustration of the pleasing harmony ; a figure often reproduced, but seldom on a scale large enough to do it justice, for Kepler insisted on representing the orbits to scale.

To the modern reader, this may seem so much nonsense ; it is difficult to see that this relationship meant any more in physical terms than did the relationships that Kepler had already rejected. But Kepler was entirely delighted, and wished for nothing more, except possibly more accurate calculations on the sizes of the orbits which would, he was sure, make the relationship even more exact. This to him was a truly physical relationship, and one moreover that came closer to relating every planet to the system of the world as a whole than anyone had done previously. To improve matters, Kepler even put the Sun in the centre of his universe (in place of making the centre of the Earth's orbit the centre of the universe, as Copernicus had done), even though this meant the reintroduction of the equant, the mathematical device which Copernicus had so hated. Even there, Kepler was to be justified ; for he was thereby led to novel speculations about the speed of planetary motions at various points in their orbits, which in turn were to permit him to abandon the equant once and for all.

The agreement between Kepler's solution of the cosmographical mystery and Copernicus' determination of the sizes of the planetary orbits was not quite as close as Kepler would have liked, in view of the manifest truth of each ; to acquire more accurate data he was willing to work with Tycho Brahe, even though it meant subordinating his own Copernicanism temporarily to the wishes of his master. Kepler hoped that Tycho would have available the information he required ; he was somewhat cast down at first to find that only the raw figures were available, and nearly all the computations were still to do, so that he could not immediately pursue his own interests. Not that he minded computing the elements of the orbit of Mars, for he had no dislike of calculation, and Mars proved rapidly to offer all sorts of interesting results. His first discovery was that the centre of the Sun and the whole of the circular orbit of Mars lay in the same plane (a point in favour of the Copernican system), even though the plane of the orbit was inclined to the ecliptic.* But more was to follow ; the longer he studied Mars, the more there seemed to be to discover.

Kepler soon became particularly disturbed by the erratic way in which the planet moved through its orbit, for its velocity varied according to no obvious law. It was certainly not uniform with respect to the Sun ; nor with respect to the centre of its orbit (a circle eccentric to the Sun) ; nor with respect to any fixed point within the orbit. True, the error involved in assuming such uniform motion was only eight minutes of arc (an error that would not have troubled Copernicus), but Kepler knew that Tycho's observations were correct to less than that, and he remained uncomfortable. As a check, he tried computing the velocity of the Earth, only to find that the Earth behaved in the same tiresome fashion. In both cases the planets moved faster when approaching the Sun, and slower when receding from it, but in neither case was there any uniformity in this change of motion. Two problems presented themselves : how to find a mathematical expression for this variation, and how to explain its existence. The solution to the first problem involved complex

* Later he determined that the planes of all planetary orbits pass through the Sun's centre.

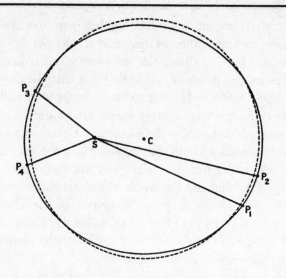

FIG. 9. KEPLER'S LAW OF AREAS

The area of triangle SP_1P_2 is equal to the area of triangle SP_3P_4, where S is the Sun, and P_1, P_2, P_3, P_4 represent positions of the planet. The dotted curve is that of a circle eccentric to S, with its centre at C (the centre of the elliptical orbit); this is obviously a close approximation to the true elliptical orbit, whose eccentricity is here slightly exaggerated even for Mars

mathematics, making use of Archimedean integration by summation of small lines and areas ; Kepler proved quite equal to its solution. The result (though its author never stated it in quite this form) was the now-familiar " second law " of Kepler : that the radius vector drawn from the Sun to the planet sweeps out equal areas in equal times (cf. figure 9).* The mathematical

* Kepler expressed it both more precisely and more tentatively : he drew his line from " the point with respect to which the eccentric is computed " or " that point, which is taken to be the centre of the world." [6] At this time Kepler was dealing with mathematical considerations only, and of course regarded the orbit as an eccentric circle.

derivation, together with several detailed proofs, occupies most of Part III of the *New Astronomy*. Here was a wonderful new mathematical discovery about the planets ; and best of all, it was susceptible of physical meaning. For the cause of this variation must lie either in the inherent properties of the planet or those of the centre of the world, and in either case Kepler thought he could find it out.

He had, already, speculated vaguely on the possible cause of the variation in planetary velocity, and wondered whether there might not be some kind of moving spirit or soul (*anima*) in the Sun, related in some way to light. But since his first speculations he had become acquainted with the concept of magnetic force, through reading Gilbert. Years later, in the Fourth Book of the *Epitome of Copernican Astronomy* (published in 1620) he was for once to defend his claim to originality by proclaiming his dependence upon others ; he wrote :

My doctrines—most of which I have taken from others— say whether love of truth or glory is mine : for I have built all Astronomy on the Copernican Hypothesis of the World ; the Observations of Tycho Brahe ; and the Magnetic Philosophy of the Englishman William Gilbert.[7]

Gilbert's magnetic philosophy now offered Kepler exactly what he required in order to explain the calculated variations in planetary velocities. Kepler developed the idea of a force (or rather " virtue ") similar to magnetism, and possessed of a power of attraction. Magnetic attraction was manifest from terrestrial experiments, and known to be capable of extending over large distances ; it could be invoked to explain why the motion of a planet being attracted to the Sun was greatest when the planet approached the Sun, least when the two bodies were farthest apart. The attraction set up by this virtue was not just a force, but a genuine ' pulling towards '—a motion, in fact. Not that this motion was exercised without restraint, for all bodies must have a resistance to motion, or it would not be necessary to explain why they move ; and all the planets, as well as the Earth, also possessed, in Kepler's view a quasi-magnetical moving virtue. The motive virtue of the Sun combined with the moving virtue of the planets to produce the peculiar variation in velocity characteristic of orbital motion. The existence of magnetic forces within each planet, varying with the size of the planet, explained the mysterious property of *gravity* : heavy bodies seek the Earth not through desire to achieve their natural place, but as a result of magnetic attraction.

This theory was to be elaborated over the years ; but as early as 1609 Kepler stated in the *New Astronomy* that the Sun, turning on its axis, emitted its magnetic virtue in much the same way as it emitted light, and in turning produced a kind of vortex. Later, in the *Epitome of Copernican Astronomy*, Kepler became more precise : he provided a diagram showing how the orientation of the magnetic poles of a planet revolving about the Sun (whose own poles were the surface and centre) necessarily explained both the orbital path and orbital velocity (see figure 10). Kepler was now secure in a physical explanation of his mathematical laws ; at the same time, his magnetic virtue was an occult force ; and the planets which were loadstones were nevertheless, as they were for Gilbert, animate bodies.

Even after finding the mathematical law of planetary motion, and a physical explanation for its existence, much remained to be done on the shape of the planetary orbit. The more he investigated, the more certain Kepler became that the planets could not be travelling in perfect circles eccentric to the Sun. Such courses might be metaphysically sound, but they were not consistent with physical fact. In the *New Astronomy* Kepler (having, as always, given the reader a summary of his views in the preface) laboriously devoted the whole of Part IV to a detailed account of the reasons which had led him to this conclusion, the mathematical

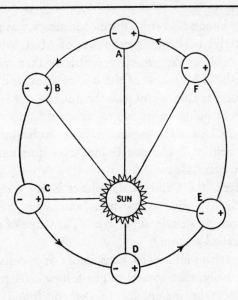

FIG. 10. KEPLER'S EXPLANATION OF THE ELLIPTICAL ORBIT IN TERMS OF MAGNETIC FORCES

As the planet revolves anti-clockwise about the Sun its two magnetic poles retain the same orientation with respect to the orbit. At A and D they are equidistant from the Sun, and the planet has no tendency to approach or recede from it. As the planet goes from A to D (as at B and C) the attractive pole is nearer the Sun and the planet therefore tends to draw closer; as it goes from D to A (as at E and F) the repelling pole is nearer the Sun and hence the planet tends to recede. The result is the ellipse

computations which had confirmed it, and the methods which he had employed to devise a substitute hypothesis and to prove that. At one stage (Kepler never spared the reader any step of his own painful progress) he even thought of abandoning the law of areas, but further tedious computation showed both that it really did hold, and that it was inconsistent with the assumption of a perfectly circular orbit. This inconsistency was particularly (and fortunately) noticeable in the case of Mars, which at best required an orbit of far greater eccentricity than was the case with the Earth.* By means of the law of areas, and with great labour, Kepler was able to compute the distance of Mars from the Sun at different points in its orbit. From this incontrovertible evidence, in the face of the ancient prejudice in favour of circular motion that neither Tycho nor Galileo ever questioned, Kepler was driven to conclude :

This is clear : the Orbit of the planet is not a circle, but passes gradually inside at the sides, and again increases to the amplitude of a circle at perigee. The shape of this kind of path is called an oval.[8]

Was Kepler perhaps influenced, consciously or unconsciously, by Tycho's suggestion that comets might follow oval paths ? Or, like Tycho, was he merely led to the conclusion by similar evidence ?

As usual, Kepler could not be content with his theoretical speculations unless they had a possible physical significance ; in this case, he thought, the deviation from circularity could be explained as the result of the influence of the magnetic force of the Sun ; this varied with distance, being greatest at perigee (when the planet was closest to the Sun), and was further combined with the tendency of the planet to rotate in a circle under the influence of its own proper motive virtue to produce an ovoid (egg-shaped) figure. The ovoid proved a most intractable curve ; after a number of attempts Kepler sighed, " if only our figure was a perfect ellipse, it could be described by the method of Archimedes." [9] But it seemed not to be anything so simple, and he struggled on, trying to calculate his ovoid by treating

---

* The eccentricity (the "flattening" of the ellipse) of the Earth's orbit is indeed quite small ; that of Mars is about five times as large, and the discrepancy in Kepler's calculations varied approximately with the square of the eccentricity.

each end as a portion of a perfect ellipse, only to find that the motions he calculated for Mars on this assumption always failed to agree with the observations. No matter what combination of ellipses he used to represent an approximation of his ovoid (which no geometry could touch), he could not handle the problem. At last, after years of effort, Kepler decided that it was the particular ovoid which he had guessed for the orbit which was at fault, and he thought it worth while to try another, though this meant a totally new set of approximations. The first ellipse he tried showed a happy numerical congruity that seemed to justify his use of the ellipse ; but he was once more defeated, this time by an arithmetical error. Once more he was in despair ; once more persistence brought relief.

So far, Kepler had treated his ellipses as mere devices of approximation : suddenly it struck him that if (as it should in accord with his "magnetic" theory) Mars librated on the diameter of an epicycle circling around the Sun (a perfectly " classical " concept) the resultant curve would be . . . an ellipse. Surprised at his own foolishness—the foolishness which prevented his seeing that his physical ideas and the geometry he had been using all led to the same inescapable conclusion—he retnrned to his calculations and found the slip in his arithmetic which had previously prevented his success. As a further reward he found that, indeed, there could be no curve *except* the perfect ellipse (with the Sun at one focus) which would fit the data, and at the same time would agree with the law of areas.[10]

Kepler's surprise, and his previous neglect of the ellipse as a possible orbit, rather than a mere calculating device, are not so unlikely as they might appear : conic sections were little studied in the sixteenth century, and as Kepler himself complained, " How many mathematicians are there who would put up with the labour of reading through the *Conics* of Apollonios of Perga ? " [11] It was a measure of the importance of Kepler's discovery that some eighty years later Newton was to advise a would-be reader of his *Principia* that Apollonios was fundamental to the comprehension of astronomical theory. Even Kepler, pleased as he was with his solution, was not quite sure that he understood the physical meaning of the ellipse as confidently as he did that of the eccentric circle, because too many of its elements were merely geometrical. True, the Sun was at one focus ; but the other focus was empty ; and so was the centre. But observation and computation could not lie ; and Kepler consoled himself by boasting that

> with extremely laborious demonstrations and by the handling of exceedingly many observations, I have discovered that the path of a Planet in the Heavens is not a circle, but an Oval route, perfectly Elliptical.[12]

And Kepler had calculated the new orbits of all the planets, including the Earth, and found them all elliptical, though mostly with far less eccentricity than was the case with Mars.

One problem still remained : what was the cause of the relative sizes of the planetary orbits ? This had been the subject of the *Cosmographical Mystery*. A quarter of a century later Kepler was to return to the study of the relationships of the velocities of the planets at various points in their orbits, the length of time it took each planet to traverse its orbit, and the average distances of the planets from the Sun. Surely there should be some interesting and relevant mathematical relationship between velocities and distances, a relationship which must, of course, be capable of physical interpretation ? And surely this relationship, when found, would once more confirm and elucidate the divine harmony and proportion of the universe.

All Kepler's later investigations on this subject are contained in *The Five Books of the Harmony of the World*, where, according to his habit, he painfully recapitulated his progress to a few certain principles, and reminded his readers of what he had previously discovered. The first book is almost entirely geometrical, being concerned with the harmonic proportions of plane figures, without any relation to astronomy. The second book (which Kepler called Architectural) deals with solid figures, and is again mainly geometrical. The third book is a Pythagorean disquisition on Harmony, explaining the mathematical proportions responsible for musical harmony. With the fourth, the physical world appears ; this is " Metaphysical, Psychological and Astrological," dealing with the souls and bodies of celestial and terrestrial bodies.

Only the fifth book, " Astronomical and Metaphysical," contains the subjects that had concerned Kepler for so long : the harmonies of celestial motions, based upon the mathematics and metaphysics of the preceding books. Here is a recapitulation of the central doctrine of the *Cosmographical Mystery*, in relation to the Copernican and Tychonic systems. But there is new doctrine as well ; in chapter III Kepler presented " a summary of the astronomical doctrines necessary for the contemplation of the harmonies of the heavens," which is really a summary of the newest of his discoveries. He described the weeks of labour spent on examining the relations between planetary motions and the size of their orbits ; he had been slow to see the truth, but at last it came and he could insist,

> But it is an absolutely certain and exact fact, that *the proportion between the periodic times of any two Planets, is precisely as the three-halves power of that of their mean distances, that is, of the orbits themselves.*[13]

Or, as in the modern expression, the squares of the periodic times of the planets are in the same ratio as the cubes of their respective mean distances from the Sun. Here, as Kepler fully realised, was a wonderful revelation ; but not even he could realise its full significance. He naturally saw its metaphysical significance far more plainly than its physical, and it immediately inspired him to further flights of computation. Elaborately he compared the proportions between the various elements of the planetary orbits, building up tables of harmonies which he then compared with the numerical proportions of various notes on the musical scale or with the geometrical proportions of the lengths of strings emitting various notes. By comparing thus astronomical, numerical and geometrical ratios he produced what he saw as a true " music of the spheres," displayed at loving length in both mathematical and musical notation. Thus triumphantly Kepler solved the ancient Pythagorean problem, and made manifest the harmony of the world.

Did Kepler really expect astronomers to discover what was to become his third law in the great morass of Pythagorean calculation ? Very probably not ; in the *Proemium* to Book v he admitted that he might have to wait a hundred years for a sympathetic reader. But he comforted himself with the thought that this was little, when God had waited six thousand years for someone to contemplate His wonders with an appreciative mind and eye. More practically, he took steps to see that his ideas had readers, even though his works were too difficult for a general audience. In the later parts of the *Epitome of Copernican Astronomy*, that peculiar mixture of elementary question and answer with advanced Keplerian astronomy, the third law was plainly stated and fully discussed, without any complications of celestial harmony since it was for an elementary audience. Not because Kepler thought the discussion of harmony in any way " non-scientific " as a modern critic might do, but because he thought it involved complex mathematics, such as the reader of the *Epitome* would not normally have ; as he had said ruefully at the beginning of the *New Astronomy*, there were few in his day who could comprehend advanced mathematical development. To its author, even the dithyramb to the Sun which concluded the *Harmony of the World* was an expression of the mathematico-physical laws of the universe, and as such intelligible only to those capable of following him on the complex mathematical road whereby he had arrived at the truth.

Kepler remains inescapably foreign to the modern world, one of the most difficult of scientists to portray accurately or to appreciate as he really was. It is not that he mixed daring new ideas with vestiges of the past ; no scientist has ever failed to do that and in many ways Galileo, his elder by only seven years, retained much more of the past than Kepler, and is nevertheless more easily comprehensible. Kepler was certainly a true Renaissance scientist, invoking the extreme past to advance the present, adopting the cosmological approach of the early Greek philosophers, and cherishing the fact that Ptolemy had written about celestial harmonies. Yet the stranger aspects of his thought were not drawn from the past, nor were they wholly of the present. There is, to be sure, something of the natural magician in Kepler ; but he is closer to the natural magic of Gilbert than to that of Porta, a strange mixture of number mysticism and passion for empirical fact. Not even Gilbert was so insistent as Kepler that mystic theory was only worth considering when it was based upon irrefutable observation and conformable with physical interpretation. There is little in Kepler of the neo-Platonic number nonsense of the late fifteenth century—entirely self-sufficient—or of the religio-philosophic pantheism of Giordano Bruno. To Kepler his newly discovered mathematical harmonies were so many laws which revealed the wonder and order of the world of God ; this was a world ruled by mathematical law, which in turn was discoverable by astronomical observation.

In many respects, indeed, Kepler was less ready to accept mathematical metaphysics than most of his contemporaries ; for his notions of harmony bore little resemblance to those of most Copernicans, like Digges and Dee. Under the influence of Tycho Brahe, Kepler totally rejected the existence of solid spheres, except for that which enclosed and held together the universe, and made it one. Under the influence of Tycho's data, Kepler had shed the time-honoured concept of the necessity for perfectly circular motion, a concept which had provided the metaphysic for the physical calculation of planetary orbits since Plato's day. Kepler's universe was thus a strange one, far more divorced from traditional astronomy than that of Copernicus. At its centre was the Sun, fixed in its place, but rotating on its axis, emitting light and magnetic virtue ; at its outer periphery the region of the fixed stars, truly at rest, bounded by an enclosing sphere. In between were scattered the planets, held in their places not by material orbs, but by the balance of motive virtues and magnetic attractions ; ceaselessly revolving about the Sun in ellipses (strange shapes!) at velocities described by the mathematical relationships of the law of areas and the law of harmonies, while the sizes of the ellipses were predetermined functions of the periodic times, and of the essential and harmonious proportions between all parts of the universe. And the clue to understanding these novelties had been the data of Tycho Brahe, interpreted by Kepler's intense and passionate conviction that mathematical harmonies governed the world, and that it was worth immense labours of calculation to find the mystic, but physical, expressions which accurately revealed what these harmonies might be. For they were as much a reflection of the work of God as the threefold division of the universe was of the three persons of the Trinity. Not the least of the wonders of Kepler's work is that with these preconceptions his discoveries were to prove exactly what was required by later natural philosophers to convert his mystic harmonies into coldly rational " mechanical " physical reasoning.

It was, however, some time before this was realised ; for in his mysticism and his daring Kepler stood equally outside the main stream of scientific development, already insisting on rationalism as its guiding principle. His scientific contemporaries read him little. He was valued by kings, princes and state officials for his skill in astrological prediction ; hence his numerous posts as mathematician to courts and senates, and the numerous offers from foreign princes, like that with which Sir Henry Wotton tried to attract him to England. Astronomers praised him, as both Maestlin and Galileo did, but none appears to have taken his ideas seriously. The first real discussion of Kepler's Laws waited until 1645, when the French astronomer Ismael Bouillaud (1605–94) in his *Astronomia Philolaica* dealt with the first two, of which he accepted only the first. The famous encyclopedic astronomy of G. B. Riccioli (1598–1671), *Almagestum Novum* (1651), mentioned the first law, only to reject it as insufficiently proved.

England had done a little better : Hariot had encouraged his pupils to read the *New Astronomy*, and to reflect upon the possibility of the truth of elliptical orbits ; but few of his pupils were real astronomers and when he died in 1621 there was no one to follow his interest. Yet nearly twenty years later Jeremiah Horrox (1619–41) wrote a thorough and informed defence of the first two laws ; this again had little immediate influence, since it was not published until some thirty years later. Seth Ward (1617–89), Savilian Professor of Astronomy at Oxford from 1649, wrote against Bouillaud in 1653, criticising his geometry ; like his adversary, he accepted only the first law. The third law was even less well known ; G. A. Borelli (1608–1679), trying to establish a system of the world based upon gravity (*Theoricae Mediceorum Planetarum*, 1666) did not accept what would have been so helpful. Fortunately not everyone ignored Kepler's best work, and some hints reached the young

Newton before the end of 1665, in time to assist him in his first formulation of the Newtonian system. After 1665 Kepler's Laws were well known and generally accepted by the best mathematical astronomers, though it was the success of Newton in using them that at last gave Kepler's discoveries the status of scientific laws.

CHAPTER XI

DEBATE AMONG THE STARS

Galileo . . ., who of late hath summoned the other worlds,
the stars to come neerer to him, & give him an account of
themselves.[1]

For half a century and more after the publication of *De Revolutionibus* the case for Copernicanism had been based on arguments
concerning harmony and probability in nature, rather than on
evidence. The followers of Copernicus were not effective
observational astronomers ; no one but Tycho had amassed any
fresh body of data that might help to settle the question of the
motion of the Earth one way or the other. Tycho's most
interesting observations—on the new star of 1572, and on comets
—had yielded arguments that were strongly anti-Aristotelian,
but not especially favourable to Copernicanism. Tycho's
destruction of the crystalline spheres could be applied to the
heliocentric system, but did nothing to render it more plausible ;
on the other hand Tycho's own system of the celestial motions
did require that the spheres be abolished. By 1610 the Tychonic
system was a powerful rival of the Copernican, at least among
those trained in science (though Galileo was to choose to ignore
it) ; its innovations, relating rather to the Aristotelian, physical
picture of the universe than to the vexed question of the Earth's
motion, did not so much prepare the way (as yet) for the still
greater innovations of Copernicus as offer instead a fresh, modern
alternative to the heliocentric doctrine. Even Kepler, in 1609,
still recognised that the two new laws of planetary motion he

had extracted from Tycho's observations could be as well applied to Tycho's system as to the Copernican, though he believed the Copernican to be the true one. In any case his discoveries, the more suspicious because they were purely mathematical, were ignored by the partisans on either side with equal impartiality. If Kepler did not fulfil Tycho's requirement that his years of calculation be devoted to proving the truth of Tycho's system, nevertheless his writings had little effect on the opposite side.

By 1610 the old arguments were growing stale. Even the literary man knew them by heart, and the great debate was declining in vigour through lack of fuel. Only fresh evidence could revive it, only some gifted writer who revelled in polemic could bring up new issues for discussion. Galileo furnished both. New evidence came from his astronomical use of the telescope; new arguments from his drawing the controversy out of the realm of mathematics into that of physics. In so doing he raised a great new issue of principle, the right of the scientific astronomer to speculate and to communicate his speculations with freedom. If, ultimately, he lost this right for himself by the hardihood of his championship, he won it for his successors.

In 1609, at the age of forty-five, Galileo was a moderately successful teacher of mathematics at Padua of seventeen years' standing, who had published nothing but a little pamphlet describing his improvement of a mathematical instrument. A year later, after the appearance of the *Sidereal Messenger*, he was world-famous and in a position to arrange his return to his native Tuscany on most honourable terms. Seldom has fame come so suddenly—or so late—in the life of a great scientist. His origins were not very different from those of Kepler. His father was a member of a decayed patrician family of Florence, a not very successful cloth merchant living in Pisa in 1564 when his eldest son was born. Though unskilful at making money Vicenzo Galilei was a cultured and accomplished musician, from whom his son evidently inherited mathematical ability. Like Kepler, Galileo was sent to a university, but it was to study medicine, not theology nor mathematics. (Tradition has it that his father forbade Galileo the art of mathematics, lest he become too entranced and drawn away from the lucrative practice of medicine.) Galileo was an even more recalcitrant pupil than Kepler, and again like Kepler he left the University of Pisa without taking a degree, having already distinguished himself in both pure and applied mathematics, to which he now applied himself energetically. Mathematical ability never made Galileo rich, but it led him to the first of a series of inventions, a new hydrostatical balance ; it brought him private pupils in Florence and Siena ; and, through his work on finding the centres of gravity of bodies, it won him the influential patronage of Guidobaldo del Monte, an authority on mechanics.

Galileo's first post, as professor of mathematics at Pisa in 1589, was neither remunerative nor agreeable, for he wrangled with the rest of the Faculty ; he was delighted to attain, with Guidobaldo's support, a similar chair at Padua, where the pay was higher, the duties lighter and the private pupils more intelligent. There he remained eighteen years, respected but inglorious, begetting illegitimate children, lamenting constantly the slenderness of his salary, the necessity for taking pupils in his house, and his exile from the Florence to which he returned each summer vacation. His interest in physics (or as he said, "philosophy") and applied mathematics resulted in no other publication than the *Geometrical and Military Compass* pamphlet of 1606, which brought him a little fame. Yet Galileo's first letter to Kepler, in 1597, shows that he had thought deeply enough on astronomy —a normal part of his mathematical teaching—to become, in private, a Copernican. Fear of ridicule (so he said) held him from avowing his conviction. Like every other professor of mathematics Galileo lectured to appreciative audiences on the famous new star (nova) of 1604; he used the occasion for a brilliantly anti-Aristotelian exegesis, which delighted his friends; but he had no novel astronomical arguments to offer as yet. Characteristically it was a problem in physics that drew him five years later into the field in which he was to win fame among contemporaries.

The change in the course of Galileo's private work was brought about by reports of the recent invention, in Holland, of a new optical device that made distant objects appear close. Less than a year afterwards Galileo wrote:

About ten months ago [May 1609] a report reached my ears that a certain Fleming had constructed a spyglass by means of which visible objects, though very distant from the eye of the observer, were distinctly seen as if nearby. Of this truly remarkable effect several stories were related, which some believed and others denied. A few days later the report was confirmed to me in a letter . . . which caused me to apply myself wholeheartedly to inquire into the means by which I might arrive at the invention of a similar instrument. This I did shortly afterwards, my basis being the theory of refraction. First I prepared a tube of lead, at the ends of which I fitted two glass lenses, both plane on one side while on the other side one was spherically convex and the other concave. Then placing my eye near the concave lens I perceived objects satisfactorily large and near, for they appeared three times closer and nine times larger [in area] than when seen with the naked eye alone. Next I constructed another one, more accurate, which represented objects as enlarged more than sixty times [in area]. Finally, sparing neither labour nor expense, I succeeded in constructing for myself so excellent an instrument that objects seen by means of it appeared nearly one thousand times larger and over thirty times closer than when regarded with our natural vision.[2]

His first conclusion was that the spyglass (*occhiale*; the word *telescope* was invented among the Lyncean Academicians in 1611) should be applied to military and naval purposes; no very original idea, as the Dutch had already so used it. But soon Galileo turned his *occhiale*, much more powerful than any yet made in Holland, to the night sky; a simple act, but one that was to revolutionise astronomy. For, and increasingly as Galileo improved upon his first attempts, it both revealed new facts and rendered all naked-eye observation obsolete.

The first object observed by Galileo, naturally enough, was the Moon. He was the first to see more than the shadows that fancy had embellished. He recognised mountains, and a little later found out how to estimate their heights from the lengths of their shadows; he saw the vast plains which he took to be seas (they are still so called). As he put it:

> if anyone wished to revive the old Pythagorean opinion that the Moon is like another Earth, its brighter part might very fitly represent the surface of the land and its darker region that of the water. I have never doubted that if our globe were seen from afar when flooded with sunlight, the land regions would appear brighter and the watery regions darker.[3]

With typical acuity he went on to explain carefully this paradoxical opinion that irregular surfaces reflect more light than smooth surfaces do, and why the edge of the Moon always looks smooth to the naked eye. At the same time he discussed earthshine (the " old Moon in the new Moon's arms ") and presented good reasons for the view (already held by some astronomers) that it was caused by sunlight reflected from the Earth to the Moon and back. All this implied that the Moon, whose celestial status no one had ever doubted, was suspiciously like the Earth itself; a telling blow at the Aristotelian twofold division of the universe into terrestrial (or sublunar) and celestial regions, and consequently an indirect argument in favour of Copernicus. If its hitherto unperceived terrestrial nature had never inhibited the revolution of the Moon, why should it be impossible for the Earth to revolve ?

Next, Galileo turned to the stars. Two facts struck him : first, that the telescope did not make the fixed stars appear larger, but only brighter ; and secondly, that he could now for the first time see so many more stars. When he turned to the planets, they revealed themselves as " globes perfectly round and definitely bounded, looking like little moons flooded all over with light," but the stars did not appear as physical bodies, even when much magnified. The telescope stripped the stars of their " sparkling rays " without enlarging them as much as other objects, or disclosing their physical nature. Thus the telescope emphasised the difference between the planets and the fixed stars. Moreover, the multitude of stars invisible to the naked eye that were seen through it showed how much of the universe had hitherto escaped observation. The mystery of the Milky Way—was it a nebulous river of light, or an aggregation of stars ?—was now solved ; it was certainly composed of a vast number of stars closely crowded together. And Galileo concluded that all the objects called " nebulae " (clouds) were similarly formed of masses of small stars.

What pleased and astonished him most, however, was the discovery of four new planets (as he called them), four satellites or moons of the planet Jupiter which he named, in honour of the ruling house of Tuscany, the Medicean stars. Their discovery was a combination of luck and persistence in observation. On 7th January 1610 Galileo, observing Jupiter with his best telescope, noticed three small stars in a line with the planet (two to the east and one to the west) that he had not seen before. He naturally took them to be fixed stars. But the next night, " happening " as he said to look at Jupiter again, he found the three stars, still in a line, now all to the west of the planet. His first assumption was that Jupiter had moved ; but the motion of Jupiter at this time was retrograde, and so the three stars, if they were fixed, should have been to the east. Further observation, continued nightly throughout the winter, convinced Galileo that the revolutions of four stars round Jupiter caused the changes he saw. This was the best evidence on behalf of the Copernican system yet discovered, or rather, it was the best evidence against the anti-Copernicans. As Galileo said :

> Here we have a fine and elegant argument for quietening the doubts of those who, while accepting with easy minds the revolutions of the planets about the Sun in the Copernican system, are mightily disturbed to have the Moon alone revolve about the Earth and accompany it in an annual revolution about the Sun. Some have believed that this structure of the universe should be rejected as impossible. But now we have not just one planet revolving about another while both run through a great orbit round the Sun ; our own eyes show us four stars which wander round Jupiter as does the Moon about the Earth, while all together trace out a grand revolution about the Sun in the space of twelve years.[4]

The *Sidereal Messenger*,* published in Latin in the spring of 1610, was a work of enormous significance. It described an optical instrument hitherto almost unknown, and never before used in astronomy ; it showed how this instrument revolutionised that science ; it made Galileo the most famous and popular astronomer in Italy ; and it spread the fame of Italian science

---

* The Latin title, *Siderius Nuncius*, is ambiguous. When, later, Galileo was accused of arrogance for proclaiming himself a messenger from the stars, he pointed out that *nuncius* means message as well as messenger. Therefore the title might be translated as *Message from the Stars*, but *Sidereal Messenger*, sanctified by long usage, embodies the most usual meaning of *nuncius*, and Galileo may have been trying to correct his critics' Latin and slip out of a tight spot at the same time.

abroad with incredible rapidity. Sir Henry Wotton, then English
ambassador to Venice, bought a copy on the day of publication,
read it with delight, and posted it off to England to entertain the
King, promising to obtain a telescope as well. His excitement
was not merely aroused by new wonders—though there was that
too—but by what he regarded as the revolutionary implications
of the little book. He wrote :

> So as upon the whole subject he [Galileo] hath first over-
> thrown all former astronomy—for we must have a new
> sphere to save the appearances—and next all astrology. For
> the virtue of these new planets must needs vary the judicial
> part, and why may there not yet be more ? . . . the author
> runneth a fortune to be either exceedingly famous or
> exceedingly ridiculous.[5]

Wotton was prescient, for Galileo was considered famous or
ridiculous depending on the astronomical doctrine favoured by
his appraiser.

Materially, the *Sidereal Messenger* was an immediate success.
Everyone who read it longed to look through the marvellous
spyglass, except a few timorous diehards who were convinced in
advance that the wonders it disclosed were in the lenses, not the
heavens (optics had so long been employed to cause scientific
illusions). No optician could grind lenses as good as those of
Galileo, and his were in great demand ; he could have made a
small fortune had he chosen to organise a workshop to make
them (as he had for his mathematical compass). He did oversee
the manufacture of a good many instruments, which he delivered
to those princes and great men who might advance his career.
Scientists, he thought, should make their own, as Kepler soon
did ; Galileo combined exasperation and scorn for astronomers
who could not find the means to see what he had seen. (We
should be inclined to forgive them, as anyone who has looked at
the heavens with low-power opera-glasses, similar in optics and

magnification to the first telescope, will agree.) But the men who tried without success to discern the new wonders in the heavens were not content to doubt—they scoffed ; and indeed, their picture of a middle-aged professor confusing the man in the moon with mountains and seas was a fit subject for ridicule.

Princes and great men saw it otherwise. The Venetian Senate promptly offered Galileo a greatly increased salary and a life-appointment, which he accepted, though he still hoped for a court position in Florence. He had been mathematical tutor to the present Grand Duke five years before ; had dedicated his book to the prince ; and had named Jupiter's satellites in his honour. Like many middle-aged professors he was tired of the routine of lectures and private teaching, endless time-wasting on dull pupils and jealous colleagues. He wanted free time, time to perfect his ideas and write his books, and, as he once disarmingly remarked to a friend, " It is impossible to obtain wages from a republic, however splendid and generous it may be, without having duties attached." [6] Besides, like many scientists since, Galileo found the prospect of communicating his discoveries to laymen most enticing. Like Leonardo long before he offered to the Duke's Secretary of State " great and remarkable things," but of more real concern to him was his personal programme of work to be done :

Two books on the system and constitution of the universe—. an immense conception full of philosophy, astronomy, and geometry. Three books on local motion—an entirely new science in which no one else, ancient or modern, has discovered any of the remarkable laws which I demonstrate. . . . Three books on mechanics . . . though other men have written on this subject, what has been done is not one-quarter of what I write, either in quantity or otherwise. I have also lesser works on physical topics, such as treatises on sound and the voice, on vision and colours, on the ocean tides, on the nature of continuous quantities, and on the motions of animals, and yet other works.[7]

Probably nothing in this huge list was more than sketched at this time. Only a part of it was ever completed, and then not according to the original design. Galileo's was to be a life not of calm contemplation, but of bitter controversy.

For the moment all was fair. Galileo secured his appointment at Florence on the terms he sought (he was to be " Philosopher" as well as "Mathematician" to the Duke, the latter alone savouring too strongly of astrology and other uncertain things); he left Padua in the summer of 1610. In Venice the Senate was furious that Galileo had broken his contract at Padua, while his friends were hardly less dismayed : reluctant to lose him, knowing that he had made enemies and foreseeing that he would make more, they judged that he was unwise to leave the security of a proud republic for the chances of an uncertain court. As Sagredo, the Venetian merchant whom Galileo was affectionately to commemorate as a most intelligent scientific virtuoso, wrote :

Where will you find freedom and self-determination as you did in Venice ? . . . At present you serve your natural prince, a great man, virtuous, young, and of singular promise ; but here you had command over those who govern and command others ; you had to serve no one but yourself ; you were as monarch of the universe.[8]

And he added ominously : " I am much disturbed by your being in a place where the authority of the friends of the Jesuits counts heavily." Galileo was too content to heed such warnings, especially now that he had convinced the philosophers at Pisa, for they had been converted, steadily if slowly, and many of the recalcitrant must by now (as Galileo sardonically remarked) have seen at first hand on their way to heaven the Moon, the Milky

Way and the Medicean stars that they had refused to glimpse through the telescope.

At Florence Galileo began his work auspiciously, with further discoveries in the skies. He had, naturally, looked to discover if all the planets had moons like the Earth and Jupiter ; only Saturn seemed promising, but his appearances were tantalisingly uncertain. For having as he thought observed two satellites, Galileo soon found that they changed their shape, while clinging closely to the planet ; cautiously—and perhaps to tease Kepler to whom he sent it—Galileo announced his discovery in a jumbled anagram, which he then publicly clarified a few months later, when he was certain. (He never knew that Saturn had in fact not " attendants," but rings.) The discovery seemed proof that not only Jupiter, but other planets had " moons." His next discovery was more important, in every respect ; again he could hardly believe his eyes, and again he announced his discovery first in an anagram, delaying its clarification until he was certain of what precisely he had seen : it was, that Venus showed phases like the Moon. It had always been an argument against the Copernican system that Venus did not vary in brightness ; now Galileo was able to show (though only his best telescope was able to demonstrate the phenomenon) that the cycle of phases occurred in such a way that Venus was " full " (and therefore brightest and biggest) only when most remote from the Earth.[9] There seemed no end to what the telescope could show.

Galileo was winning converts everywhere, even in Rome. Because of what came later, it is difficult to remember that Galileo's relations with the Church were, at first, peculiarly cordial. The Jesuit mathematicians at their College in Rome were among the most competent in Italy ; one of them, Clavius (1537–1612) (he had long ago been responsible for the final computations of the Gregorian calendar), soon accepted Galileo's telescopic discoveries, and others followed suit. Indeed, they welcomed all additions to astronomical knowledge, though they were committed to the geostatic system. Galileo wished to re-establish relationships with Clavius, whom he had known long before ; and he had hopes of influencing the highest ecclesiastical officials, since the Church had always been interested in astronomy. Possibly Galileo was even toying with the idea of trying to convince the Pope that the Church should accept the Copernican system, on which it had, after all, never officially pronounced. His visit to Rome in the spring of 1611 was a triumphant success. The Jesuits were cordial ; the head of their College, Cardinal Bellarmine (1542–1621), after his mathematicians had assured him that there was no doubt of the truth of Galileo's observations, was friendly. At the same time the anti-clerical Lyncean Academy admitted him as a member, and christened his spyglass " telescope "; Galileo was immensely proud of his new status, and called himself " Lyncean Academician " ever afterwards.

After all these triumphs Galileo returned to Florence to find himself embroiled in controversy, a controversy which threatened to become endless. He enjoyed polemic, he did not suffer fools gladly, and as a professional scientist he needed to defend his scientific reputation : but above all, this was a cosmological controversy. Even before he had publicly taken up the cudgels for Copernicus he had begun the necessarily concomitant attack on Aristotle ; and now the two aspects of the debate drew closer together. First came an anti-Aristotelian debate over floating bodies, which began at the Grand Duke's dinner table with a discussion of why ice floats in a cooling drink ; after the publication of Galileo's *Discourse on Floating Bodies* (1612) it continued with renewed acerbity because of Galileo's insistence on writing in Italian, so that he could appeal to the educated classes over the heads of the traditional men of learning.

Next came (in 1613) Galileo's *History and Demonstration Con-*

cerning *Sunspots and their Phenomena*, a tract in the form of letters provoked by a book in which a German Jesuit, Christopher Scheiner (1573–1650), claimed to be the discoverer of these phenomena. Galileo claimed priority, because he had observed them three weeks before Scheiner (in fact large sunspots had been observed for centuries with the naked eye, and a German astronomer, Johannes Fabricius, was the first to publish an account of telescopic observations of sunspots in 1611) ; but the question of priority soon gave way to the more interesting one of interpretation. Scheiner found the best (and least disturbing) conclusion to be that these were small bodies moving about the Sun, which then might be said to have satellites like other planets. Galileo denied this conclusion, with its anti-Copernican implications, and insisted (correctly) that the spots were on the Sun, substantiating his claim by computations based upon the laws of mathematical optics. He never guessed their exact nature, but believed that they were indeed " spots " or imperfections, whose existence falsified Aristotle's theory of celestial perfection. He also used the apparent motion of the spots as an argument in favour of the Sun's rotation upon its axis. Once again he possessed arguments which could be used with effect against Aristotle —or rather against his followers ; for he remarked (as he was often to do later) that Aristotle himself was too intelligent to have been capable of accepting the ideas put forward in his name ! Elated by the cogency of his own arguments, Galileo chose this as the place in which to announce his discoveries about Venus and Saturn, with the conclusion :

> And perhaps this planet also, no less than horned Venus, harmonizes to perfection with the great Copernican system, to the universal revelation of which doctrine propitious breezes are now seen to be directed towards us, leaving little fear of clouds or crosswinds.[10]

For Galileo was determined to continue discussing the Copernican system, and in his own special way : arguing from observation, and appealing to the intelligence and common sense of the educated Italian public. If, as well, he could discredit all anti-Copernicans by ridicule and logical reasoning, that made his task so much the easier.

For the moment this seemed a perfectly safe course. The Jesuits at Rome continued to accept his new discoveries, while rejecting his interpretation of them, even though his old friend Clavius was now dead. There were certainly attacks, clerical and academic, upon Galileo and his dangerous doctrines, but they did no more than arouse interest. As the lay public became better informed, inevitably the question of the relation of Copernicanism and the Bible, of science and revelation, began to be debated. After one of his pupils had been involved in a debate on the matter at Court, Galileo supplied a long and thorough analysis of the problem, and a cogent defence of the independence of scientific investigation, together with arguments on the compatibility of Copernicanism and Scripture ; after expansion and re-writing, it became the *Letter to the Grand Duchess Christina*. Galileo wrote with conviction, though one of his chief arguments was based upon the somewhat flippant epigram of Cardinal Baronius, " the intention of the Holy Ghost is to teach us how to go to heaven, not how heaven goes." [11] He argued that the Bible was not a scientific text, so we need not take its casual remarks as scientific statements. Further, he remarked, sensibly enough, that if a scientific theory is false, it may be refuted by demonstration ; and if it can be refuted by demonstration, it cannot be dangerous. (But Galileo failed to see that the inability of the scientist to imagine a contrary hypothesis need not argue the truth of his initial explanation.) Finally, Galileo used the old and perfectly orthodox argument, that since nature and Scripture are two divine texts, they must give the intelligent reader the same conclusions ; but where his opponents relied in every case upon the evidence of Scripture, Galileo preferred that of the senses. Galileo's friends and the Tuscan court were pleased with the essay. But it was not a wise one to have written, even for private circulation ; for here Galileo was on theological ground, and his scientific reputation did not give him the right to compete with established clerical authority.

Indeed, partly as a result of the *Letter* (which circulated in manuscript), the attacks on Galileo were growing sharper. A Dominican preacher, who had been expounding the book of Joshua (with its anti-Copernican text), delivered an impassioned sermon in which he attacked Galileo, Copernicus and all mathematicians as inimical to the Christian faith, and subversive to the State. Very shortly after, a copy of Galileo's letter on science and religion was sent to the Holy Office. Galileo countered by improving on his original version ; seemingly he was unperturbed by the fact that to anti-Copernicans he had become the leading Copernican spokesman, and the most dangerous. Galileo's Roman friends tried to urge him to a less open belligerence ; Prince Cesi, patron of the Lyncean Academy, wrote :

> Those enemies of knowledge who take it upon themselves to disturb you in your heroic and most useful inventions and writings, are such perfidious and rabid beings as can never rest, and the best way to demolish them altogether is to pay no heed to them and to attend to your health, so that you may complete all your books and give them to the world in spite of their efforts.[12]

Which was a tactful way of reminding Galileo that he had not done much about that great series of treatises which he had promised to give to the world. And, as Cesi reminded him, Cardinal Bellarmine, receptive to new ideas though he was, had always maintained that Copernicanism was contrary to Scripture, though an interesting mathematical hypothesis. Too much boldness might make Bellarmine and others think that the open discussion which the Church had so far permitted might present a threat to the faith. Indeed, Cardinal Barberini had told a friend of Galileo that it was as well not to try to improve upon Copernicus : it was better to treat astronomy mathematically than to try to convert theologians. And all the more so since what began as sound scientific arguments often became distorted in the process of popularisation. This was the first hint that Galileo was regarded as a particular menace because he wrote in Italian, for the non-learned, who did not always know how far it was safe to carry scientific conclusions.

As if to confirm the fact that Galileo's work provided the best arguments on the Copernican side, a fact that automatically made Galileo the most dangerous Copernican from a clerical point of view, there was published in Naples an essay by a Carmelite friar, Father Foscarini, in which the author used Galileo's observations as proof of the truth of the Copernican doctrine ; at the same time Foscarini argued that Copernicanism was not contrary to Scripture. Foscarini asked Bellarmine for his opinion ; the Cardinal quickly replied that any discussion of these matters was acceptable *provided* that the discussion was couched in hypothetical or purely mathematical terms.* As Bellarmine put it :

> to say that assuming the Earth moves and the Sun stands still saves all the appearances better than epicycles and eccentrics is to speak well ; this has no danger in it, and suffices for mathematicians. But to seek to affirm that the Sun is really fixed in the centre of the heavens and merely turns upon itself without travelling from east to west, and that the Earth is situated in the third sphere and revolves very swiftly around the Sun, is a very dangerous thing, not only because

* As he, and everyone but such ardent Copernicans as Galileo and Kepler who had detected that Osiander was the real author of the Preface to *De Revolutionibus*, thought Copernicus himself had done.

it irritates all the theologians and scholastic philosophers, but because it injures our holy faith and makes sacred Scripture false.

Admittedly, if the Copernican hypothesis could be *proved*, then Scripture could and must be re-interpreted. " But," he declared firmly,

I do not think there is any such demonstration, since none has been shown to me. To demonstrate that the appearances are saved by assuming the Sun at the centre and the Earth in the heavens is not the same thing as demonstrating that in truth the Sun *is* in the centre and the Earth *is* in the heavens. I believe that the first demonstration may exist, but I have very grave doubts about the second ; and in case of doubt one may not abandon the Holy Scriptures as expounded by the holy Fathers.[13]

It seemed to Galileo that there was nothing new in this, and he need fear nothing ; he was even mildly indignant that anyone should think that he had been meddling with theology. After all, he had but followed doctrines set forth in a book accepted by the Church ; was it fair that, doing so, he should be accused by " ignorant philosophers " and preachers of saying things contrary to the faith ? All he wished was to convince everyone that these things were *not* contrary to faith. Determined on this path, which he could not see to be a dangerous one, he went to Rome, where he had a pleasant few months debating and discussing with gusto and success, enjoying the way all contrary arguments collapsed before the cogency of his controversial skill.★

But in fact this was not the way to please the authorities. Pope Paul was no friend to scientists or literary men, disliked

★ At the same time, he wrote an essay on his theory of the tides, arguing that they were caused not by the attraction of the Moon, but by the double motion of the Earth ; this he thought an irrefutable confirmation of the Copernican system. It was read by only a few, until it appeared in 1632 as part of the *Dialogue*.

ingenious subtleties, and was inclined to think that Galileo's opinions must be pernicious and heretical, since they were scientific, literary and ingenious. Cardinals Bellarmine and Barberini could not approve of the way in which Galileo had ignored their friendly cautions. The more Galileo's cleverness won him friends, the more it also won him enemies. And many ecclesiastics were seriously concerned, as Bellarmine had been for some time, over the consequences of what Galileo wrote, since those who read him carried his arguments to extremes. The Holy Office considered Foscarini's book, and necessarily Galileo's work as well. The Congregation of the Index completed its deliberations in March, 1616 : the opinion that the Sun is the centre of the world, and immovable, they declared " foolish and absurd, philosophically false and formally heretical "; the opinion that the Earth is not the centre, but moves, both by rotation and revolution, they declared equally false in philosophy, and " at least erroneous in faith." Foscarini's book was prohibited ; those of Copernicus and Didacus à Stunica were placed on the Index until corrected (very minor corrections sufficed). As for Galileo, Bellarmine was instructed to admonish him not to hold or defend these (Copernican) opinions, which was duly done.[14]

Though his enemies claimed Galileo had been forced to recant, he felt he had not done badly. For security, he asked Cardinal Bellarmine for a certificate that he was cleared, and had suffered nothing ; and he wrote home :

As may be seen from the very nature of the business, I am not in the least concerned, nor would I have been involved had it not been for my enemies, as I have said before. What I have done may always be seen from my writings (which I keep so that I may always silence the malevolent), and I can show that my activity in this matter has been such that not even a saint could have dealt more reverently or more zealously with the holy Church than I. This is perhaps not equally true of my enemies, who have not scrupled to scheme, slander, and make diabolical suggestions.[15]

Bold words ; it was true that many within the Church regretted the whole affair, but nevertheless it was serious. Galileo still felt rebellious ; his mood must have been similar to that which made him, many years later, write in the margin of the *Dialogue* :

In the matter of introducing novelties. And who can doubt that it will lead to the worst disorders when minds created free by God are compelled to submit slavishly to an outside will ? When we are told to deny our senses and subject them to the whim of others ? When people devoid of any competence whatsoever are made judges over experts and are granted authority to treat them as they please ? Those are the novelties which are apt to bring about the ruin of commonwealths and the subversion of the state.[16]

But for the moment—and the more so as his health was persistently bad—there was nothing to do but wait.

Though he could not appear publicly as a Copernican, there was nothing to prevent his appearing openly as an anti-Aristotelian, and controversy soon flared up again with the appearance of a series of comets in 1618. Ironically, for once Galileo did not have observation on his side (for he was ill in bed) and, perhaps as a consequence, his views were scientifically unsound. The affair was complex ; early in 1619 there appeared an anonymous pamphlet, soon known to be written by a Jesuit, Father Grassi (1583-1654), adopting Tycho's view that comets were heavenly bodies located beyond the Moon. Grassi supported this concept with arguments about parallax and telescopic appearance. Galileo had never been sympathetic to Tycho's ideas, partly because they offered a sound alternative to the Copernican doctrine, and he treated Grassi's account as if it were a direct attack on the Copernican system. Not that he could say so openly; but he could discredit Grassi's scientific argument by an ingenious counter-theory of his own. This he did in a work ostensibly written by one of his pupils, Mario Guiducci (1585-1646), but actually mainly written by himself : a *Discourse on Comets*, read by Guiducci before the Florentine Academy, and then published. Rather unfairly, the *Discourse* began by attacking the Aristotelian theory of comets (unfairly, since Grassi had said nothing of the nature of comets, and because, as Galileo well knew, he would be obliged as a Jesuit to defend Aristotle even though he disagreed with some Aristotelian views). Aristotle had held that comets shine because of friction set up as they move through the air,* and an attack on this view could only have been intended to provoke Grassi to a reply. Galileo's own notions about comets—that they were the result of an earthly exhalation rising towards the Sun, shining by refracted light, illusions like haloes—permitted him to discount Grassi's optical and parallactic arguments, since these would not hold if the comets were not solid bodies.

If Galileo had wished to provoke Grassi, he succeeded ; the result was a violent attack on Galileo, and a fervid defence of Aristotle, under the title *The Astronomical and Philosophical Balance, On which the Opinions of Galileo Galilei Regarding Comets are Weighed, as Well as Those Presented in the Florentine Academy by Mario Guiducci and Recently Published*. Here Grassi tried to show that comets were real and solid bodies, having a circular path about the Sun (a theory more nearly correct than Galileo's) ; he also tried to defend the Aristotelian theory of heat in all its ramifications, some of them absurd. Hence the detailed analysis of the nature of heat that characterised the best part of *The Assayer* (*Il Saggiatore*, 1623), and the merciless drubbing Galileo

---

* Carried away by the heat of controversy, Galileo adopted an equally untenable view : that substances grow hot with friction only when they are soft enough so that some material can be rubbed off and " consumed."

gave his hapless adversary, in the course of which the weak side of his own argument was obscured.* Indeed, Grassi's rejoinder, *A Reckoning of Weights for the Balance* was too dull and heavy-handed to count for much.

Besides the joy of battle, Galileo had other reasons for feeling that he had won an important advance in the campaign for the right to discuss scientific theories freely. Though it was published under the sponsorship of the Lyncean Academy, Galileo was able to dedicate the *Assayer* to the new Pope, Urban VIII, that Cardinal Barberini who had always befriended him and who now expressed himself as delighted at the wit with which Galileo overcame his Jesuit opponent. The situation seemed promising to Galileo ; now that an intellectual was Pope, one moreover who was a personal friend and no great protector of narrow-minded orthodoxy, he thought the moment right to try the effect of personal diplomacy. In the spring of 1624 Galileo went to Rome to try to secure more freedom for discussion of the Copernican system. As he had done thirteen years before, he carried a new scientific instrument—a compound microscope —which he used to reveal new wonders in the living world, as the telescope had revealed new wonders in the celestial world.† This attracted interest and proved him still to be a creative scientist, all to the good when scientific prestige was needed to back up his arguments.

His optimism was, it seemed, fully justified. He saw the Pope

---

* Cf. ch. VII, pp. 261-3 above. Kepler, sympathetic to Galileo's position, was yet too loyal to Tycho not to defend his master against Galileo's attacks. This he did in an appendix to a defence of Tycho he was then publishing, *Tychonis Brahei Dani Hyperaspites* (*The Shieldbearer of Tycho*).

† Microscopes had been known for some time, though not, apparently, in Rome, but their zoological use was new. Galileo gave an instrument to Cesi; in 1625 the Lynceans christened it with its modern name, and one of them, Francesco Stelluti, produced an account of the anatomy of the bee as revealed by the microscope. (Cf. p. 242 above.)

several times, was received cordially, given a number of Agnus Dei medals, and much good advice. Best of all, there was opportunity for long discussions about the problem of the Copernican system. Galileo inquired about the effect of further physical arguments in favour of the Copernican system ; suppose, for example, he could show that if the Earth were assumed to move, this would explain the tides without recourse to " occult " attractions between the Moon and the sea. Would this be admitted as a strong, even conclusive, proof of the Copernican system ? Presumably it was at this time that the Pope pointed out to Galileo that he should never, especially in questions bordering on theology, forget that the weak and fallible intellect of man could not always understand the ways in which God chose to work. Even if to human reason it seemed that there was only one way of constructing the universe, nevertheless it did not follow that God *had* constructed the universe in this fashion. God could not be constrained by the limits of human reason ; because a man thought he had irrefutable proof of the Earth's motion, it did not follow that God had chosen to make the Earth move. However well inclined the Pope might be towards free scientific discussion, as Supreme Pontiff he was responsible for the safety of men's souls, and he judged that open support of the Copernican system was dangerous because it might cast doubt upon the infallibility of Holy Scripture. Especially was this true of discussions in Italian which treated the matter in a popular and easy vein, for this was readily open to misconstruction. It was all too easy to think that Galileo said the Bible spoke untruth, when all he said was that it needed re-interpretation ; the ordinary lay reader, unfamiliar with theology or mathematics, was apt to assume that if the astronomers differed from the Bible, and claimed to be right, the Bible was no longer to be trusted. The Copernican system was a useful mathematical device ; it was best to leave it there, and never to discuss it except as a scientific hypothesis. To Galileo, this was permission enough, and he returned to Florence determined to take up the defence of the Copernican system once more.

As long ago as 1610, Galileo had promised the world a great cosmological treatise which he temporarily called *The System of the World*. His theory of the tides was intended as the clinching argument in favour of the Earth's motion, which was to be supported by his own telescopic observations, and his own discoveries about motion. Now he began to write seriously ; in Italian, of course, and in dialogue form. In Galileo's youth literary men had enjoyed writing literary and philosophic dialogues in the manner of Plato ; and the dialogue form gave him greater freedom to express what he believed to be true without announcing it as his own conviction. For this could be no gay and slashing attack on the anti-Copernicans ; he must prove the truth of the Copernican system beyond doubt, for, as he had written long before,

> The most expeditious and safe way for me to prove that the Copernican position is not contrary to Scripture will be to demonstrate by a thousand truths that it is true, and that the contrary cannot be sustained in any way. Whence, since two truths cannot be in contradiction, it necessarily follows that this and Scripture are perfectly in accord.[17]

Although Galileo could contemplate the possibility that science might compel theologians to admit that the Bible was not literally true, at the same time he had to remember the Pope's argument, that no scientific proof had sufficient force to permit men to say that God must have done things in this way and no other. No easy task ; and after a year Galileo stopped writing, and turned instead to magnetic experiments ; he had presumably arrived at the point where he wished to make use of Gilbert's discovery of the magnetic nature of the Earth, and wanted to confirm Gilbert's experiments. But once laid aside he did not find it easy to pick the work up again ; it was not until the autumn of 1629 that he once more began writing. This time he finished the book in a few months.

As usual, Galileo hoped to have his work published in Rome with the assistance of the Lyncean Academy, and Cesi agreed to see it through the press. But it needed a licence, of course, and permission from the Pope was a desirable precaution. Once again, Galileo went to Rome, and had audience with the Pope ; as before, the Pope was friendly, and assured him he was free to discuss the Copernican system provided he avoided theological controversy, and provided he did not claim to have proved the truth of what was only an hypothesis. So the title must be one which gave a sure indication of the contents ; *Dialogue on the Two Chief Systems of the World* (*Dialogo sopra i due Massimi Sistemi del Mondo, Tolemaico e Copernicano*) was chosen. And there must be a preface in which the hypothetical character of the pro-Copernican arguments must be clearly stated. On the strength of this, Galileo submitted the book to the censors ; they were worried, but when the Pope had given permission, what could they do ? Only ask for minor adjustments to preface and conclusion, which they did ; indeed the preface was only released after the rest of the book was completed. All should have been plain sailing, but there were vexatious delays : Cesi died ; there was plague ; it was difficult to oversee the work from afar ; when it was decided to print in Florence there were new censors to be satisfied. At last the book appeared early in 1632.

The *Dialogue* was an important and sensational work. It was nearly ten years since Galileo had published anything, and there had been a dearth of Italian books on astronomy. Here was a lively discussion of the Copernican system, convincingly pro-Copernican yet couched in such terms that one could read it with a clear conscience. It was, as Galileo explained in the preface, intended to reveal all the experimental arguments in favour of the Earth's mobility ; secondly, all the celestial arguments ; and thirdly, as " an ingenious speculation," his theory of the motion of the Earth as a cause of the tides. And all this, ostensibly, at least, not to prove that the Earth moves, but rather to show that, in spite of what other nations had said after 1616, Italians were free to speculate upon Copernican problems, and could do so as well as any other nation. To which end, as Galileo declared " I have taken the Copernican side in the discourse, proceeding as with a pure mathematical hypothesis and striving by every artifice to represent it as superior to supposing the earth motionless—not, indeed, absolutely, but as against the arguments of some professed Peripatetics [Aristotelians]." [18] And this was done in a vivid literary style ; the three interlocutors—Salviati the Copernican, Sagredo the intelligent sceptic and Copernican convert, Simplicio the Aristotelian—were all modelled upon real men, though Simplicio was not named after a contemporary Italian, but after the sixth-century commentator upon Aristotle so often quoted by his followers.

To the talent for polemic which Galileo had displayed so many times he here gave full play, softened only by the fiction that this was a polite conversation between three gentlemen. Nevertheless, Galileo employed every device he knew to the confusion of poor Simplicio who, like a disciple of Socrates, is continually and ruthlessly led to expose his ignorance and lack of comprehension of the Aristotelian tenets he thinks he knows so well. Logic, telescopic discovery, the new Galilean dynamics, the theory of the tides, and all contemporary anti-Aristotelian physics were pressed into service to support the Copernican side.* The

* The scientific arguments, particularly the application of Galilean dynamics to terrestrial and celestial physics, were to be of far-reaching importance for the development of science. Because they are not relevant to the *Dialogue* as an instrument in the Copernican debate, but rather relate to the development of seventeenth-century physics, they will not be discussed here. A detailed

conversation was allowed to range freely from Earth to the celestial regions and back again, until every anti-Aristotelian and pro-Copernican argument had been expressed and supported, while the inherent weakness of the anti-Copernican system became more and more apparent. Poor Simplicio was beaten at every turn; it is not therefore particularly convincing when, on the very last page, his protest,

I know that if asked whether God in His infinite power and wisdom could have conferred upon the watery element its observed reciprocating motion using some other means than moving its containing vessels, both of you would reply that He could have, and that He would have known how to do this in many ways that are unthinkable to our minds. From this I forthwith conclude that, this being so, it would be excessive boldness for anyone to limit and restrict the Divine power and wisdom to some particular fancy of his own,

was met by Salviati's polite but hardly fervent rejoinder,

An admirable and angelic doctrine, and well in accord with another one, also Divine, which, while it grants to us the right to argue about the constitution of the universe (perhaps in order that the working of the human mind shall not be curtailed or made lazy) adds that we cannot discover the work of His hands.[19]

Particularly when Sagredo's only comment was to say how much he looked forward to Salviati's promised exposition of " our Academician's [Galileo's] new science of natural and constrained local motions." It is dutiful; it is correct; it is what the Pope had told Galileo to say; but it lacks conviction. Warmth and zeal were expended on the Copernican side of the argument; obedience alone marked the conclusion.

---

analysis will be found in chapter 2 of volume III in this series, *From Galileo to Newton*.

No wonder that the Pope was hurt and dismayed : he had allowed, even encouraged Galileo to write, under certain conditions ; and his benevolence had been abused. The Preface was dubiously sincere ; the conclusion, the Pope's own ingenious argument about the omnipotence of God, was put into the mouth of Simplicio, the simpleton, who throughout the rest of the book continually got the worst of the argument. The Pope could not but agree with those who suggested that Galileo was laughing at both argument and Pontiff. The whole book was a mistake, and should not have been licensed ; as no one wished to blame the censors, it must be that Galileo had somehow misled them. Had he perhaps altered the text ? The preface, last of the book to be printed, was in a different type face from the rest ; perhaps it had been added after the censors had seen the rest. The Jesuits were now thoroughly anti-Galilean, and the Pope in no mood to defend his former favourite. Sale of the *Dialogue* was suspended, and Galileo's friends became seriously alarmed. The Florentine Ambassador obtained an interview with the Pope and reported to the alarmed Grand Duke * that

His Holiness broke out in great anger, and on the spur of the moment said to me that all the same Galileo had dared to enter where he ought not, and into the most grave and dangerous matters that could be raised at this time.[20]

The Pope was both angry and alarmed lest, in deserting mathematics for Scriptural and religious controversy, Galileo had strayed from the Faith. He asked the Holy Office to consider the affair, and summoned Galileo peremptorily to Rome. Galileo was shocked ; he was used to better treatment, and he thought he had the Pope's support. Besides he was ill. Yet he could still hope that it would be possible " to justify myself fully and make plain my innocence and holy zeal towards the Church." [21]

* This was no longer Cosimo II, who had died in 1621, but his successor Ferdinand II, equally loyal to Galileo.

For once, Galileo did not want to go to Rome and conduct the affair in person ; no one wished to deal more closely than he need with the Inquisition, and he was genuinely ill. The Pope thought he was unduly arrogant, hardly credited his pleas of ill health, reasonable as they were in a man of nearly seventy, and at last threatened that if Galileo did not come immediately, he would be brought as a prisoner. Yet when he did arrive in Rome, it was two months before he was finally examined by the Inquisitors. To his surprise, he found that the chief offence attributed to him was that he had disobeyed the injunction of 1616 ; the Inquisitors seemed to claim that he had been told that he must not defend, hold, teach or discuss the Copernican system, when his recollection was that he had merely been told not to hold or defend it, which would leave him free to discuss it, as in the *Dialogue*.[22] How could he prove his recollection except by his sworn word and by inference from the certificate that Bellarmine had given him ? For Bellarmine was now dead. To make things worse his judges appeared to have evidence to the contrary. Galileo never, of course, saw the evidence ; it was in fact an unsigned, unofficial minute found in the files for 1616. It purported to record as happening what would have taken place had Galileo not immediately agreed to accept Bellarmine's warning.[23] With this document in existence, however, it was impossible not to feel that Galileo should never have written the *Dialogue*, nor tried to get it published. Yet there may have been some doubt about the document, which should have been in a more official form ; certainly Galileo came off well in his interrogations, and was kindly treated, though the strain of imprisonment by the Holy Office while the interrogations were conducted was great, despite the attention paid to his personal comfort. Indeed, after two examinations, Galileo was released to the custody of the Florentine Ambassador, and deliberations continued. There were a certain number of private interviews ; Galileo must be made to realise that, in the eyes of the Church, he had been guilty of contumacy, and could not expect leniency unless he would submit to the will of the Church without further constraint, and show his appreciation of the gentleness with which he had been treated.

Finally, four months after his arrival in Rome, the decision was reached : late in June 1633, Galileo was once more summoned to the Holy Office to receive the decree of the Inquisitors. He was censured for disobedience ; the *Dialogue* was prohibited ; he was to abjure his errors and confess his disobedience, after which he was to be detained in the prisons of the Holy Office during the Pope's pleasure. (In fact, the sentence of imprisonment was immediately commuted to house arrest in one of the Roman residences of the Medici.) But on two points Galileo won : he begged that he should not be required to say that he was not a good Catholic, nor that he had ever knowingly deceived anyone in publishing his book.[24] What, in fact, he abjured was having appeared to hold opinions which the Church regarded as heretical ; and what he promised was never to hold them again in such a fashion that he could be again suspected of possible heresy. There was nothing to which he could not swear with a clear conscience, as a loyal Catholic ; but he felt the social disgrace of the sentence and subsequent imprisonment keenly, perhaps more keenly than the prohibition of the *Dialogue*, which he had expected.

Many legends have arisen about Galileo's trial and sentence. In the nineteenth century, when freedom of thought seemed secure and rationalism triumphant, the whole affair seemed incomprehensible : either Galileo must have been broken by torture, or else he must have recanted with tongue in cheek. But Galileo was an Italian of 1633, not a nineteenth-century North European Protestant. He thought the Holy Office, even the Pope, misguided, as individuals ; but as representatives of the

Church he obeyed them as it was his duty to do. He thought his punishment severe, and never ceased trying to have it mitigated even when, within the year, he was allowed to return (under restriction) to his villa at Arcetri, near Florence. He mourned his failure to use the pressure of educated Italian opinion to change the prejudice of the ecclesiastical hierarchy. He shared the opinion of his friends, many of them clerics, who blamed it all on the enmity of the Jesuits. But it did not occur to him to rebel.

Ironically, the decree of the Church and the trial and conviction of Galileo did more to promote Copernicanism than to discourage it. A Latin edition of the *Dialogue* appeared in Strasbourg in 1635, together with Latin and Italian editions of *The Letter to the Grand Duchess*. Italians might be forced to silence, or to apply their Copernican theories only to Jupiter and its satellites, but Catholic scientists in France and elsewhere cheerfully ignored the decree. Descartes was unique among French scientists in never openly supporting Copernicanism when he believed it to be the true system of the world. Father Mersenne, a devout Churchman, translated much of Galileo's work on mechanics into French, and became a convert to Copernicanism. Gassendi (1592–1655), also a cleric, not only continued to correspond with Galileo after 1633, but continued to defend the Copernican system. Indeed, there were many more serious scientists who supported the Copernican position after 1633 than before 1610. Galileo had provided telescopic information that showed how little the ancients had known about the universe and that fitted far better with the Copernican than with the Ptolemaic system. And his trial and condemnation had made the choice between the Copernican and Ptolemaic systems a matter of conscience ; no one could be content to wait and see. For this was one of the most serious scientific issues of the day ; and those who chose not to be Copernicans had at least to be

Tychonians if they wished to be serious scientists. Galileo had won, while seeming to lose ; for he had convinced men that the debate was not about theology, but about the heavens. Acceptance or rejection of Copernicanism was no longer a matter related to religious belief ; it rested upon the evidence of the stars, such as Galileo had provided.

EPILOGUE

Galileo's trial marks the climax of the great debate on cosmology, and the end of the long search for a new astronomy begun by Peurbach. Galileo demonstrated the road which astronomy was now to follow, for it was only through Galilean dynamics that the Newtonian synthesis came into being. The dynamics Galileo had used as an incidental argument in the *Dialogue* was to be elucidated at length in the *Discourses on Two New Sciences* which he indomitably completed, in spite of confinement and blindness. That book was smuggled out of Italy and printed in Holland in 1638, as if to show that nothing could stop the new bent which Galileo had given to natural philosophy. Kepler, dying in 1630, never saw the *Dialogue*, never knew of Galileo's new dynamics, nor that Galileo had ignored all his own elaborate calculations. Yet it was a combination of Galilean dynamics with Keplerian mathematical astronomy that made possible the ultimate triumph of the new astronomy.

Astronomy matured earlier than the natural sciences. Yet in a sense Harvey's work also marks a point of triumph, and of completion. The fifteenth-century attempts to capture human anatomy and physiology through the eyes of Galen had led first to independent study, then to new concepts and new knowledge. Finally, these in turn had led to the overthrow of the central pillar of Galenic physiology by means of the doctrine of the circulation of the blood. Though this discovery was still rejected by many in 1630, few could doubt that modern physicians knew more than the ancients, nor that experimental methods were as well suited to the investigation of the frame of man as to any other part of the natural world. Even the ancillary sciences of botany and zoology were well on the way to independent existence.

One of the most noticeable changes in the period between 1450 and 1630 is the change in attitude towards the ancients. In 1450 men attempted no more than comprehension of what the ancients had discovered, certain that this was the most that could be known ; by 1630 things had so changed that the works of the ancients were available in various vernacular translations, and even the barely literate who read these versions were aware that the authority of the Greek and Roman past was under attack. Ancient learning was increasingly old-fashioned ; what had been new in 1500 was outmoded by 1600, so relatively rapidly had ideas changed. In 1536 Petrus Ramus as a wildly daring young man could, perhaps prematurely, publicly defend the thesis that everything Aristotle had taught was false ; forty years later Aristotle's philosophy was still a university subject, but bright undergraduates like Francis Bacon were already saying that the study of Aristotle was a great waste of time. By 1630 it was obvious that the way was clear for a new physics, as it was for a new cosmology ; only Aristotle's zoological work still, precariously, survived.

In 1450 the scientist was either a classical scholar or dangerously close to a magician. By 1630 he was either a new kind of learned man or a technical craftsman. As ancient authority declined and self-confidence in the ability of the moderns grew, the necessity for a classical education grew less, though every scientist was still expected to read and write Latin competently. The sheer success of science and the steady advance of rationalism generally meant the end of the magical tradition. Mathematician no longer meant astrologer ; the word chemistry replaced alchemy as a new science was born ; the number mysticism Kepler loved gave way to number theory, such as Fermat (1601–65) explored ; natural magic was about to be replaced by experimental science and the mechanical philosophy. Science and rationalism were to become synonymous, cemented together by Descartes's *Discourse on Method* (1637).

Of very practical significance to the individual scientist was the changed position of science in the learned world. Peurbach and Regiomontanus had lectured on literature, not mathematics nor astronomy ; Vesalius was appointed lecturer in surgery, not anatomy ; in 1500 there were few university posts in science, and no scientist could expect respect from the learned world unless he were also a humanist. By 1600 things were very different. There were chairs of mathematics at all major universities, and many minor ones (like Graz) ; these supported cosmographers, astronomers and applied mathematicians in considerable numbers. Their pay and prestige was at first lower than those of the corresponding chairs in the older faculty of medicine, but even this began to change after 1600, as the experience of Galileo was to show. Harvey found the Lumleian chair profitable in terms of pay and opportunity for research, and chairs of anatomy, botany and even chemistry became indispensable to good medical faculties. New scientific chairs were founded : the Savilian chairs at Oxford, the Lucasian chair at Cambridge were well-paid, well-regarded posts which could draw men away from Gresham College in London. They were founded often by wealthy amateurs, receptive to the progress of science, and aware of its potentialities. As the content of science became more technical, there was a greater demand for textbooks and manuals, first in Latin, later in the vernaculars. In 1550 those who knew no Latin were expected to be interested in little beyond elementary mathematics, pure and applied. In the early years of the seventeenth century Galileo showed that the most novel and complex ideas could be presented in common language. Galileo's example was followed more and more, except for very technical works, though of course all important books in English, French or Italian were regularly translated into Latin for the benefit of the learned everywhere. The sheer volume of scientific books published reflects the growth of science itself, and the growth in the size of the audience capable of appreciating them. Increase in numbers of the men engaged in scientific work was not yet great enough to warrant widespread formation of scientific societies, but only a generation separates the Lyncean Academy from the formal societies of enduring importance, the Royal Society in England and the Académie des Sciences in France.

Science became more self-assured partly as it became more useful, though its utility was limited as yet, and its practical potentialities could not be predicted with assurance. Anatomy helped the surgeon, though only up to a point; he was incompetent to deal with internal disorders. Better understanding of plant structure did nothing to advance medicine; new plants from strange lands provided the physician with more drugs, but they were not necessarily better for being exotic. The discovery of the circulation of the blood paradoxically led to more blood-letting, not less. Chemistry added new drugs to the pharmacopoeia; whether this was pernicious or beneficial is as debatable now as it was then. Purges and emetics were now both cheaper and more violent than they had been a century earlier; the death rate remained unvaryingly high, though wound-surgery was perhaps marginally more effective. The chemist learned more from the craftsman as yet than he could hope to teach in return. In contrast, astronomy and applied mathematics were immediately and genuinely useful. Astronomy satisfied many needs: through astrology it offered man certainty for the future, reassuring if not quietening his mind; through calendrical computation it gave a more certain date for Easter, quietening man's soul; through navigation it protected men's bodies on perilous voyages through the oceans. New methods, new instruments, new ideas were all tried ; though many remained impracticable, and some of the best inventions were merely empirical, yet the success of applied astronomy was indubitable. No wonder that Bacon saw science as of practical benefit, nor that others besides Bacon were optimistic about the potential usefulness of science. Part of the assurance of the later seventeenth century on this point came from the undoubted success of the sixteenth century, a somewhat premature triumph that could not immediately be continued.

Interest in the useful application of science meant interest in technical problems and encouragement of craftsman and engineer. The fifteenth century saw the military engineer of the Middle Ages concern himself with a host of civilian problems. This, together with the increase in demand for astronomical and surveying instruments, encouraged the development of a new profession, that of instrument-maker and mathematical practitioner. Quadrant, cross-staff, back-staff, sector, logarithmic and navigational scales, magnetic compass, theodolite, declinometer—a host of new instruments appeared, demanding some mathematical knowledge and much mechanical skill for their construction. Science enormously increased the stock-in-trade of the mathematical practitioner. First came the navigational instruments and charts ; then astronomical instruments, such as those Tycho invented and taught others to construct and use ; finally, in the early seventeenth century, various optical devices. Galileo turned the spectacle maker into a telescope maker. As the seventeenth century wore on, it was to become common for scientific inventions to provide new wares for the craftsman as well as new tools of scientific investigation for the scientist himself.

Though the sixteenth century saw increased interest in science, and its spread among the relatively unlearned, it paradoxically did not see a parallel influence on the learned world. Once science had been a part of every man's learned education, as it was to be again; science had formed a part of the university curriculum in the Middle Ages, when every clerk had read Aristotle *On the Heavens*. Turning away from the scholasticism of the universities with which Aristotelian science was indissolubly associated, the new humanism preferred literature and philology to natural philosophy. Mathematics fared best, regarded as the training for the mind advocated by Plato, whose doctrines provided a convenient alternative to those of Aristotle. But the very success of the new science left the non-scientific philosopher far behind; how could he accept a repudiation of ancient learning coupled with a tendency, however faint as yet, to believe that modern man might know more than the ancients knew, when he was still evaluating the various doctrines of the ancients? Particularly was this true as the astronomical revolution gathered strength and upset the fundamental human belief in an earth-centred cosmos made for man. Astronomy, once the most commonly understood science, had burst all bounds to become both highly mathematical and highly abstract; and as the astronomer's universe became vast in extent and strange in appearance the non-scientific intellectual often took refuge not in rebellion (as the poets did) but in indifference. The philosopher did not feel the need to apprehend the prodigious changes in the universe brought about by science; like Montaigne he was content to assume that it was all a play with hypotheses, which could matter little. Only a few saw it, as yet, otherwise; both Bruno and Bacon showed, in very different ways, what use philosophy could make of speculations about the physical universe. Soon, no philosopher could afford to ignore the new cosmos invented by scientists and strangely made real by their, fascinatingly novel methods.

BIBLIOGRAPHY AND NOTES

Wherever possible, all discussions have been based on original sources, as far as these have been available to me. Where these sources exist in modern editions (which often contain excellent introductory material) they have been listed here. Also listed are reasonably available works which I have found useful, and which may profitably be consulted for further details. Some of these contain fairly extensive bibliographies. For this reason, no attempt has been made to be exhaustive, nor to give many references to scholarly articles.

There are two modern works dealing with the period of this book : Lynn Thorndike, *Science and Thought in the Fifteenth Century* (New York, 1929), by a mediaevalist who found the fifteenth century a disappointment ; and George Sarton, *Six Wings : Men of Science in the Renaissance* (Bloomington, Indiana, 1957), mainly biographical and bibliographical. For technology there is the recent authoritative *A History of Technology*, ed. C. Singer, E. J. Holmyard, A. R. Hall and T. I. Williams (Oxford, 1954–8) ; volume III deals with the Renaissance, and includes discussion of various aspects of applied science. A collective volume *La Science au Seizième Siècle*, Colloque international de Royaumont (Paris, 1960) deals with many interesting topics.

CHAPTER I: THE TRIUMPH OF OUR NEW AGE

The standard work on the literary and philological aspects of humanism is still J. E. Sandys, *A History of Classical Scholarship*, II, London, 1908 ; George Sarton, *The Appreciation of Ancient and Medieval Science during the Renaissance* (Philadelphia, 1955) is full of relevant information, E. G. R. Taylor, *The Haven-Finding Art* (New York, 1957) is a delightful, lively and authoritative history of navigation. J. Bensaude, *L'Astronomie Nautique au Portugal à l'époque des grandes découvertes* (Bern, 1912) is valuable for its original sources. The best history of astronomy is still J. L. E. Dreyer, *A History of Planetary Systems from Thales to Kepler* (Cambridge, 1906) ; 2nd ed. as *A History of Astronomy* (New York, 1953). Lloyd A. Brown, *The Story of Maps* (Boston, 1950) is the best history of cartography for this period ; E. L. Stevenson, *Portolan Charts* (New York, 1911) deals with the earlier sea charts.

Notes for Chapter I (pp. 17–49)

1. Jean Fernel, *De Abditis Rerum Causis*, 1548 ; quoted by Sherrington, *The Endeavour of Jean Fernel*, p. 136 (see below, ch. v).

2. Sandys, p. 73.

3. Taylor, p. 159.

4. *Merchant of Venice*, act v, scene 1, Lorenzo to Jessica.

5. *De Revolutionibus*, Preface (see below, ch. III).

6. Printed by George Sarton in *Osiris*, 5, 1938, " The Scientific Literature Transmitted through the Incunabula," 41–247.

CHAPTER II: THE PLEASURE AND DELIGHT OF NATURE

Agnes Arber, *Herbals* (2nd ed., Cambridge, 1953) and Wilfrid Blunt, *The Art of Botanical Illustration* (London, 1950) contain excellent and authoritative accounts of the botanical side. C. E. Raven, *English Naturalists from Neckham to Ray* (Cambridge, 1947) deals with both botanists and zoologists, but is difficult reading. E. J. Cole, *A History of Comparative Anatomy* (London, 1944 and 1949) is extremely interesting and reliable. Hariot's account of Virginia and John White's paintings are reproduced in Stefan Lorant, *The New World* (New York, 1946) ; Oviedo y Valdes, *Natural History of the West Indies*, ed. S. A. Stoudemire (Chapel Hill, N. Carolina, 1959), E. Nordenskiold, *The History of Biology* (New York, 1928, 1946) and Charles Singer, *A History of Biology* (2nd ed. New York, 1950) contain good general accounts ; more detailed is É. Callot, *La Renaissance des Sciences de la Vie au XVIᵉ Siècle* (Paris, 1951). H. B. Adelmann, *The Embryological Treatises of Hieronymus Fabricius of Aquapendente* (Ithaca, N.Y., 1942) contains text and commentary.

Notes for Chapter II (pp. 50–67)

1. *De Historia Stirpium*, Basle, 1542, Preface, sig. 2v. Quoted in Arber, p. 67.

2. *De Historia Animalium*, IV, 1558, De Aquatalibus.

3. Adam Zaluziansky von Zaluzian, *Methodi Herbariae Libri Tres*, Prague, 1592, quoted in Arber, p. 144.

CHAPTER III: THE COPERNICAN REVOLUTION

The best modern translation of the Preface and Book 1 of *De Revolutionibus* is by J. F. Dobson and Selig Brodetsky, *Occasional Notes*, Royal Astronomical Society, no. 10, 1947 (available as a pamphlet). *Commentariolus* and the *Narratio Prima* of Rheticus have been translated by Edward Rosen, *Three Copernican Treaties* (New York, 1939, 2nd ed. with identical pagination, New York, 1959). Much of the first book of Copernicus is printed with commentary in Thomas S. Kuhn, *The Copernican Revolution* (Cambridge, Mass.,

1956). Dreyer, *History of Astronomy* is helpful ; Angus Armitage, *The World of Copernicus* (New York, 1951) is the most reliable brief life. There is a stimulating discussion by A. Koyré in R. Taton, ed., *L'Histoire Générale des Sciences*, t. II, *La Science Moderne* (1450–1800) (Paris, 1959) ; Koyré has also translated the Preface and Book I into French, with commentary (Paris, 1934).

Notes for Chapter III (pp. 68–89)

1. Rheticus, *Narratio Prima*, in Rosen, *Three Copernican Treatises*, p. 109.
2. *De Revolutionibus*, pp. 4–5. (Quotations have sometimes been modified.)
3. *ibid.*, p. 3.
4. *ibid.*, p. 6.
5. They are printed at the end of the edition used here.
6. *De Revolutionibus*, p. 5.
7. *ibid.*, sect. 8, p. 15.
8. *Three Copernican Treatises*, pp. 58–59.
9. *De Revolutionibus*, sect. 9, p. 16.
10. *ibid.*, sect. 10, p. 19.
11. *ibid.*, sect. 8, p. 14.
12. The preface is translated by Rosen in *Three Copernican Treatises*, pp. 24–25.

CHAPTER IV: THE GREAT DEBATE

There are detailed presentations in Kuhn, *op. cit.* ; in Dorothy Stimson, *The Gradual Acceptance of the Copernican Theory of the Universe* (New York, 1917) ; and, limited to England, but the best general account of English astronomy in the period, F. R. Johnson, *Astronomical Thought in Renaissance England* (Baltimore, 1937), a pioneer work. A. D. White, *A History of the Warfare of Science with Theology in Christendom* (New York, 1899) is an interesting example of the nineteenth-century rationalist-secular view, with useful quotations. Tycho Brahe's *Astronomiae Instauratae Mechanica* (1598) has been translated as *Tycho Brahe's Description of his Instruments and Scientific Work*, ed. H. Raeder, E. Stromgren and B. Stromgren (Copenhagen, 1946) ; and his cosmological description has been translated in " Tycho Brahe's System of the World " by Marie Boas and A. R. Hall, *Occasional Notes*, Royal Astronomical Society, 3, no. 21, 1959, pp. 253–63. J. L. E. Dreyer, *Tycho Brahe* (Edinburgh, 1890) is the standard biography by the astronomer who was also the modern editor of *Tychonis Brahe Opera Omnia* (15 vols., Copenhagen, 1913–29). D. W. Singer, *Giordano Bruno his Life and Thought* (New York, 1950) is a useful biography and contains a translation of *On the Infinite Universe and World*. William Gilbert's *De Magnete* is most pleasantly available in the handsome edition translated and edited by Sylvanus P. Thompson (London, 1900, facsimile edition, New York, 1958). Frances A. Yates, *The French*

Academies of the Sixteenth Century (London, 1947) is an interesting and authoritative account of a little-known intellectual society. A. Koyré, *From the Closed World to the Infinite Universe* (Baltimore, 1957) is a stimulating discussion of a cosmological-philosophical problem intensified by the Copernican debate.

Notes for Chapter IV (pp. 90–128)

1. Galileo, *Dialogue Concerning the Two Chief World Systems*, tr. by Stillman Drake (Berkeley, 1953), p. 128 (The Second Day; Sagredo to Simplicio).

2. *Tycho Brahe's Description of his Instruments*, pp. 45–6, and p. 110.

3. Dreyer, *History of Astronomy*, pp. 318, 345. Reinhold's edition of Peurbach went through several editions.

4. Tycho Brahe, *op. cit.*, p. 107.

5. Stimson, *Gradual Acceptance of the Copernican Universe*, p. 44, cited from V. de La Fuente, *Historia de las Universidades . . . de España*, 1884.

6. Quoted at length by F. R. Johnson, pp. 127–8. (Spelling here modernised.)

7. Yates, *French Academies*, p. 96.

8. *ibid.*, pp. 96–97.

9. Dreyer, *History of Astronomy*, p. 350.

10. Johnson, p. 183.

11. Fourth Day. Quoted from the English translation by Joshua Sylvester, who somewhat heightened the effect.

12. *Universae Naturae Theatrum*, Book V, section 2, quoted by Stimson, pp. 45–47.

13. *Essays*, Book II, chapter 12, " An Apologie of Raymonde Sebonde," Florio's translation.

14. *An Anatomy of the World*, 1611.

15. " To the Reader," a prefatory introduction to *A Perfit Description of the Coelestiall Orbes*, appended to *A Prognostication Everlasting* (London, 1576), unpaged. There were seven editions of the combined publication between 1576 and 1605. There are long quotations in Johnson.

16. Slightly modified from Digges' own Latinate construction. The diagram has been often reproduced, for example by Johnson (p. 166). That this infinite universe is theological, not purely physical, was first pointed out by Koyré in *From the Closed World to the Infinite Universe*, where the diagram is reproduced on p. 37.

17. *Tycho Brahe's Description of his Scientific Instruments*, pp. 106–8.

18. *ibid.*, p. 108.

19. *ibid.*, p. 110.

20. *ibid.*, p. 117.

21. "Tycho Brahe's System of the World," p. 255. Chapter x of *Recent Phenomena*; *Opera Omnia*, t. IV., p. 222.

22. *ibid.*, p. 258; Chapter 8 of *Recent Phenomena*, from which the subsequent quotations are also drawn.

23. *On the Magnet*, Book VI, chapter 3, p. 215.

24. *ibid.*, chapter 5, p. 226.

25. *ibid.*, chapter 9, p. 240; chapter 3, pp. 214–215.

26. *ibid.*, Book V, chapter 12, p. 210.

27. Singer, *Giordano Bruno*, p. 250.

28. *ibid.*, p. 229.

29. Quoted in Armitage, *The World of Copernicus*, p. 94.

30. White, *History of the Warfare of Science with Theology*, pp. 126–7.

31. Calvin has often been (erroneously) cited as an outspoken critic of Copernicus; the history of this myth has been examined by Edward Rosen, "Calvin's Attitude Toward Copernicus," *Journal of the History of Ideas*, 21, 1960, pp. 431–41.

CHAPTER V: THE FRAME OF MAN AND ITS ILLS

Much of the source material is now available in recent translations, most of which contain useful and authoritative commentary. Galen, *On Anatomical Procedures*, ed. by Charles Singer, Publication no. 7 of the Wellcome Historical Medical Museum (London, 1959); *De Usu Partium*, French tr. by C. Daremberg, in *Œuvres Anatomiques, Physiologiques et Médicales de Galien* (Paris, 1854). Mondino de' Liucci, *Anatomia* (Florence, 1930) in Italian. A. Benivieni, *De Abditis Nonnullis ac Mirandis Morborum et Sanationum Causis* (*The Hidden Causes of Disease*, 1506), ed. by C. Singer (Springfield, Ill., 1954). *Leonardo da Vinci on the Human Body*, ed. C. D. O'Malley and J. B. de C. M. Saunders (New York, 1952). Berengario da Carpi, *A Short Introduction to Anatomy*, tr. L. R. Lind (Chicago, 1959). *The Epitome of Andreas Vesalius*, tr. L. R. Lind (New York, 1949); *A Prelude to Modern Science . . . the "Tabulae Anatomicae Sex" of Vesalius*, ed. C. Singer and C. Rabin (Cambridge, 1946, with valuable introduction); *The Illustrations from the Works of Andreas Vesalius*, ed. C. D. O'Malley and J. B. de C. M. Saunders (New York, 1950). *Hieronimij Fracastorii de Contagione et Contagionis Morbis et Eorum Curatione*, tr. W. C. Wright (New York and London, 1930). Ambroise Paré, *Apologie and Treatise*, ed. G. Keynes (London, 1951). William Clowes, *Selected Writings*, ed. F. N. L. Poynter (London, 1948).

On Fernel, there is the enthusiastic account by Sir Charles Sherrington, *The Endeavour of Jean Fernel* (Cambridge, 1946). The best general discussion is in Charles Singer, *The Evolution of Anatomy* (London, 1925), 2nd ed. *A Short History of Anatomy & Physiology from the Greeks to Harvey* (New York, 1957).

On the techniques of preparing chemical medicines, there is R. J. Forbes, *A Short History of the Art of Distillation* (Leiden, 1948); on their use, George Urdang, "How Chemicals entered the Official Pharmacopoeias," *Archives Internationales d'Histoire des Sciences*, 7, 1954, 303-14. There is an excellent account of Beguin's life and work by T. S. Patterson, "Jean Beguin and his *Tyrocinium Chemicum*," *Annals of Science*, 11, 1937.

Notes for Chapter V (pp. 129-165)

1. *On Anatomical Procedures*, pp. 33-4.
2. *Epitome*, p. xxxv.
3. *De Abditis Nonnullis* . . ., case 32, pp. 79-80.
4. *On Anatomical Procedures*, p. 2.
5. Usually known as *Isagogae Breves*.
6. *Six Tables* in *A Prelude to Modern Science*, p. 2. The originals were published in 1538 and intended for " professors and students of medicine."
7. *A Prelude to Modern Science*, p. xxxv.
8. *On Anatomical Procedure*, Book v, chapter 13.
9. *Short Introduction to Anatomy*, p. 35.
10. *ibid.*, p. 147.
11. *De Fabrica*, in *Opera Omnia* (2 vols., Leyden, 1725), Book III, chapter 1, p. 305.
12. *ibid.*, p. 306.
13. *ibid.*, Book v, chapter 1, pp. 412-13.
14. *ibid.*, Book III, chapter 12, p. 340.
15. *ibid.*, Book vi, chapter 7, p. 511. The statement in the first edition differs slightly in wording, but the meaning is identical.
16. *ibid.*, Book vi, chapter 15, p. 516.
17. *ibid.*
18. *ibid.*, p. 535.
19. From Guillaume Plancy's *Life of Fernel* (1607) in Sherrington, *Endeavour of Jean Fernel*, pp. 150-70.
20. *De contagione*, p. 7.
21. Quoted in Forbes, *Short History of the Art of Distillation*, p. 108.
22. Quoted by Urdang in " How Chemicals entered the Official Pharmacopoeias," p. 309. The decree was emended in 1613, when the new edition of the Pharmacopoieas listed these drugs.
23. *A Briefe Aunswere of Josephus Quercetanus Armeniacus, Doctor of Physic, concerning the Original and Causes of Mettales, Set foorth against Chemists. Concerning the Spagericall Preparations, and use of Minerall, Animall and Vegitable Medicines.* With additions by John Hester. (London, 1591), p. lv.
24. *Basilica Chymica & Praxis Chymiatricae, or Royal and Practical Chemistry*

. . . *Being a Translation of Oswald Crollius his Royal Chemistry. Augmented and Inlarged by John Hartman. As also the Practice of Chymistry of John Hartman.* (London, 1670), p. 1.

25. English translation (London, 1669), p. 1.

CHAPTER VI: RAVISHED BY MAGIC

The great work on magic in this period is Lynn Thorndike, *A History of Magic and Experimental Science* (New York, 1923–41); volumes 4–6 cover the fifteenth and sixteenth centuries; this is indispensable as a work of reference, but is not interpretative. D. C. Allen, *The Star-Crossed Renaissance* (Durham, N. C., 1941), treats of astrology; J. M. Stillman, *The Story of Early Chemistry* (New York, 1924), deals with the rational and E. J. Holmyard, *Alchemy* (Penguin, 1957), with the irrational aspects of chemistry in this period. For Paracelsus there are *Selected Writings*, ed. J. Jacobi (London, 1951); A. E. Waite, *Hermetical and Alchemical Writings of Paracelsus* (London, 1894); T. P. Sherlock, "The Chemical Work of Paracelsus," *Ambix*, III, 1948, 33–63, a sane and balanced analysis; and, most recently, Walter Pagel, *Paracelsus: an Introduction to Philosophical Medicine* (Basle and New York, 1958); for the anti-Paracelsan point of view, A. R. Hall, "'Paracelsus' Again," *The Cambridge Journal*, VI, 1953, 301–10. Thomas Norton's *Ordinall of Alchimy* was edited by E. J. Holmyard from a seventeenth-century edition (London, 1928). "Basil Valentine" may be read in *The Triumphal Chariot of Antimony with the Commentary of Theodore Kerckringus*, a translation of the Latin edition of 1685 by A. E. Waite (London, 1893). Porta's *Natural Magick* has been reprinted in a facsimile of the London edition of 1658; the edition used here, the second (1669), has the same pagination. There are several editions of Cardan's autobiographical *The Book of My Life*.

Notes for Chapter VI (pp. 166–196)

1. Francis Bacon, *The Advancement of Learning*, in *Philosophical Works*, p. 57 (see below, for chapter 8).

2. Christopher Marlowe, *The Tragical History of Dr. Faustus*.

3. Sherrington, *Endeavour of Jean Fernel*, p. 153.

4. *New Star*, p. 8.

5. *Tycho Brahe's Description of his Instruments*, p. 117.

6. *New Star*, p. 13.

7. *Description*, pp. 117–18.

8. *ibid.*

9. Norton, *Ordinall of Alchimy*, p. 13.

10. *Pirotechnia*, tr. and ed. C. S. Smith and M. T. Gnudi, 2nd ed. (New York, 1959), pp. 35–6.

11. *Pirotechnia*, pp. 336–7.

12. *Selected Writings*, p. 211.

13. *ibid.*, pp. 133–4.

14. Quoted by Sherlock, p. 41.

15. *Selected Writings*, p. 164.

16. *Triumphal Chariot of Antimony*, p. 75.

17. *ibid.*, pp. 76–7.

18. *ibid.*, pp. 78–9. The preparation of the arcanum is given on pp. 113–15 ; the best description of the " star " on pp. 175–7.

19. Sherrington, p. 42.

20. *Hermetical and Alchemical Writings*, I, p. 17.

21. *Selected Writings*, pp. 196–7.

22. *Ordinall of Alchimy*, p. 21.

23. Henry Cornelius Agrippa, *The Vanity of Arts and Sciences* (London, 1694), p. 109.

24. *ibid.*, p. 110.

25. *Natural Magick* (2nd ed., London, 1669), pp. 1–2. Substantially the same definition appeared in the original version of 1558.

26. Henry Billingsley's English translation of Euclid (London, 1570), un-paged ; spelling modernised.

27. *De Augmentis Scientiarum*, Book III, chapter V ; *Philosophical Works*, p. 474.

28. *Natural Magick*, p. 211.

29. *On the Magnet*, " Preface to the Candid Reader."

30. *ibid.*, Book IV, chapter 2, pp. 155–9.

31. *ibid.*, Book V, chapter 8, p. 200.

32. *ibid.*, Book II, chapter 4, p. 65.

33. *ibid.*, Book V, chapter 12, p. 209.

CHAPTER VII: THE USES OF MATHEMATICS

On the art of perspective, see Erwin Panofsky, *Albrecht Dürer* (2nd ed. Princeton, 1945). For navigation, E. G. R. Taylor, *op. cit.* ; *Tudor Geography* (London, 1930) ; *Late Tudor and Early Stuart Geography, 1584-1650* (London, 1934), for Dee and Wright ; and *Mathematical Practitioners of Tudor and Stuart England* (Cambridge, 1954), an invaluable work of reference. D. W. Waters, *The Art of Navigation in Tudor and Stuart England* (London, 1959), is a very complete survey, with long bibliography. For mechanics, Marshall Clagett, *The Science of Mechanics in the Middle Ages* (Madison, Wisconsin, 1959), a source book with commentary, is indispensable for the background. Galileo's *On Motion* and *On Mechanics* are available in one volume (University of Wisconsin Publications in Mediaeval Science no. 5, Madison, Wisconsin,

1960), ed. I. E. Drabkin and Stillman Drake. Stevin's work is available in English in *The Principal Works of Simon Stevin*, vol. I, *Mechanics*, ed. E. J. Dijksterhuis (Amsterdam, 1955), with a useful summary of statical methods. For the development of sixteenth-century dynamics, A. Koyré, *Études Galiléennes*, 3 vols. (Paris, 1939), is detailed, penetrating and stimulating ; especially relevant are vols. I (*A L'Aube de la Science Classique*) and III (*La Loi de la Chute des Corps*). There are many useful histories of mathematics. A good summary of algebra in this period is A. Koyré in *La Science Moderne*. General summaries are D. E. Smith, *A History of Mathematics*, 2 vols. (Boston and London, 1923 ; reprinted New York, 1959) ; and, still very useful, Montucla, *Histoire des Mathématiques* (Paris, 1758)—vol. I covers the fifteenth and sixteenth centuries. Henry Morley, *The Life of Girolamo Cardano* (2 vols., London, 1854), is invaluable for the Cardan-Tartaglia dispute, for which it provides source material in translation ; for Tartaglia's dynamics, A. Koyré, in *La Science au Seizième Siècle*, pp. 93–116.

Notes for Chapter VII (pp. 197–237)

1. John Dee's Preface to Billingsly's Euclid.

2. From Panofsky, *Albrecht Dürer*, I, pp. 247–60, esp. p. 254. Dürer's work covered much besides perspective ; he was interested in the "geometrical" construction of calligraphy, and in aspects of conic sections and regular polygons.

3. Robert Hues, *Tractatus de Globis et eorum usu*, Hakluyt Society (London, 1889), from the English edition of 1638.

4. "The Compendious Rehearsall of John Dee . . . Anno 1592. Nov. 9," *Chetham Society Miscellanies*, I, XXIV, 1851, p. 5.

5. *Certaine Errors in Navigation* (London, 1599), Part I, chapter 1, unpaged.

6. *ibid.*, Part II, chapter 2, spelling modernised.

7. Dee's Preface to Euclid.

8. *Theatre des instrumens* (Lyon, 1579), dedication.

9. Stevin wrote in Dutch not merely to make his work available to his unlearned compatriots, but because he had convinced himself that Dutch was one of the oldest European languages, and that it was singularly well-suited for expressing scientific concepts in simple (which to him meant short) words ; the first edition of the *Art of Weighing* had a long preface *On the Worth of the Dutch Language*. Naturally, Stevin's works were not well known until they appeared in Latin, French or English, which a few did instantly. His work on mechanics was available in Latin translation in *Hypomnemata Mathematica*, 1608.

10. *Works*, I, p. 95.

11. *On Motion & On Mechanics*, p. 159.

12. *ibid.*, p. 67.

13. Clagett, *Science of Mechanics*, p. 350.

14. *Quesiti et Inventioni Diverse* (Venice, 1546), Book I, Question 3, quoted from Henry Crew, *The Rise of Modern Physics* (London, 1928), p. 75. In his *Nova Scientia* (1537) he still held that natural and violent motion could not mix, except at the very end of the trajectory followed under forced motion.

15. *On Motion*, p. 35.

16. *Works*, I, p. 511.

17. *ibid.*, pp. 65–66.

18. *Opere*, Edizione Nazionale, ed. A. Favaro, X, p. 115; quoted in French by A. Koyré, *La Loi de la Chute des Corps*, pp. 4–25, with penetrating commentary.

19. Tartaglia gave his version of the story in Book IX of *Quesiti et Inventioni Diverse* ; quoted in Morley, *Life of Cardan*, I, pp. 222 ff.

20. Quoted in E. G. R. Taylor, *The Haven Finding Art*, p. 211.

21. Napier calculated his tables so that $\log_N a = 10^7 \log_e \frac{10^7}{a}$. Or, if the whole sine be taken for 10^7, $N = e^{-1}$. Hence Napier's base is the reciprocal of the base of the usual modern " Napierian " logarithms.

22. Quoted in E. G. R. Taylor, *Late Tudor and Early Stuart Geography*, footnote 3, p. 67.

CHAPTER VIII: THE ORGANISATION AND REORGANISATION OF SCIENCE

Sixteenth- and early seventeenth-century societies have been little discussed, and that often inaccurately. Frances Yates, *French Academies*, is reliable. For Mersenne there is the monumental *Mersenne ou la Naissance du Mécanisme* of Robert Lenoble (Paris, 1943). Peiresc's correspondence has been edited by P. Tamisay de Larroque (7 vols., Paris, 1888–98). For Raleigh and his friends, M. C. Bradbrook, *The School of Night* (Cambridge, 1936), gives a fascinating literary exegesis. John Ward, *Lives of the Professors of Gresham College* (London, 1740), is the only real source, though there is some material in F. R. Johnson, *Astronomical Thought in Renaissance England*. A convenient edition of Bacon's works is *The Philosophical Works of Francis Bacon*, ed. J. M. Robertson from the edition of Ellis, Spedding and Heath (London, 1905). A convenient study and summary is Fulton H. Anderson, *Philosophy of Francis Bacon* (Chicago, 1948). Benjamin Farrington, *Francis Bacon Philosopher of Industrial Science* (New York, 1949), is over-concerned with Bacon's interest in scientific utility.

Notes for Chapter VIII (pp. 238–264)

1. Bacon, *The Great Instauration : Proemium, Works*, p. 241.
2. *The Compendious Rehearsall* (cf. chapter 8, note 4), pp. 7–8.
3. *Advancement of Learning*, p. 60.

4. *Novum Organum*, Book I, aph. lxxiv.

5. *ibid.*, aph. lxxx.

6. *Advancement of Learning*, p. 92.

7. *ibid.*, p. 95.

8. *ibid.*, p. 80.

9. *ibid.*, p. 81. Cf. *Descriptio Globi Intellectualis*, chapter II.

10. *Novum Organum*, Book I, aph. lxi.

11. *ibid.*, aph. lxx.

12. *ibid.*, aph. xix.

13. *ibid.*, Book II, aph. xxxix.

14. *ibid.*

15. *Descriptio Globi Intellectualis* (written about 1612 and published post-humously), chapter V, p. 681.

16. *ibid.*

17. *Advancement of Learning*, p. 94.

18. *Novum Organum*, Book II, aph. ii.

19. *ibid.*, aph. iii.

20. *ibid.*, aph. xx.

21. *ibid.*, aph. lii.

22. Stillman Drake, *Discoveries and Opinions of Galileo* (New York, 1957), p. 274 (see below, ch. 11).

23. *ibid.*, p. 276-7.

24. *ibid.*, p. 272.

25. *ibid.*, p. 277.

26. *Journal tenu par Isaac Beeckman de 1604 à 1634*, ed. Cornelis de Waard (3 vols., The Hague, 1939-45), I, 216 (1618).

CHAPTER IX: CIRCLES APPEAR IN PHYSIOLOGY

Charles Singer, *The Discovery of the Circulation of the Blood* (London, 1956, originally printed 1920) is a good brief account. C. D. O'Malley, *Michael Servetus, A Translation of his Geographical, Medical and Astrological Writings* (Philadelphia, 1953), is invaluable as a source. John F. Fulton, *Michael Servetus, Humanist and Martyr* (New York, 1953), is briefly biographical and biblio-graphical. Michael Foster, *Lectures on the History of Physiology during the Sixteenth, Seventeenth and Eighteenth Centuries* (Cambridge, 1924), though out of date in some respects is useful, especially for its translation of relevant passages. Hieronymus Fabricius, *De Venarum Ostiolis 1603*, facsimile edition with introduction, translation and notes by K. J. Franklin (Springfield, Illinois, 1933). There are many editions of Harvey's *De Motu Cordis* ; the most recent (with Latin and English texts) is that of K. J. Franklin (Oxford, 1957) which I have quoted here. The best study is H. P. Bayon, " William Harvey,

Physician and Biologist," *Annals of Science*, III, IV, 1938–9 ; the latest biography, disappointingly florid in style, is Louis Chauvois, *William Harvey* (Paris and London, 1957, in both French and English) ; this draws heavily on D'Arcy Power, *William Harvey* (London, 1897).

Notes for Chapter IX (pp. 265–286)

1. *De Motu Cordis*, Dedication.

2. O'Malley, *Michael Servetus*, p. 203. The relevant passage covers about six printed pages in English translation.

3. Juan Valverde published a Spanish paraphrase of Vesalius's *Fabrica* with the title *Historia de la Composicion del Cuerpo Humano* (Rome, 1556); the preface is dated 1554. He there mentioned the pulmonary circulation with the remark, "nobody before me has ever said this." As he was at one time a pupil of Realdus Columbus, he may have heard him lecture on the subject ; but the date when Columbus first did so is unknown. As Valverde's work was in Spanish, it can hardly have had much influence in Italian medical circles—where the original *Fabrica* would surely have been preferred. See Fulton, *Servetus*, p. 41.

4. Translated in Foster, *Lectures on the History of Physiology*, pp. 29 ff., from Book VII of *De Re Anatomica* (pp. 326–7 of the Frankfurt edition of 1593).

5. *De Motu Cordis*, p. 11 (Introduction) : " some . . . deny, in opposition to Colombo's view, that the lungs either make or retain spirits."

6. Foster, *op. cit.*, pp. 33–4.

7. This fact is noted by one of Cesalpino's most ardent partisans, Arturo Castiglioni, in his excellent *History of Medicine* (2nd ed. in English, New York, 1947), p. 440. For some reason, however, historians of medicine are extremely *parti pris* on the whole circulatory problem and their conclusions often bear little relation to their evidence.

8. *De Venarum Ostiolis*, pp. 53–4 (translation modified) and p. 74 (p. 4 of the original Latin edition).

9. *ibid.*, p. 47 (p. 1 of original).

10. *De Motu Cordis*, chapter 1, " I have, however [published] the more gladly in that Hieronymus Fabricius of Aquapendente, after having dealt carefully and learnedly in a special treatise with almost all the parts of animals, left only the heart untouched," p. 24.

11. *ibid.*, introduction, pp. 15–16.

12. " Praelectiones anatomicae universalis," quoted by Chauvois, *William Harvey*, from D'Arcy Power's transcription and translation, pp. 62–9.

13. Chauvois, *op. cit.*, p. 106.

14. *De Motu Cordis*, To His Colleagues, p. 7.

15. *ibid.*, chapter 7, p. 55.

16. *ibid.*, chapter 17, p. 111.

17. *ibid.*, chapter 6, p. 44.

18. *ibid.*, chapter 5, p. 41.

19. *ibid.*, chapter 6, p. 44.

20. *ibid.*, chapter 13, p. 82. Harvey further remarked that "the discoverer of the valves did not understand their real function, and others went no farther."

21. *ibid.*, chapter 14, p. 87.

22. *ibid.*, chapter 15, p. 88.

23. *ibid.*, chapter 8, pp. 58–9.

24. *ibid.*, chapter 16, p. 93.

CHAPTER X: CIRCLES VANISH FROM ASTRONOMY

Johannes Kepler, *Gesammelte Werke*, ed. Max Caspar (Munich, 1938 ff.), has replaced the older edition edited by Frisch ; all references are to the Caspar edition. Most of Kepler's major works were translated into German by Caspar. Caspar's biography of Kepler (English translation by C. D. Hellman, London and New York, 1959) is Germanic and thorough on the facts of his life, but is over-laudatory of Kepler, over-detailed, and not at all analytical. Carola Baumgardt, *Johannes Kepler : Life and Letters* (New York, 1951), is very brief ; the translations of the letters are rather free, and the biographical material hagiographic. A most useful short survey is A. Koyré, "L'Œuvre astronomique de Kepler," *XVIIᵉ Siècle* (Bulletin de la Société d'Étude du XVIIᵉ Siècle, Paris, 1956, no. 30), pp. 69–109. Kepler's reasons for believing that the universe was finite are discussed in Koyré, *From the Closed World to the Infinite Universe*, chapter 3. The account in Dreyer, *History of Astronomy* is reliable and fairly technical. Koyré, *La révolution astronomique* (Paris, 1961), indispensable.

Notes for Chapter X (pp. 287–312)

1. Kepler to Maestlin, 10 December 1601, *Gesammelte Werke*, IV, p. 203.

2. *Mysterium Cosmographicum*, Preface to the Reader, *Gesammelte Werke*, I, p. 9.

3. *ibid.* cf. *Epitome of Copernican Astronomy*, Book IV, Part I, *Gesammelte Werke*, VII, p. 258.

4. *Nova Astronomia*, chapter XIX, *Gesammelte Werke*, III, p. 178.

5. *Mysterium Cosmographicum*, p. 13.

6. *Nova Astronomia*, chapter XXXIII, p. 236.

7. *Epitome*, p. 254.

8. *Nova Astronomia*, chapter XLIV, p. 286.

9. *Nova Astronomia*, chapter XLVII, p. 297.

10. *Nova Astronomia*, chapter LIX, pp. 367 *et seq.*

11. *Nova Astronomia*, Introduction, p. 18.

12. *Nova Astronomia*, p. 35.

13. *Harmonices Mundi*, *Gesammelte Werke*, VI, Book V, chapter 3, p. 302.

CHAPTER XI: DEBATE AMONG THE STARS

The standard edition is the Edizione Nazionale, *Opere*, ed. A. Favaro (Firenze, 1890–1909, and 1929–39). *The Sidereal Messenger* is available in the English translations of E. S. Carlos (London, 1880 and 1960) and of Stillman Drake, the latter in *Discoveries and Opinions of Galileo* (New York, 1957), which contains *The Letter to the Grand Duchess Christina*, part of *The Assayer*, and some biographical material. *The Controversy on the Comets of 1618*, tr. Stillman Drake and C. D. O'Malley (Philadelphia, 1960), contains English versions of all the relevant documents. *The Discourse on Bodies in Water*, a facsimile of the 1665 translation by T. Salusbury, has been re-edited by Stillman Drake (Urbana, Illinois, 1960). There are two English editions of the *Dialogue* : that edited by Giorgio de Santillana from a seventeenth-century English translation, *Dialogue on the Great World Systems* (Chicago, 1953) and that by Stillman Drake, *Dialogue Concerning the Two Chief World Systems* (Berkeley, 1953) : quotations in the text are from the latter. Giorgio de Santillana, *The Crime of Galileo* (Chicago, 1955 ; there are later English and French editions) is a complete, impassioned and thoroughly documented account of the trial. Edward Rosen, *The Naming of the Telescope* (New York, 1947), is a very scholarly little work. There is no satisfactory biography.

Notes for Chapter XI (pp. 313–343)

1. John Donne, *Ignatius his Conclave* (1610).

2. *Sidereal Messenger*, in *Discoveries & Opinions*, pp. 28–9. Cf. substantially the same account given 29 August 1609 in a letter to his brother-in-law Landucci, quoted in J. J. Fahie, *Galileo his Life and Work* (London, 1903), pp. 77–78, though there he says " two months ago." The first inventor of the telescope is uncertain ; many years later (1634), Isaac Beeckmann recorded in his journal that the son of one of the claimants said his father had, in fact, copied an instrument previously made in Italy. But there is no other evidence, and certainly it was unknown in Venice in 1609. Hariot had a " perspective glass " before 1590, but its exact nature had never been determined. He may have used it to study the heavens before 1609, but he certainly announced no discoveries.

3. *Sidereal Messenger*, p. 34.

4. *ibid.*, p. 57.

5. Letter to the Earl of Salisbury, 13 March 1610, in Logan Pearsall Smith, *The Life and Letters of Sir Henry Wotton* (Oxford, 1927), I, pp. 486–7.

6. *Opere*, X, p. 233, quoted in *Discoveries and Opinions*, p. 65.

7. *Discoveries and Opinions*, p. 63.

8. *Opere*, IX, pp. 170–2 ; *Discoveries and Opinions*, p. 67.

9. These discoveries were publicly announced in 1612, in *Letters on Sunspots* (*Discoveries and Opinions*, pp. 87–144).

10. *Letters on Sunspots*, p. 144.

11. *Letter to the Grand Duchess*, *Discoveries and Opinions*, p. 186.

12. *Opere*, XII, pp. 128–30, letter of 12 January 1615.

13. *Discoveries and Opinions*, pp. 163–4.

14. The whole formality is discussed in detail in *The Crime of Galileo*, chapters 5 and 6.

15. *Opere*, XII, pp. 243–5 ; *Discoveries and Opinions*, p. 219.

16. Note in Galileo's hand in his copy of the *Dialogue* ; quoted in Santillana's edition, p. x.

17. *Opere*, XII, pp. 183–5 ; *Discoveries and Opinions*, p. 166.

18. *Dialogue*, Drake's edition, pp. 5–6.

19. *ibid.*, p. 465.

20. *Opere*, XIII, pp. 383–5, letter of Niccolini, 5 September 1632.

21. *Opere*, XIII, p. 411, letter of 16 October.

22. *Opere*, XV, p. 55, letter of Niccolini, 27 February 1633.

23. Much historical controversy has raged over the provenance of this document. At one time it was thought to be an interpolation into the files of 1616 at a later date ; it is now regarded as genuine in date, but the work of some character in the drama who was displeased with the leniency with which Galileo was then treated and foresaw that there might arise another occasion when he would be on trial. *The Crime of Galileo*, chapter 13, pp. 261–74, discusses the matter fully.

24. *The Crime of Galileo*, chapter 15, p. 311, from the diary of a friend of Galileo, Buonamici.

INDEX

Accademia dei Lincei, 242-3, 324, 327, 333 n., 336, 347

Achillini, Alessandro (1463-1512), 133-4

Acosta, Jose d' (d. 1600), 63-4

Agricola, Georg (1490-1555), 158-9, 173, 188, 210

Agrippa von Nettesheim, Cornelius (c. 1486-c. 1534), 183

Alchemy, 21, 166-7, 171-81

Aldrovandi, Ulysse (1522-1605), 58, 61

al-Kwarizmi (9th cent. A.D.), 228

Alpino, Prospero (1553-1617), 55

Anatomy, 67, 129-53, 245, 265-86 passim, 347 ; see also Illustration

Apian, Peter (1495-1552), 113, 201, 202

Apollonios (b. c. 262 B.C.), 25, 47, 226, 227, 305, 306

Archimedes (287-212 B.C.), 25, 74 n., 103, 211-12, 214-15, 222, 223, 226, 227, 240, 300, 304

Aristarchos of Samos (fl. 280 B.C.), 74 n., 95, 99, 103

Aristotle (384-322 B.C.), 26, 27, 65-6, 87, 88, 120, 149, 198, 211, 215 f., 250, 258, 273-5, 278, 284, 325
attacks on, 27, 43, 96-7, 99-100,

101, 113, 119, 194, 195, 222, 240, 254, 313, 316, 324, 332, 337-8, 345, 349
cosmology, 31 n., 40, 42, 75-6, 111, 119, 317
physics, 78-80, 99, 215, 216-18, 221, 223
zoology, 51, 52-3, 59-60, 61-2, 149, 151, 345

Astrology and astrologers, 20-1, 42-3, 166-71, 267, 290, 345

Astronomy, 39-49, 68-128, 201, 207-312, 347-8, 349
calendrical, 21, 42, 48, 70, 71, 323
Jupiter's satellites, 318-19, 323
navigational, 34-9
novae (new stars), 20, 101, 110-11, 170, 313, 316
observation of comets, 20, 101, 113, 118, 169, 313
parallax, 82, 116
phases of Venus, 323, 325
planetary conjunctions, 20, 39, 110, 297
planetary theory, 40-9, 75-85, 113-117, 298-307
sunspots, 325 ; see also Planetary Tables and Instruments

Augustine, St. (354–430), 51, 124

Avicenna (979–1037), 130, 176

Bacon, Francis (1561–1626), 185–6, 195, 238, 247–60, 262, 279, 286, 345, 348, 349
philosophy of science, 248–57
theory of matter, 257–60, 264

Baïf, Jean Antoine de (1532–89), 97–8

Barbarini, Cardinal, later Pope Urban VIII (1568–1644), 328, 330, 333–335, 336, 339–41

Barbaro, Ermalao (1453–93), 53

Bartas, Guillaume du (1544–90), 101–103

Basil Valentine, see Thölde

Bauhin, G. (1560–1624), 55

Bauhin, J. (1541–1612), 55

Beeckmann, Isaac (1588–1637), 263–4, 366

Beguin, Jean (c. 1550–1620), 165

Behaim, Martin (1459–1507), 34, 37

Bellarmine, Cardinal (1542–1621), 324, 327–31, 340

Belon, Pierre (1517–64), 59–60, 61

Benedetti, G. B. (1530–90), 99, 215, 216, 221, 222, 224

Benivieni, Antonio (c. 1440–1502), 130

Berengario, Jacopo da Carpi (c. 1460– c. 1530), 140, 142–3, 145

Besson, Jacques (fl. 1550–70), 189, 202, 210, 211

Biringuccio, Vanoccio (fl. c. 1540), 173–4, 210

Blaeu, Willem (1571–1638), 209–10

Bodin, Jean (1520–96), 103–4

Borelli, G. A. (1608–79), 311

Borough, William (1537–98), 235

Bostocke, Richard (fl. 1585), 99–100

Botany, 53–6, 62–7
botanic gardens, 56, 58
herbaria, 56; see also Illustration

Bouillaud, Ismael (1605–94), or Boulliaud or Bullialdus, 311

Bourne, William, 203

Boyle, Robert (1626–91), 163 n., 276

Bracciolini, Poggio (1380–1459), 23

Branca, G. (1571–1640), 210

Briggs, Henry (1561–1630), 207 n., 236, 246

Brues, Guy de (fl. 1557), 97–8

Brunfels, Otto (1488–1534), 54

Bruno, Giordano (c. 1548–1600), 122–125, 177, 294, 309, 349

Brunschwygk, Hieronymus (c. 1450–1512), 160, 173

Caius, John (1510–73), 19, 60, 136–7, 154

Calvin, Jean (1509–64), 126, 268, 357

Camerarius, Joachim (1534–98), 55

Canano, Giambattista (1515–79), 267 n.

Cardan, Jerome (1501–76), 188, 194, 199

Cardan, Jerome (*contd.*)
  as astrologer, 169
  as magician, 186–7
  as mathematician, 231–2
Cartography, 21–2, 30–4, 205–10
  globes, 34, 206
  plane charts, 205
  portolans, 31
  projection, 33–4, 205–10
Celsus (*fl.* A.D. 14–27), 23, 25, 26
Cesalpino, Andreas (1519–1603), 65–67, 274–6
Cesi, Federigo, Duke of Aquasparta (1585–1630), 242–3, 327, 333 n., 336
Chaucer, Geoffrey (1328–1400), 159, 174
Chemistry, 173–4, 178, 180–1, 347
  chemical medicine, 159–65 ; *see also* Alchemy
Chrysoloras, Manuel (*c.* 1355–1415), 23
Cicero (106–43 B.C.), 19, 23
Clavius, Christopher (1537–1612), 323, 326
Clowes, William (1544–1604), 156–7
Clusius, Charles (1526–1609), 55
Coiter, Volcher (1543–76), 61
College of Physicians, 25, 154, 164, 245, 279
Collège Royale, 198, 201, 245
Columbus, Christopher (*c.* 1446–1506), 31 n., 34, 39, 192
Columbus, Realdus (d. 1559), 152, 272–4, 279, 364

Commandino, Federigo (1509–75), 189, 212 n., 214, 226
Constantinople, fall of, 24
Copernicus, Nicholas (1473–1543), 30, 42, 43, 48, 68–89, 90–3, 97–100, 115, 121, 129, 146, 166, 198, 235, 256, 289, 298, 299
  Commentariolus, 71, 73, 76–7, 80–2, 91
  De Revolutionibus, 69, 70, 73, 74, 76–7, 91, 92, 313, 330
  system of, 77, 78–89, 93–4, 294, 295, 307, 313–43 *passim*
  acceptance and rejection of, 90–128, 257, 288, 326–43
  as observational astronomer, 92
Cordus, Valerius (1515–44), 55
Croll, Oswald (1580–1609), 163

Davis, John (1552–1605), 203
Dee, John (1527–1608), 93, 105–6, 107, 111, 120, 122, 167, 184–5, 190, 197, 199, 206–7, 209, 226, 241, 244, 245, 310
Descartes, René (1596–1650), 217, 224, 230, 238, 248, 264, 286, 342, 346
Didacus à Stunica (*c.* 1610–80), 124–5, 330
Digges, Leonard (d. *c.* 1571), 105, 169, 200
Digges, Thomas (*c.* 1543–75), 105–109, 111, 120, 121, 122, 241, 288, 310

Diophantos (c. A.D. 250), 226

Dioscorides (fl. c. A.D. 50), 29, 50, 51–53, 55, 64

Dodoens, R. (1517–85), 55

Donis projection (of Donnus Nicolaus Germanus), 34

Donne, John (1573–1631), 104, 293

Dürer, Albrecht (1471–1528), 137, 200–1

Dryander, Johannes (d. 1560), 143

Eccentrics, 41, 43–5, 75, 86, 117, 300n.
defined, 43

Ens, Gaspar, 189

Epicuros (341–270 B.C.), 98, 108

Epicycles, 41, 44–5, 75, 86, 103, 115, 117
defined, 44

Equant, 44–5, 86, 115, 117, 298
defined, 45

Erasmus (1465–1536), 27, 28 n., 122

Estienne, Charles (1504–64), 143, 276 n.

Euclid (fl. c. 300 B.C.), 26, 93, 197, 199, 225, 240, 245

Eudoxos (fl. c. 365 B.C.), 75, 103, 108

Eustachio, Bartolomeo (1520–72), 30, 152

Fabricius, Hieronymus of Aquapendente (1537–1619), 152, 276

Fabricius, Hieronymus of Aquapendente (contd.)
embryology, 61–2
valves in the veins, 276–8

Fabricius, Johannes (fl. 1610), 325

Fallopius, Gabriel (1523–62), 152

Feild, John (1525 ?–1587), 93

Fermat, Pierre (1601–65), 346

Fernel, Jean (d. 1557), 153–4, 168, 181–2, 199, 206

Finé, Oronce (1494–1555), 201, 202, 206, 245

Fludd, Robert (1574–1637), 189

Foscarini, Paolo Antonio (fl. 1615), 328, 330

Fracastoro, Girolamo (1484–1553), 75–6, 155 n., 157–8, 187

Frisius, Gemma (1508–55), 202, 205, 206

Fuchs, Leonhard (1501–66), 51, 54–6, 57, 135, 267

Galen (A.D. 129–199), 25, 26, 65, 129–138, 141–2, 146–54 passim, 162, 164, 169, 176, 265–84 passim, 344
De juvamentis, 131–2
On Anatomical Procedures, 135–7, 144, 153, 272
On the Motion of Muscles, 135
On the Natural Faculties, 134–5
On the Use of the Parts, 131–4, 138, 153, 265, 270

Galileo Galilei (1564–1642), 73, 96,

Galileo Galilei (*contd.*)
116, 170, 220, 240, 242–3, 248,
304, 309, 311, 313–43, 344, 346,
348
as a Copernican, 290, 315, 323–43
astronomical observations, 256, 292,
317–9, 323, 325
on physics, 214–16, 221–5, 344
theory of matter, 260–3, 264
Gassendi, Pierre (1592–1655), 342
Gellibrand, Henry (1597–1636), 246–
247
Gerard, John (1545–1607), 55
Gesner, Conrad (1516–65), 51, 57–8,
60, 160 n.
Ghini, Luca (d. 1556), 56
Gilbert, William (1540–1603), 120–
122, 124, 159, 254, 256, 257, 285,
288, 301, 302, 335
as natural magician, 190–6, 309
cosmology of, 120–2, 195, 301
on navigation, 192–4
Giovanni da Vigo (1460–1525), 156,
160
Grassi, Horatio (1583–1654), 331–3
Gresham, Sir Thomas (1519 ?–79),
246
Gresham College, 236, 246–7, 286,
346
Guarino of Verona (1370–1460), 23, 25
Guiducci, Mario (1585–1646), 332
Guinther, Johannes of Andernach
(1487–1574), 135, 144, 145, 266,
267
Gunter, Edmund (1581–1626), 246

Hariot, Thomas (1560–1621), 64, 207,
232, 242, 311, 366
Hartman, Johann (1568–1631), 163
Harvey, William (1578–1657), 67,
240, 251 n., 274, 276, 277, 278–
286, 344, 346
Henry the Navigator (1394–1460), 36
Heraclides of Pontus (*fl.* 4th cent.
B.C.), 94, 116
Herbals, 51, 52, 54–6
Hero of Alexandria (*fl.* A.D. 63), 25,
189–90
Hipparchos (*fl. c.* 146–127 B.C.), 110,
233, 287
Hippocrates (b. *c.* 460 B.C.), 149, 158
Hondius, Jodocus (1563–1611), 206–8
Hood, Thomas (*fl.* 1580), 245–6
Horrox, Jeremiah (1619–41), 311
Hues, Robert (1553–1632), 202–3, 207
Humanism and humanists, 18–19, 22–
28, 43, 49, 51, 57, 66, 68–9, 88,
134–7, 345
defined, 18

Illustration,
anatomical 29, 138–41, 143, 144–5
botanical, 29, 52, 53–5
engineering, 53
Instruments, 37, 109–10, 209, 241, 244
astrolabe, 33 n., 37
backstaff, 203, 210, 348
cross-staff, 37–8, 348
declinometer, 193, 348

Instruments (*contd.*)
  Gunter's sector, 204, 348
  log, 203, 210
  magnetic compass, 17, 36, 348
  microscope, 242, 255–6, 333
  quadrant, 37, 38, 205, 348
  telescope, 242, 243, 256, 314, 316–320, 323, 324, 348

Jacopo Angelo (*fl.* 1406), 23–4, 32
Jan Stephen van Calcar (1499–*c.* 1550), 145, 145–6 n.
Jordanus Nemorarius (*fl. c.* 1200), 214

Kelly, Edward (1555–95), 175
Kepler, Johann (1571–1630), 45, 74, 89, 95–6, 112, 120, 124, 127, 170, 233, 240, 285, 287–312, 313–14, 315, 328 n., 333 n., 344, 345
  as geometer, 227, 292

Leonard of Pisa (*c.* 1180–*c.* 1250), 228
Leonardo da Vinci (1452–1519), 30, 137–9, 142, 201, 212, 215, 224, 321
Leurechon, Jean, 189
Linacre, Thomas (*c.* 1460–1524), 19, 25, 134, 135, 201
Lobelius, Matthias (1538–1616), 55

Lucretius (*c.* 99–55 B.C.), 23, 97, 101, 108, 123
Lull, Raymond (*c.* 1235–1315), 122
Lusitanus, Amatus (1511–68), 276 n.
Luther, Martin (1483–1546), 73, 126

Maestlin, Michael (1550–1631), 95–6, 111, 288–90, 311
Magnus, Olaus (d. 1568), 155
Manilius (1st cent. B.C.), 48
Massa, Niccolo (*c.* 1489–1569), 142 n., 143
Mathematics, 21, 197–237, 349
  algebra, 228–32
  algorism, 22, 225, 227
  arithmetic, 22, 227–8, 232–3
  geometry, 226–7, 240
  logarithms, 235–7, 348, 362
  trigonometry, 48, 77, 225, 232–5
Mattioli, Pietro Andrea (1501–77), 55
Maurolyco, Francesco (1494–1575), 30, 227
Medicine, 19–20, 153, 157–65, 285, 346, 347
  chemical, 159–65, 347
  magical, 181–2 ; *see also* Anatomy, College of Physicians *and* Surgery
Melanchthon, Philip (1497–1560), 126
Mercator, Gerard (1512–94), 205–6, 208, 210
Mersenne, Marin (1588–1648), 215, 243–4, 342
Monardes (1493–1588), 62–3

Mondino de Luzzi (*c.* 1275–1326), 131–6, 138, 140, 142–3, 240

Montaigne, Michel (1533–92), 91, 104, 349

Monte, Guidobaldo del (1545–1607), 214, 315

Mouffet, Thomas (1553–1604), 60

Münster, Sebastian (1489–1552), 201

Napier, John (1550–1614), 235–6, 246

Natural History, 50–60, 62–4; *see also* Herbals

Natural Magic, 21, 166–7, 183–96

Navigation, 21, 30–9, 201–5, 246–7, 348
latitude determination, 37–9
longitude determination, 39; *see also* Cartography *and* Instruments

Newton, Isaac (1642–1727), 71, 94, 312, 344

Niccolo da Reggio (*fl.* 1322), 132, 135

Nicholas of Cusa (1401–64), 45–6, 70, 123, 124

Norman, Robert (*fl.* 1560–96), 193, 202

Norton, Thomas (*fl.* 1477), 172–3, 182–3

Norwood, Richard (1590–1675), 204

Nuñez, Pedro (1502–78), 205–6, 208

Oresme, Nicole (*c.* 1323–82), 124, 219–20

Osiander, Andreas (1498–1552), 72, 88, 106, 125, 328 n.

Oviedo y Valdes (1478–1557), 62

Pacioli, Luca (d. *c.* 1510), 227, 228, 230

Pappus (3rd cent. A.D.), 226

Paracelsus (1493–1541), 99–100, 158, 159, 161–2, 165, 176–9, 182

Paré, Ambroise (1510–90), 156

Paul V, Pope, 324, 329–30

Peiresc, Nicolas Claude Fabri de (1580–1637), 244

Pélerin, Jean, 200

Penny, Thomas (1530–88), 57, 60

Peurbach, George (1423–69), 19, 44–45, 47–8, 69, 70, 92, 234, 240, 344, 346

Philolaus (5th cent. B.C.), 74, 103

Physics, 211–25
Aristotelian, *see* Aristotle
dynamics, 215–25, 321, 337 n., 344
theory of matter, 257–64

Piccolomini, Aeneas Sylvius (1405–1464), 24

Pico della Mirandola (1463–94), 42, 168, 169

Planetary Tables,
Alphonsine, 42, 92, 93, 110
Ephemerides, 20, 48, 92–3, 169
Prutenic, 92, 93, 110
Rudolphine, 292, 293

Plantin, Christopher (1514–88), 55

Plato (429–348 B.C.), 24, 27, 43, 88,
    149, 151, 169, 198, 310, 335, 349
Pléiade, 97–8, 243, 244
Pliny the Elder (A.D. 23–79), 50, 51–3
Popularisation, 22–3, 28, 199–201,
    346–7
Porta, G. B. della (d. 1615), 183–4,
    187–9, 191, 242–3, 244, 309
Printing press, 17, 28–30, 47–8
Proclus (410–85), 201
Ptolemy, 26, 40, 43, 46, 65, 68, 78, 79,
    85, 87, 89, 93, 233, 287, 309
    Almagest, 28, 47–8, 68, 74, 76–7
    Geography, 24, 28 n., 29, 30–4, 267
    system, 42–5, 48–9, 69, 74–5, 80,
    81 n., 84, 86, 96, 113, 119, 288,
    295
Pythagorean cosmology, 69, 74, 86–7,
    89, 103, 107

Quercetanus, Joseph (c. 1544–1609),
    162–3

Raleigh, Walter (1552 ?–1618), 64,
    242
Ramelli, Agostino (1531–90), 210–11
Ramus, Petrus (1515–42), 238, 257,
    345
Recorde, Robert (1510 ?–58), 23, 94–
    95, 230, 245
Regiomontanus, Johann Müller (1436–

1476), 19, 20, 28, 47–9, 68, 169,
    189 n., 212 n., 226, 234–5, 346
Reinhold, Erasmus (1511–53), 92–3,
    95, 235
Rete mirabile, 141–2, 143
Reymers, Nicolas (*fl.* 1588), 119 n.
Rheticus, Georg Joachim (1514–76),
    71–2, 91, 93, 96, 126, 235
Riccioli, G. B. (1598–1671), 311
Rondelet, Guillaume (1507–66), 58,
    59–60, 61
Ronsard, Pierre de (1524–85), 97–8
Rothmann, Christopher (*fl.* 1587), 96
Ruel, Jean (1474–1537), 55
Ruini, Carlo (c. 1530–1598), 60–1

Sacrobosco (John of Holywood, c.
    1200–50), 47, 240
Sagredo, Giovan Francesco (1571–
    1620), 322, 337–8
Scheiner, Christopher (1573–1650),
    325
Scholasticism, 18, 27, 96
Schrick, Michael Puff von (d. 1472),
    160
Sendivogius, Michael (1556–1636 or
    1566–1646), 175–6
Sennert, Daniel (1572–1637), 164,
    263 n.
Servetus, Michael (1509–53), 169,
    266–71, 273
Seton, Alexander "the Cosmopo-
    lite" (d. 1604), 175–6

Stelluti, Francesco (1577–1653), 242, 333 n.

Stevin, Simon (1548–1620), 222 n., 361
  on mathematics, 230
  on mechanics, 212–14
  on navigation, 192, 202

Strabo (c. 64 B.C.–A.D. 21), 25

Surgery, 154–7

Sylvius (1478–1553), 276

Syphilis, 19, 155–6

Tartaglia, Niccolo (1500–57), 212, 215, 221, 222, 231–2, 362

Thabit ibn Qurra (c. 830–901), 42

Theophrastos (c. 372–c. 288 B.C.), 51, 52–3

Thölde, Johann (fl. 1600), 163, 179–81

Translations and translators, 24–6, 52, 132, 134–6, 189, 197, 201, 211–12, 226, 346–7

Turner, William (c. 1510–68), 55, 57

Tyard, Pontus de (c. 1521–1605), 93, 98–9

Tycho Brahe (1546–1601), 92, 93, 96, 109–19, 121, 123, 126, 127, 209, 240, 241, 246, 287–96 passim, 298, 299, 301, 304, 310, 313, 331, 348
  Tychonic system, 116–9, 257, 295, 307, 313–14
  and alchemy, 110, 171–2
  and astrology, 109, 111, 170–1

Universities, 19, 93–4, 132, 239–41, 346, 349
  Bologna, 56, 58, 70, 131, 140
  Cambridge, 204, 246, 257, 278, 346
  Cracow, 68, 70
  Ferrara, 70
  Heidelberg, 201
  Leipzig, 109
  Louvain, 143, 144, 202, 205, 206
  Marburg, 163
  Montpellier, 59
  Naples, 122
  Oxford, 202, 218, 219, 246, 311, 346
  Padua, 56, 70, 144, 152, 214, 272, 276–7, 278, 290, 314, 315, 322
  Paris, 59, 135, 144, 154, 168, 218, 219, 267, 276
  Pisa, 65, 272, 315
  Salamanca, 93
  Toulouse, 267
  Tübingen, 95, 133 n., 288–9
  Vienna, 19, 47
  Wittenberg, 71, 91, 93

Valla, Giorgio (d. 1499), 25

Veranzio, Faust, 210

Vergil (70–19 B.C.), 19, 85

Verrochio (1435–88), 137

Vesalius, Andreas (1514–64), 129–30, 135, 136, 137, 141, 142, 143–52, 166, 240, 265–6, 270, 271, 272, 276 n., 346, 364
  Epitome, 145

Vesalius, Andreas (*contd.*)
Fabrica, 145–52, 153
Tabulae Sex, 144–5
Vicary, Thomas (d. 1561), 154
Viète, François (1540–1603), 230, 232

Waghenaer, Lucas Janszoon (b. *c.* 1540), 209
Walton, Isaak (1593–1683), 59
Ward, Seth (1617–89), 311
Wateson, George, 159
Weiditz, Hans (*fl.* 1520), 54
White, John (*fl.* 1585), 64

Wotton, Edward (1492–1555), 60
Wotton, Sir Henry (1568–1639), 311, 320
Wotton, William (1666–1727), 271
Wright, Edward (1558–1615), 120, 204, 207–9, 210, 236

Zacuto, Abraham (*fl.* 1473), 37, 205
Zaluziansky, Adam (*c.* 1500–*c.* 1610), 64–5
Zoology, 53–4, 56–62, 64, 67
Zonca, V. (1568–1602), 210

harper 🔥 torchbooks

HUMANITIES AND SOCIAL SCIENCES

American Studies: General

THOMAS C. COCHRAN: The Inner Revolution. *Essays on the Social Sciences in History*　TB/1140

EDWARD S. CORWIN: American Constitutional History. *Essays edited by Alpheus T. Mason and Gerald Garvey* △　TB/1136

CARL N. DEGLER, Ed.: Pivotal Interpretations of American History　TB/1240, TB/1241

A. HUNTER DUPREE: Science in the Federal Government: *A History of Policies and Activities to 1940*　TB/573

A. S. EISENSTADT, Ed.: The Craft of American History: *Recent Essays in American Historical Writing*
Vol. I　TB/1255;　Vol. II　TB/1256

OSCAR HANDLIN: This Was America: *As Recorded by European Travelers in the Eighteenth, Nineteenth and Twentieth Centuries. Illus.*　TB/1119

MARCUS LEE HANSEN: The Atlantic Migration: 1607-1860. *Edited by Arthur M. Schlesinger*　TB/1052

MARCUS LEE HANSEN: The Immigrant in American History.　TB/1120

JOHN HIGHAM, Ed.: The Reconstruction of American History △　TB/1068

ROBERT H. JACKSON: The Supreme Court in the American System of Government　TB/1106

JOHN F. KENNEDY: A Nation of Immigrants. △ *Illus.*　TB/1118

RALPH BARTON PERRY: Puritanism and Democracy　TB/1138

ARNOLD ROSE: The Negro in America　TB/3048

MAURICE R. STEIN: The Eclipse of Community. *An Interpretation of American Studies*　TB/1128

W. LLOYD WARNER and Associates: Democracy in Jonesville: *A Study in Quality and Inequality* ¶　TB/1129

W. LLOYD WARNER: Social Class in America: *The Evaluation of Status*　TB/1013

American Studies: Colonial

BERNARD BAILYN, Ed.: Apologia of Robert Keayne: *Self-Portrait of a Puritan Merchant*　TB/1201

BERNARD BAILYN: The New England Merchants in the Seventeenth Century　TB/1149

JOSEPH CHARLES: The Origins of the American Party System　TB/1049

LAWRENCE HENRY GIPSON: The Coming of the Revolution: 1763-1775. † *Illus.*　TB/3007

LEONARD W. LEVY: Freedom of Speech and Press in Early American History: *Legacy of Suppression*　TB/1109

PERRY MILLER: Errand Into the Wilderness　TB/1139

PERRY MILLER & T. H. JOHNSON, Eds.: The Puritans: *A Sourcebook of Their Writings*
Vol. I　TB/1093;　Vol. II　TB/1094

EDMUND S. MORGAN, Ed.: The Diary of Michael Wigglesworth, 1653-1657: *The Conscience of a Puritan*　TB/1228

EDMUND S. MORGAN: The Puritan Family: *Religion and Domestic Relations in Seventeenth-Century New England*　TB/1227

RICHARD B. MORRIS: Government and Labor in Early America　TB/1244

KENNETH B. MURDOCK: Literature and Theology in Colonial New England　TB/99

WALLACE NOTESTEIN: The English People on the Eve of Colonization: 1603-1630. † *Illus.*　TB/3006

LOUIS B. WRIGHT: The Cultural Life of the American Colonies: 1607-1763. † *Illus.*　TB/3005

American Studies: From the Revolution to 1860

JOHN R. ALDEN: The American Revolution: 1775-1783. † *Illus.*　TB/3011

MAX BELOFF, Ed.: The Debate on the American Revolution, 1761-1783: *A Sourcebook* △　TB/1225

RAY A. BILLINGTON: The Far Western Frontier: 1830-1860. † *Illus.*　TB/3012

EDMUND BURKE: On the American Revolution: *Selected Speeches and Letters. ‡ Edited by Elliott Robert Barkan*　TB/3068

WHITNEY R. CROSS: The Burned-Over District: *The Social and Intellectual History of Enthusiastic Religion in Western New York, 1800-1850*　TB/1242

GEORGE DANGERFIELD: The Awakening of American Nationalism: 1815-1828. † *Illus.*　TB/3061

CLEMENT EATON: The Freedom-of-Thought Struggle in the Old South. *Revised and Enlarged. Illus.*　TB/1150

CLEMENT EATON: The Growth of Southern Civilization: 1790-1860. † *Illus.*　TB/3040

LOUIS FILLER: The Crusade Against Slavery: 1830-1860. † *Illus.*　TB/3029

DIXON RYAN FOX: The Decline of Aristocracy in the Politics of New York: 1801-1840. ‡ *Edited by Robert V. Remini*　TB/3064

FELIX GILBERT: The Beginnings of American Foreign Policy: *To the Farewell Address*　TB/1200

FRANCIS GRIERSON: The Valley of Shadows: *The Coming of the Civil War in Lincoln's Midwest: A Contemporary Account*　TB/1246

FRANCIS J. GRUND: Aristocracy in America: *Social Class in the Formative Years of the New Nation*　TB/1001

ALEXANDER HAMILTON: The Reports of Alexander Hamilton. ‡ *Edited by Jacob E. Cooke*　TB/3060

THOMAS JEFFERSON: Notes on the State of Virginia. ‡ *Edited by Thomas P. Abernethy*　TB/3052

---

† The New American Nation Series, edited by Henry Steele Commager and Richard B. Morris.

‡ American Perspectives series, edited by Bernard Wishy and William E. Leuchtenburg.

* The Rise of Modern Europe series, edited by William L. Langer.

¶ Researches in the Social, Cultural, and Behavioral Sciences, edited by Benjamin Nelson.

§ The Library of Religion and Culture, edited by Benjamin Nelson.

Σ Harper Modern Science Series, edited by James R. Newman.

º Not for sale in Canada.

△ Not for sale in the U. K.

JAMES MADISON: The Forging of American Federalism: Selected Writings of James Madison. Edited by Saul K. Padover   TB/1226

BERNARD MAYO: Myths and Men: Patrick Henry, George Washington, Thomas Jefferson   TB/1108

JOHN C. MILLER: Alexander Hamilton and the Growth of the New Nation   TB/3057

RICHARD B. MORRIS, Ed.: The Era of the American Revolution   TB/1180

R. B. NYE: The Cultural Life of the New Nation: 1776-1801. † Illus.   TB/3026

FRANCIS S. PHILBRICK: The Rise of the West, 1754-1830. † Illus.   TB/3067

TIMOTHY L. SMITH: Revivalism and Social Reform: American Protestantism on the Eve of the Civil War   TB/1229

FRANK THISTLETHWAITE: America and the Atlantic Community: Anglo-American Aspects, 1790-1850 TB/1107

A. F. TYLER: Freedom's Ferment: Phases of American Social History from the Revolution to the Outbreak of the Civil War. 31 illus.   TB/1074

GLYNDON G. VAN DEUSEN: The Jacksonian Era: 1828-1848. † Illus.   TB/3028

LOUIS B. WRIGHT: Culture on the Moving Frontier   TB/1053

American Studies: The Civil War to 1900

THOMAS C. COCHRAN & WILLIAM MILLER: The Age of Enterprise: A Social History of Industrial America TB/1054

W. A. DUNNING: Essays on the Civil War and Reconstruction. Introduction by David Donald   TB/1181

W. A. DUNNING: Reconstruction, Political and Economic: 1865-1877   TB/1073

HAROLD U. FAULKNER: Politics, Reform and Expansion: 1890-1900. † Illus.   TB/3020

HELEN HUNT JACKSON: A Century of Dishonor: The Early Crusade for Indian Reform. ‡ Edited by Andrew F. Rolle   TB/3063

ALBERT D. KIRWAN: Revolt of the Rednecks: Mississippi Politics, 1876-1925   TB/1199

ROBERT GREEN MCCLOSKEY: American Conservatism in the Age of Enterprise: 1865-1910   TB/1137

ARTHUR MANN: Yankee Reformers in the Urban Age: Social Reform in Boston, 1880-1900   TB/1247

WHITELAW REID: After the War: A Tour of the Southern States, 1865-1866. ‡ Edited by C. Vann Woodward   TB/3066

CHARLES H. SHINN: Mining Camps: A Study in American Frontier Government. ‡ Edited by Rodman W. Paul   TB/3062

VERNON LANE WHARTON: The Negro in Mississippi: 1865-1890   TB/1178

American Studies: 1900 to the Present

RAY STANNARD BAKER: Following the Color Line: American Negro Citizenship in Progressive Era. ‡ Illus. Edited by Dewey W. Grantham, Jr.   TB/3053

RANDOLPH S. BOURNE: War and the Intellectuals: Collected Essays, 1915-1919. ‡ Edited by Carl Resek   TB/3043

A. RUSSELL BUCHANAN: The United States and World War II. † Illus.   Vol. I TB/3044; Vol. II TB/3045

ABRAHAM CAHAN: The Rise of David Levinsky: a documentary novel of social mobility in early twentieth century America. Intro. by John Higham   TB/1028

THOMAS C. COCHRAN: The American Business System: A Historical Perspective, 1900-1955   TB/1080

FOSTER RHEA DULLES: America's Rise to World Power: 1898-1954. † Illus.   TB/3021

JOHN D. HICKS: Republican Ascendancy: 1921-1933. † Illus.   TB/3041

SIDNEY HOOK: Reason, Social Myths, and Democracy   TB/1237

ROBERT HUNTER: Poverty: Social Conscience in the Progressive Era. ‡ Edited by Peter d'A. Jones   TB/3065

WILLIAM L. LANGER & S. EVERETT GLEASON: The Challenge to Isolation: The World Crisis of 1937-1940 and American Foreign Policy
    Vol. I TB/3054; Vol. II TB/3055

WILLIAM E. LEUCHTENBURG: Franklin D. Roosevelt and the New Deal: 1932-1940. † Illus.   TB/3025

ARTHUR S. LINK: Woodrow Wilson and the Progressive Era: 1910-1917. † Illus.   TB/3023

GEORGE E. MOWRY: The Era of Theodore Roosevelt and the Birth of Modern America: 1900-1912. † Illus.   TB/3022

RUSSEL B. NYE: Midwestern Progressive Politics: A Historical Study of Its Origins and Development, 1870-1958   TB/1202

WALTER RAUSCHENBUSCH: Christianity and the Social Crisis. ‡ Edited by Robert D. Cross   TB/3059

JACOB RIIS: The Making of an American. ‡ Edited by Roy Lubove   TB/3070

PHILIP SELZNICK: TVA and the Grass Roots: A Study in the Sociology of Formal Organization   TB/1230

IDA M. TARBELL: The History of the Standard Oil Company: Briefer Version. ‡ Edited by David M. Chalmers   TB/3071

GEORGE B. TINDALL, Ed.: A Populist Reader ‡   TB/3069

TWELVE SOUTHERNERS: I'll Take My Stand: The South and the Agrarian Tradition. Intro. by Louis D. Rubin, Jr., Biographical Essays by Virginia Rock   TB/1072

WALTER E. WEYL: The New Democracy: An Essay on Certain Political Tendencies in the United States. ‡ Edited by Charles B. Forcey   TB/3042

Anthropology

JACQUES BARZUN: Race: A Study in Superstition. Revised Edition   TB/1172

JOSEPH B. CASAGRANDE, Ed.: In the Company of Man: Twenty Portraits of Anthropological Informants. Illus.   TB/3047

W. E. LE GROS CLARK: The Antecedents of Man: Intro. to Evolution of the Primates. o △ Illus.   TB/559

CORA DU BOIS: The People of Alor. New Preface by the author. Illus.   Vol. I TB/1042; Vol. II TB/1043

RAYMOND FIRTH, Ed.: Man and Culture: An Evaluation of the Work of Bronislaw Malinowski ¶ o △   TB/1133

DAVID LANDY: Tropical Childhood: Cultural Transmission and Learning in a Puerto Rican Village ¶   TB/1235

L. S. B. LEAKEY: Adam's Ancestors: The Evolution of Man and His Culture. △ Illus.   TB/1019

ROBERT H. LOWIE: Primitive Society. Introduction by Fred Eggan   TB/1056

EDWARD BURNETT TYLOR: The Origins of Culture. Part I of "Primitive Culture." § Intro. by Paul Radin TB/33

EDWARD BURNETT TYLOR: Religion in Primitive Culture. Part II of "Primitive Culture." § Intro. by Paul Radin   TB/34

W. LLOYD WARNER: A Black Civilization: A Study of an Australian Tribe. ¶ Illus.   TB/3056

Art and Art History

WALTER LOWRIE: Art in the Early Church. Revised Edition. 452 illus.   TB/124

EMILE MÂLE: The Gothic Image: Religious Art in France of the Thirteenth Century. § △ 190 illus.   TB/44

MILLARD MEISS: Painting in Florence and Siena after the Black Death: The Arts, Religion and Society in the Mid-Fourteenth Century. 169 illus.   TB/1148

ERICH NEUMANN: The Archetypal World of Henry Moore. △ 107 illus.   TB/2020

DORA & ERWIN PANOFSKY: Pandora's Box: The Changing Aspects of a Mythical Symbol. Revised Edition. Illus.   TB/2021

ERWIN PANOFSKY: Studies in Iconology: Humanistic Themes in the Art of the Renaissance. △ 180 illustrations   TB/1077

2

ALEXANDRE PIANKOFF: The Shrines of Tut-Ankh-Amon. *Edited by N. Rambova. 117 illus.*   TB/2011

JEAN SEZNEC: The Survival of the Pagan Gods: *The Mythological Tradition and Its Place in Renaissance Humanism and Art. 108 illustrations*   TB/2004

OTTO VON SIMSON: The Gothic Cathedral: *Origins of Gothic Architecture and the Medieval Concept of Order.* △ *58 illus.*   TB/2018

HEINRICH ZIMMER: Myth and Symbols in Indian Art and Civilization. *70 illustrations*   TB/2005

Business, Economics & Economic History

REINHARD BENDIX: Work and Authority in Industry: *Ideologies of Management in the Course of Industrialization*   TB/3035

GILBERT BURCK & EDITORS OF FORTUNE: The Computer Age: *And Its Potential for Management*   TB/1179

THOMAS C. COCHRAN: The American Business System: *A Historical Perspective, 1900-1955*   TB/1080

THOMAS C. COCHRAN: The Inner Revolution: *Essays on the Social Sciences in History*   TB/1140

THOMAS C. COCHRAN & WILLIAM MILLER: The Age of Enterprise: *A Social History of Industrial America* TB/1054

ROBERT DAHL & CHARLES E. LINDBLOM: Politics, Economics, and Welfare: *Planning and Politico-Economic Systems Resolved into Basic Social Processes*   TB/3037

PETER F. DRUCKER: The New Society: *The Anatomy of Industrial Order* △   TB/1082

EDITORS OF FORTUNE: America in the Sixties: *The Economy and the Society*   TB/1015

ROBERT L. HEILBRONER: The Great Ascent: *The Struggle for Economic Development in Our Time*   TB/3030

FRANK H. KNIGHT: The Economic Organization   TB/1214

FRANK H. KNIGHT: Risk, Uncertainty and Profit   TB/1215

ABBA P. LERNER: Everybody's Business: *Current Assumptions in Economics and Public Policy*   TB/3051

ROBERT GREEN MC CLOSKEY: American Conservatism in the Age of Enterprise, 1865-1910 △   TB/1137

PAUL MANTOUX: The Industrial Revolution in the Eighteenth Century: *The Beginnings of the Modern Factory System in England* ○ △   TB/1079

WILLIAM MILLER, Ed.: Men in Business: *Essays on the Historical Role of the Entrepreneur*   TB/1081

RICHARD B. MORRIS: Government and Labor in Early America △   TB/1244

HERBERT SIMON: The Shape of Automation: *For Men and Management*   TB/1245

PERRIN STRYKER: The Character of the Executive: *Eleven Studies in Managerial Qualities*   TB/1041

PIERRE URI: Partnership for Progress: *A Program for Transatlantic Action*   TB/3036

Contemporary Culture

JACQUES BARZUN: The House of Intellect △   TB/1051

CLARK KERR: The Uses of the University   TB/1264

JOHN U. NEF: Cultural Foundations of Industrial Civilization △   TB/1024

NATHAN M. PUSEY: The Age of the Scholar: *Observations on Education in a Troubled Decade*   TB/1157

PAUL VALÉRY: The Outlook for Intelligence △   TB/2016

RAYMOND WILLIAMS: Culture and Society, 1780-1950 ○ △   TB/1252

RAYMOND WILLIAMS: The Long Revolution.○ △ *Revised Edition*   TB/1253

Historiography & Philosophy of History

JACOB BURCKHARDT: On History and Historians. △ *Introduction by H. R. Trevor-Roper*   TB/1216

WILHELM DILTHEY: Pattern and Meaning in History: *Thoughts on History and Society.* ○ △ *Edited with an Introduction by H. P. Rickman*   TB/1075

J. H. HEXTER: Reappraisals in History: *New Views on History & Society in Early Modern Europe* △ TB/1100

H. STUART HUGHES: History as Art and as Science: *Twin Vistas on the Past*   TB/1207

RAYMOND KLIBANSKY & H. J. PATON, Eds.: Philosophy and History: *The Ernst Cassirer Festschrift. Illus.*   TB/1115

GEORGE H. NADEL, Ed.: Studies in the Philosophy of History: *Selected Essays from History and Theory*   TB/1208

JOSE ORTEGA Y GASSET: The Modern Theme. *Introduction by Jose Ferrater Mora*   TB/1038

KARL R. POPPER: The Open Society and Its Enemies △
  Vol. I: *The Spell of Plato*   TB/1101
  Vol. II: *The High Tide of Prophecy: Hegel, Marx and the Aftermath*   TB/1102

KARL R. POPPER: The Poverty of Historicism ○ △ TB/1126

G. J. RENIER: History: Its Purpose and Method △ TB/1209

W. H. WALSH: Philosophy of History: *An Introduction* △   TB/1020

History: General

L. CARRINGTON GOODRICH: A Short History of the Chinese People. △ *Illus.*   TB/3015

DAN N. JACOBS & HANS H. BAERWALD: Chinese Communism: *Selected Documents*   TB/3031

BERNARD LEWIS: The Arabs in History △   TB/1029

History: Ancient and Medieval

A. ANDREWES: The Greek Tyrants △   TB/1103

ADOLF ERMAN, Ed.: The Ancient Egyptians: *A Sourcebook of Their Writings. New material and Introduction by William Kelly Simpson*   TB/1233

MICHAEL GRANT: Ancient History ○ △   TB/1190

SAMUEL NOAH KRAMER: Sumerian Mythology   TB/1055

NAPHTALI LEWIS & MEYER REINHOLD, Eds.: Roman Civilization. Sourcebook I: *The Republic*   TB/1231

NAPHTALI LEWIS & MEYER REINHOLD, Eds.: Roman Civilization. Sourcebook II: *The Empire*   TB/1232

History: Medieval

P. BOISSONNADE: Life and Work in Medieval Europe: *The Evolution of the Medieval Economy, the 5th to the 15th Century.* ○ △ *Preface by Lynn White, Jr.* TB/1141

HELEN CAM: England before Elizabeth △   TB/1026

NORMAN COHN: The Pursuit of the Millennium: *Revolutionary Messianism in Medieval and Reformation Europe* △   TB/1037

G. G. COULTON: Medieval Village, Manor, and Monastery   TB/1022

CHRISTOPHER DAWSON, Ed.: Mission to Asia: *Narratives and Letters of the Franciscan Missionaries in Mongolia and China in the 13th and 14 Centuries* △   TB/315

HEINRICH FICHTENAU: The Carolingian Empire: *The Age of Charlemagne* △   TB/1142

F. L. GANSHOF: Feudalism △   TB/1058

DENO GEANAKOPLOS: Byzantine East and Latin West: *Two Worlds of Christendom in the Middle Ages and Renaissance* △   TB/1265

EDWARD GIBBON: The Triumph of Christendom in the Roman Empire (Chaps. XV-XX of "Decline and Fall," J. B. Bury edition). § △ *Illus.*   TB/46

W. O. HASSALL, Ed.: Medieval England: *As Viewed by Contemporaries* △   TB/1205

DENYS HAY: The Medieval Centuries ○ △   TB/1192

J. M. HUSSEY: The Byzantine World △   TB/1057

FERDINAND LOT: The End of the Ancient World and the Beginnings of the Middle Ages. *Introduction by Glanville Downey*   TB/1044

G. MOLLAT: The Popes at Avignon: 1305-1378 △ TB/308

CHARLES PETIT-DUTAILLIS: The Feudal Monarchy in France and England: *From the Tenth to the Thirteenth Century* ○ △   TB/1165

HENRI PIRENNE: Early Democracies in the Low Countries: *Urban Society and Political Conflict in the Middle Ages and the Renaissance. Introduction by John H. Mundy*   TB/1110

STEVEN RUNCIMAN: A History of the Crusades. △
Volume I: *The First Crusade and the Foundation of the Kingdom of Jerusalem. Illus.* TB/1143
Volume II: *The Kingdom of Jerusalem and the Frankish East, 1100-1187. Illus.* TB/1243

FERDINAND SCHEVILL: Siena: *The History of a Medieval Commune. Intro. by William M. Bowsky* TB/1164

SULPICIUS SEVERUS et al.: The Western Fathers: *Being the Lives of Martin of Tours, Ambrose, Augustine of Hippo, Honoratus of Arles and Germanus of Auxerre. △ Edited and trans. by F. O. Hoare* TB/309

HENRY OSBORN TAYLOR: The Classical Heritage of the Middle Ages. *Foreword and Biblio. by Kenneth M. Setton* TB/1117

F. VAN DER MEER: Augustine The Bishop: *Church and Society at the Dawn of the Middle Ages △* TB/304

J. M. WALLACE-HADRILL: The Barbarian West: *The Early Middle Ages, A.D. 400-1000 △* TB/1061

History: Renaissance & Reformation

JACOB BURCKHARDT: The Civilization of the Renaissance in Italy. *△ Intro. by Benjamin Nelson & Charles Trinkaus. Illus.* Vol. I TB/40; Vol. II TB/41

JOHN CALVIN & JACOPO SADOLETO: A Reformation Debate. *Edited by John C. Olin* TB/1239

ERNST CASSIRER: The Individual and the Cosmos in Renaissance Philosophy. *△ Translated with an Introduction by Mario Domandi* TB/1097

FEDERICO CHABOD: Machiavelli and the Renaissance △ TB/1193

EDWARD P. CHEYNEY: The Dawn of a New Era, 1250-1453. * *Illus.* TB/3002

G. CONSTANT: The Reformation in England: *The English Schism, Henry VIII, 1509-1547 △* TB/314

R. TREVOR DAVIES: The Golden Century of Spain, 1501-1621 ᵒ △ TB/1194

DESIDERIUS ERASMUS: Christian Humanism and the Reformation: *Selected Writings. Edited and translated by John C. Olin* TB/1166

WALLACE K. FERGUSON et al.: Facets of the Renaissance TB/1098

WALLACE K. FERGUSON et al.: The Renaissance: *Six Essays. Illus.* TB/1084

JOHN NEVILLE FIGGIS: The Divine Right of Kings. *Introduction by G. R. Elton* TB/1191

JOHN NEVILLE FIGGIS: Political Thought from Gerson to Grotius: *1414-1625: Seven Studies. Introduction by Garrett Mattingly* TB/1032

MYRON P. GILMORE: The World of Humanism, 1453-1517. * *Illus.* TB/3003

FRANCESCO GUICCIARDINI: Maxims and Reflections of a Renaissance Statesman (Ricordi). *Trans. by Mario Domandi. Intro. by Nicolai Rubinstein* TB/1160

J. H. HEXTER: More's Utopia: *The Biography of an Idea. New Epilogue by the Author* TB/1195

HAJO HOLBORN: Ulrich von Hutten and the German Reformation TB/1238

JOHAN HUIZINGA: Erasmus and the Age of Reformation. *△ Illus.* TB/19

JOEL HURSTFIELD, Ed.: The Reformation Crisis △ TB/1267

ULRICH VON HUTTEN et al.: On the Eve of the Reformation: *"Letters of Obscure Men." Introduction by Hajo Holborn* TB/1124

PAUL O. KRISTELLER: Renaissance Thought: *The Classic, Scholastic, and Humanist Strains* TB/1048

PAUL O. KRISTELLER: Renaissance Thought II: *Papers on Humanism and the Arts* TB/1163

NICCOLÒ MACHIAVELLI: History of Florence and of the Affairs of Italy: *from the earliest times to the death of Lorenzo the Magnificent. Introduction by Felix Gilbert* TB/1027

ALFRED VON MARTIN: Sociology of the Renaissance. *Introduction by Wallace K. Ferguson △* TB/1099

GARRETT MATTINGLY et al.: Renaissance Profiles. *△ Edited by J. H. Plumb* TB/1162

MILLARD MEISS: Painting in Florence and Siena after the Black Death: *The Arts, Religion and Society in the Mid-Fourteenth Century. △ 169 illus.* TB/1148

J. E. NEALE: The Age of Catherine de Medici ᵒ △ TB/1085

ERWIN PANOFSKY: Studies in Iconology: *Humanistic Themes in the Art of the Renaissance. △ 180 illustrations* TB/1077

J. H. PARRY: The Establishment of the European Hegemony: *1415-1715: Trade and Exploration in the Age of the Renaissance △* TB/1045

J. H. PLUMB: The Italian Renaissance: *A Concise Survey of Its History and Culture △* TB/1161

A. F. POLLARD: Henry VIII. ᵒ △ *Introduction by A. G. Dickens* TB/1249

A. F. POLLARD: Wolsey. ᵒ △ *Introduction by A. G. Dickens* TB/1248

CECIL ROTH: The Jews in the Renaissance. *Illus.* TB/834

A. L. ROWSE: The Expansion of Elizabethan England. ᵒ △ *Illus.* TB/1220

GORDON RUPP: Luther's Progress to the Diet of Worms ᵒ △ TB/120

FERDINAND SCHEVILL: The Medici. *Illus.* TB/1010

FERDINAND SCHEVILL: Medieval and Renaissance Florence. *Illus.* Volume I: *Medieval Florence* TB/1090
Volume II: *The Coming of Humanism and the Age of the Medici* TB/1091

G. M. TREVELYAN: England in the Age of Wycliffe, 1368-1520 ᵒ △ TB/1112

VESPASIANO: Renaissance Princes, Popes, and Prelates: *The Vespasiano Memoirs: Lives of Illustrious Men of the XVth Century. Intro. by Myron P. Gilmore* TB/1111

History: Modern European

FREDERICK B. ARTZ: Reaction and Revolution, 1815-1832. * *Illus.* TB/3034

MAX BELOFF: The Age of Absolutism, 1660-1815 △ TB/1062

ROBERT C. BINKLEY: Realism and Nationalism, 1852-1871. * *Illus.* TB/3038

ASA BRIGGS: The Making of Modern England, 1784-1867: *The Age of Improvement ᵒ △* TB/1203

CRANE BRINTON: A Decade of Revolution, 1789-1799. * *Illus.* TB/3018

D. W. BROGAN: The Development of Modern France. ᵒ △
Volume I: *From the Fall of the Empire to the Dreyfus Affair* TB/1184
Volume II: *The Shadow of War, World War I, Between the Two Wars. New Introduction by the Author* TB/1185

J. BRONOWSKI & BRUCE MAZLISH: The Western Intellectual Tradition: *From Leonardo to Hegel △* TB/3001

GEOFFREY BRUUN: Europe and the French Imperium, 1799-1814. * *Illus.* TB/3033

ALAN BULLOCK: Hitler, A Study in Tyranny. ᵒ △ *Illus.* TB/1123

E. H. CARR: The Twenty Years' Crisis, 1919-1939: *An Introduction to the Study of International Relations ᵒ △* TB/1122

GORDON A. CRAIG: From Bismarck to Adenauer: *Aspects of German Statecraft. Revised Edition* TB/1171

WALTER L. DORN: Competition for Empire, 1740-1763. * *Illus.* TB/3032

FRANKLIN L. FORD: Robe and Sword: *The Regrouping of the French Aristocracy after Louis XIV* TB/1217

CARL J. FRIEDRICH: The Age of the Baroque, 1610-1660. * *Illus.* TB/3004

RENÉ FUELOEP-MILLER: The Mind and Face of Bolshevism: *An Examination of Cultural Life in Soviet Russia. New Epilogue by the Author* TB/1188

M. DOROTHY GEORGE: London Life in the Eighteenth Century ᵒ △ TB/1182

LEO GERSHOY: From Despotism to Revolution, 1763-1789. * *Illus.* TB/3017

C. C. GILLISPIE: Genesis and Geology: *The Decades before Darwin* § TB/51

4

ALBERT GOODWIN: The French Revolution △ TB/1064

ALBERT GUÉRARD: France in the Classical Age: *The Life and Death of an Ideal* △ TB/1183

CARLTON J. H. HAYES: A Generation of Materialism, 1871-1900. * *Illus.* TB/3039

J. H. HEXTER: Reappraisals in History: *New Views on History and Society in Early Modern Europe* △ TB/1100

STANLEY HOFFMANN et al.: In Search of France: *The Economy, Society and Political System in the Twentieth Century* TB/1219

A. R. HUMPHREYS: The Augustan World: *Society, Thought, & Letters in 18th Century England* ○ △ TB/1105

DAN N. JACOBS, Ed.: The New Communist Manifesto and Related Documents. *Third edition, revised* TB/1078

HANS KOHN: The Mind of Germany: *The Education of a Nation* △ TB/1204

HANS KOHN, Ed.: The Mind of Modern Russia: *Historical and Political Thought of Russia's Great Age* TB/1065

FRANK E. MANUEL: The Prophets of Paris: Turgot, Condorcet, Saint-Simon, Fourier, and Comte TB/1218

KINGSLEY MARTIN: French Liberal Thought in the Eighteenth Century: *A Study of Political Ideas from Bayle to Condorcet* TB/1114

L. B. NAMIER: Personalities and Powers: *Selected Essays* △ TB/1186

L. B. NAMIER: Vanished Supremacies: *Essays on European History, 1812-1918* ○△ TB/1088

JOHN U. NEF: Western Civilization Since the Renaissance: *Peace, War, Industry, and the Arts* TB/1113

FREDERICK L. NUSSBAUM: The Triumph of Science and Reason, 1660-1685. * *Illus.* TB/3009

JOHN PLAMENATZ: German Marxism and Russian Communism. ○△ *New Preface by the Author* TB/1189

RAYMOND W. POSTGATE, Ed.: Revolution from 1789 to 1906: *Selected Documents* TB/1063

PENFIELD ROBERTS: The Quest for Security, 1715-1740. * *Illus.* TB/3016

PRISCILLA ROBERTSON: Revolutions of 1848: *A Social History* TB/1025

LOUIS, DUC DE SAINT-SIMON: Versailles, The Court, and Louis XIV. ○ △ *Introductory Note by Peter Gay* TB/1250

ALBERT SOREL: Europe Under the Old Regime. *Translated by Francis H. Herrick* TB/1121

N. N. SUKHANOV: The Russian Revolution, 1917: *Eyewitness Account.* △ *Edited by Joel Carmichael* Vol. I TB/1066; Vol. II TB/1067

A. J. P. TAYLOR: From Napoleon to Lenin: *Historical Essays* ○△ TB/1268

A. J. P. TAYLOR: The Habsburg Monarchy, 1809-1918: *A History of the Austrian Empire and Austria-Hungary* ○△ TB/1187

G. M. TREVELYAN: British History in the Nineteenth Century and After: *1782-1919.* ○△ *Second Edition* TB/1251

H. R. TREVOR-ROPER: Historical Essays ○△ TB/1269

JOHN B. WOLF: The Emergence of the Great Powers, 1685-1715. * *Illus.* TB/3010

JOHN B. WOLF: France: 1814-1919: *The Rise of a Liberal-Democratic Society* TB/3019

Intellectual History & History of Ideas

HERSCHEL BAKER: The Image of Man: *A Study of the Idea of Human Dignity in Classical Antiquity, the Middle Ages, and the Renaissance* TB/1047

R. R. BOLGAR: The Classical Heritage and Its Beneficiaries: *From the Carolingian Age to the End of the Renaissance* △ TB/1125

RANDOLPH S. BOURNE: War and the Intellectuals: *Collected Essays, 1915-1919.* △ ‡ *Edited by Carl Resek* TB/3043

J. BRONOWSKI & BRUCE MAZLISH: The Western Intellectual Tradition: *From Leonardo to Hegel* △ TB/3001

ERNST CASSIRER: The Individual and the Cosmos in Renaissance Philosophy. △ *Translated with an Introduction by Mario Domandi* TB/1097

NORMAN COHN: The Pursuit of the Millennium: *Revolutionary Messianism in Medieval and Reformation Europe* △ TB/1037

C. C. GILLISPIE: Genesis and Geology: *The Decades before Darwin* § TB/51

G. RACHEL LEVY: Religion Conceptions of the Stone Age and Their Influence upon European Thought. △ *Illus. Introduction by Henri Frankfort* TB/106

ARTHUR O. LOVEJOY: The Great Chain of Being: *A Study of the History of an Idea* TB/1009

FRANK E. MANUEL: The Prophets of Paris: Turgot, Condorcet, Saint-Simon, Fourier, and Comte TB/1218

PERRY MILLER & T. H. JOHNSON, Editors: The Puritans: *A Sourcebook of Their Writings* Vol. I TB/1093; Vol. II TB/1094

MILTON C. NAHM: Genius and Creativity: *An Essay in the History of Ideas* TB/1196

ROBERT PAYNE: Hubris: *A Study of Pride. Foreword by Sir Herbert Read* TB/1031

RALPH BARTON PERRY: The Thought and Character of William James: *Briefer Version* TB/1156

GEORG SIMMEL et al.: Essays on Sociology, Philosophy, and Aesthetics. ¶ *Edited by Kurt H. Wolff* TB/1234

BRUNO SNELL: The Discovery of the Mind: *The Greek Origins of European Thought* △ TB/1018

PAGET TOYNBEE: Dante Alighieri: *His Life and Works. Edited with Intro. by Charles S. Singleton* TB/1206

ERNEST LEE TUVESON: Millennium and Utopia: *A Study in the Background of the Idea of Progress.* ¶ *New Preface by the Author* TB/1134

PAUL VALÉRY: The Outlook for Intelligence △ TB/2016

PHILIP P. WIENER: Evolution and the Founders of Pragmatism. △ *Foreword by John Dewey* TB/1212

BASIL WILLEY: Nineteenth Century Studies: *Coleridge to Matthew Arnold* △ TB/1261

BASIL WILLEY: More Nineteenth Century Studies: *A Group of Honest Doubters* ○ △ TB/1262

Literature, Poetry, The Novel & Criticism

JAMES BAIRD: Ishmael: *The Art of Melville in the Contexts of International Primitivism* TB/1023

JACQUES BARZUN: The House of Intellect △ TB/1051

W. J. BATE: From Classic to Romantic: *Premises of Taste in Eighteenth Century England* TB/1036

RACHEL BESPALOFF: On the Iliad TB/2006

R. P. BLACKMUR et al.: Lectures in Criticism. *Introduction by Huntington Cairns* TB/2003

JAMES BOSWELL: The Life of Dr. Johnson & The Journal of a Tour to the Hebrides with Samuel Johnson LL.D.: *Selections.* ○ △ *Edited by F. V. Morley. Illus. by Ernest Shepard* TB/1254

ABRAHAM CAHAN: The Rise of David Levinsky: *a documentary novel of social mobility in early twentieth century America. Intro. by John Higham* TB/1028

ERNST R. CURTIUS: European Literature and the Latin Middle Ages △ TB/2015

GEORGE ELIOT: Daniel Deronda: *a novel. Introduction by F. R. Leavis* TB/1039

ADOLF ERMAN, Ed.: The Ancient Egyptians: *A Sourcebook of Their Writings. New Material and Introduction by William Kelly Simpson* TB/1233

ÉTIENNE GILSON: Dante and Philosophy TB/1089

ALFRED HARBAGE: As They Liked It: *A Study of Shakespeare's Moral Artistry* TB/1035

STANLEY R. HOPPER, Ed.: Spiritual Problems in Contemporary Literature § TB/21

A. R. HUMPHREYS: The Augustan World: *Society, Thought and Letters in 18th Century England* ○ △ TB/1105

ALDOUS HUXLEY: Antic Hay & The Giaconda Smile. ○ △ *Introduction by Martin Green* TB/3503

ALDOUS HUXLEY: Brave New World & Brave New World Revisited. ⁰ △ *Introduction by Martin Green*   TB/3501

HENRY JAMES: The Tragic Muse: *a novel. Introduction by Leon Edel*   TB/1017

ARNOLD KETTLE: An Introduction to the English Novel. △
Volume I: *Defoe to George Eliot*   TB/1011
Volume II: *Henry James to the Present*   TB/1012

RICHMOND LATTIMORE: The Poetry of Greek Tragedy △   TB/1257

J. B. LEISHMAN: The Monarch of Wit: *An Analytical and Comparative Study of the Poetry of John Donne* ⁰ △   TB/1258

J. B. LEISHMAN: Themes and Variations in Shakespeare's Sonnets ⁰ △   TB/1259

ROGER SHERMAN LOOMIS: The Development of Arthurian Romance △   TB/1167

JOHN STUART MILL: On Bentham and Coleridge. △ *Introduction by F. R. Leavis*   TB/1070

KENNETH B. MURDOCK: Literature and Theology in Colonial New England   TB/99

SAMUEL PEPYS: The Diary of Samuel Pepys. ⁰ *Edited by O. F. Morshead. Illus. by Ernest Shepard*   TB/1007

ST.-JOHN PERSE: Seamarks   TB/2002

V. DE S. PINTO: Crisis in English Poetry, 1880–1940 ⁰   TB/1260

GEORGE SANTAYANA: Interpretations of Poetry and Religion §   TB/9

C. K. STEAD: The New Poetic: *Yeats to Eliot* △   TB/1263

HEINRICH STRAUMANN: American Literature in the Twentieth Century. △ *Third Edition, Revised*   TB/1168

PAGET TOYNBEE: Dante Alighieri: *His Life and Works. Edited with Intro. by Charles S. Singleton*   TB/1206

DOROTHY VAN GHENT: The English Novel: *Form and Function*   TB/1050

E. B. WHITE: One Man's Meat. *Introduction by Walter Blair*   TB/3505

BASIL WILLEY: Nineteenth Century Studies: *Coleridge to Matthew Arnold* △   TB/1261

BASIL WILLEY: More Nineteenth Century Studies: *A Group of Honest Doubters* ⁰ △   TB/1262

RAYMOND WILLIAMS: Culture and Society, 1780–1950 ⁰ △   TB/1252

RAYMOND WILLIAMS: The Long Revolution. ⁰ △ *Revised Edition*   TB/1253

MORTON DAUWEN ZABEL, Editor: Literary Opinion in America   Vol. I TB/3013;   Vol. II TB/3014

Myth, Symbol & Folklore

JOSEPH CAMPBELL, Editor: Pagan and Christian Mysteries *Illus.*   TB/2013

MIRCEA ELIADE: Cosmos and History: *The Myth of the Eternal Return* § △   TB/2050

MIRCEA ELIADE: Rites and Symbols of Initiation: *The Mysteries of Birth and Rebirth* § △   TB/1236

C. G. JUNG & C. KERÉNYI: Essays on a Science of Mythology: *The Myths of the Divine Child and the Divine Maiden*   TB/2014

DORA & ERWIN PANOFSKY: Pandora's Box: *The Changing Aspects of a Mythical Symbol.* △ *Revised edition. Illus.*   TB/2021

ERWIN PANOFSKY: Studies in Iconology: *Humanistic Themes in the Art of the Renaissance.* △ *180 illustrations*   TB/1077

JEAN SEZNEC: The Survival of the Pagan Gods: *The Mythological Tradition and its Place in Renaissance Humanism and Art.* △ *108 illustrations*   TB/2004

HELLMUT WILHELM: Change: *Eight Lectures on the I Ching* △   TB/2019

HEINRICH ZIMMER: Myths and Symbols in Indian Art and Civilization. △ *70 illustrations*   TB/2005

Philosophy

G. E. M. ANSCOMBE: An Introduction to Wittgenstein's Tractatus. ⁰ △ *Second Edition, Revised*   TB/1210

HENRI BERGSON: Time and Free Will: *An Essay on the Immediate Data of Consciousness* ⁰ △   TB/1021

H. J. BLACKHAM: Six Existentialist Thinkers: *Kierkegaard, Nietzsche, Jaspers, Marcel, Heidegger, Sartre* ⁰ △   TB/1002

CRANE BRINTON: Nietzsche. *New Preface, Bibliography and Epilogue by the Author*   TB/1197

ERNST CASSIRER: The Individual and the Cosmos in Renaissance Philosophy. △ *Translated with an Introduction by Mario Domandi*   TB/1097

ERNST CASSIRER: Rousseau, Kant and Goethe. *Introduction by Peter Gay*   TB/1092

FREDERICK COPLESTON: Medieval Philosophy ⁰ △   TB/376

F. M. CORNFORD: Principium Sapientiae: *A Study of the Origins of Greek Philosophical Thought. Edited by W. K. C. Guthrie*   TB/1213

F. M. CORNFORD: From Religion to Philosophy: *A Study in the Origins of Western Speculation* §   TB/20

WILFRID DESAN: The Tragic Finale: *An Essay on the Philosophy of Jean-Paul Sartre*   TB/1030

A. P. D'ENTRÈVES: Natural Law: *An Historical Survey* △   TB/1223

HERBERT FINGARETTE: The Self in Transformation: *Psychoanalysis, Philosophy and the Life of the Spirit* ¶   TB/1177

PAUL FRIEDLÄNDER: Plato: *An Introduction* △   TB/2017

ÉTIENNE GILSON: Dante and Philosophy   TB/1089

WILLIAM CHASE GREENE: Moira: *Fate, Good, and Evil in Greek Thought*   TB/1104

W. K. C. GUTHRIE: The Greek Philosophers: *From Thales to Aristotle* ⁰ △   TB/1008

F. H. HEINEMANN: Existentialism and the Modern Predicament △   TB/28

ISAAC HUSIK: A History of Medieval Jewish Philosophy   JP/3

EDMUND HUSSERL: Phenomenology and the Crisis of Philosophy. *Translated with an Introduction by Quentin Lauer*   TB/1170

IMMANUEL KANT: The Doctrine of Virtue, *being Part II of the Metaphysic of Morals. Trans. with Notes & Intro. by Mary J. Gregor. Foreword by H. J. Paton*   TB/110

IMMANUEL KANT: Groundwork of the Metaphysic of Morals. *Trans. & analyzed by H. J. Paton*   TB/1159

IMMANUEL KANT: Lectures on Ethics. § △ *Introduction by Lewis W. Beck*   TB/105

IMMANUEL KANT: Religion Within the Limits of Reason Alone. § *Intro. by T. M. Greene & J. Silber*   TB/67

QUENTIN LAUER: Phenomenology: *Its Genesis and Prospect*   TB/1169

GABRIEL MARCEL: Being and Having: *An Existential Diary.* △ *Intro. by James Collins*   TB/310

GEORGE A. MORGAN: What Nietzsche Means   TB/1198

PHILO, SAADYA GAON, & JEHUDA HALEVI: Three Jewish Philosophers. *Ed. by Hans Lewy, Alexander Altmann, & Isaak Heinemann*   TB/813

MICHAEL POLANYI: Personal Knowledge: *Towards a Post-Critical Philosophy* △   TB/1158

WILLARD VAN ORMAN QUINE: Elementary Logic: *Revised Edition*   TB/577

WILLARD VAN ORMAN QUINE: From a Logical Point of View: *Logico-Philosophical Essays*   TB/566

BERTRAND RUSSELL et al.: The Philosophy of Bertrand Russell. *Edited by Paul Arthur Schilpp*
Vol. I TB/1095;   Vol. II TB/1096

L. S. STEBBING: A Modern Introduction to Logic △ TB/538

ALFRED NORTH WHITEHEAD: Process and Reality: *An Essay in Cosmology* △   TB/1033

PHILIP P. WIENER: Evolution and the Founders of Pragmatism. *Foreword by John Dewey*   TB/1212

WILHELM WINDELBAND: A History of Philosophy
Vol. I: *Greek, Roman, Medieval*   TB/38
Vol. II: *Renaissance, Enlightenment, Modern*   TB/39

LUDWIG WITTGENSTEIN: The Blue and Brown Books ⁰   TB/1211

Political Science & Government

JEREMY BENTHAM: The Handbook of Political Fallacies: Introduction by Crane Brinton   TB/1069

KENNETH E. BOULDING: Conflict and Defense: A General Theory   TB/3024

CRANE BRINTON: English Political Thought in the Nineteenth Century   TB/1071

EDWARD S. CORWIN: American Constitutional History: Essays edited by Alpheus T. Mason and Gerald Garvey   TB/1136

ROBERT DAHL & CHARLES E. LINDBLOM: Politics, Economics, and Welfare: Planning and Politico-Economic Systems Resolved into Basic Social Processes   TB/3037

JOHN NEVILLE FIGGIS: The Divine Right of Kings. Introduction by G. R. Elton   TB/1191

JOHN NEVILLE FIGGIS: Political Thought from Gerson to Grotius: 1414-1625: Seven Studies. Introduction by Garrett Mattingly   TB/1032

F. L. GANSHOF: Feudalism △   TB/1058

G. P. GOOCH: English Democratic Ideas in the Seventeenth Century   TB/1006

J. H. HEXTER: More's Utopia: The Biography of an Idea. New Epilogue by the Author   TB/1195

SIDNEY HOOK: Reason, Social Myths and Democracy △   TB/1237

ROBERT H. JACKSON: The Supreme Court in the American System of Government △   TB/1106

DAN N. JACOBS, Ed.: The New Communist Manifesto and Related Documents. Third Edition, Revised   TB/1078

DAN N. JACOBS & HANS BAERWALD, Eds.: Chinese Communism: Selected Documents   TB/3031

ROBERT GREEN MC CLOSKEY: American Conservatism in the Age of Enterprise, 1865-1910   TB/1137

KINGSLEY MARTIN: French Liberal Thought in the Eighteenth Century: Political Ideas from Bayle to Condorcet △   TB/1114

ROBERTO MICHELS: First Lectures in Political Sociology. Edited by Alfred de Grazia ¶ °   TB/1224

JOHN STUART MILL: On Bentham and Coleridge. △ Introduction by F. R. Leavis   TB/1070

BARRINGTON MOORE, JR.: Political Power and Social Theory: Seven Studies ¶   TB/1221

BARRINGTON MOORE, JR.: Soviet Politics—The Dilemma of Power: The Role of Ideas in Social Change ¶   TB/1222

BARRINGTON MOORE, JR.: Terror and Progress—USSR: Some Sources of Change and Stability in the Soviet Dictatorship ¶   TB/1266

JOHN B. MORRALL: Political Thought in Medieval Times △   TB/1076

JOHN PLAMENATZ: German Marxism and Russian Communism. ° △ New Preface by the Author   TB/1189

KARL R. POPPER: The Open Society and Its Enemies △
Vol. I: The Spell of Plato   TB/1101
Vol. II: The High Tide of Prophecy: Hegel, Marx and the Aftermath   TB/1102

HENRI DE SAINT-SIMON: Social Organization, The Science of Man, and Other Writings. Edited and Translated by Felix Markham   TB/1152

JOSEPH A. SCHUMPETER: Capitalism, Socialism and Democracy △   TB/3008

CHARLES H. SHINN: Mining Camps: A Study in American Frontier Government. ‡ Edited by Rodman W. Paul   TB/3062

Psychology

ALFRED ADLER: The Individual Psychology of Alfred Adler. △ Edited by Heinz L. and Rowena R. Ansbacher   TB/1154

ALFRED ADLER: Problems of Neurosis. Introduction by Heinz L. Ansbacher   TB/1145

ANTON T. BOISEN: The Exploration of the Inner World: A Study of Mental Disorder and Religious Experience   TB/87

HERBERT FINGARETTE: The Self in Transformation: Psychoanalysis, Philosophy and the Life of the Spirit ¶   TB/1177

SIGMUND FREUD: On Creativity and the Unconscious: Papers on the Psychology of Art, Literature, Love, Religion. § △ Intro. by Benjamin Nelson   TB/45

C. JUDSON HERRICK: The Evolution of Human Nature   TB/545

WILLIAM JAMES: Psychology: The Briefer Course. Edited with an Intro. by Gordon Allport   TB/1034

C. G. JUNG: Psychological Reflections △   TB/2001

C. G. JUNG: Symbols of Transformation: An Analysis of the Prelude to a Case of Schizophrenia. △ Illus.
Vol. I TB/2009; Vol. II TB/2010

C. G. JUNG & C. KERÉNYI: Essays on a Science of Mythology: The Myths of the Divine Child and the Divine Maiden   TB/2014

JOHN T. MC NEILL: A History of the Cure of Souls   TB/126

KARL MENNINGER: Theory of Psychoanalytic Technique   TB/1144

ERICH NEUMANN: Amor and Psyche: The Psychic Development of the Feminine △   TB/2012

ERICH NEUMANN: The Archetypal World of Henry Moore. △ 107 illus.   TB/2020

ERICH NEUMANN: The Origins and History of Consciousness △   Vol. I Illus. TB/2007; Vol. II TB/2008

C. P. OBERNDORF: A History of Psychoanalysis in America   TB/1147

RALPH BARTON PERRY: The Thought and Character of William James: Briefer Version   TB/1156

JEAN PIAGET, BÄRBEL INHELDER, & ALINA SZEMINSKA: The Child's Conception of Geometry ° △   TB/1146

JOHN H. SCHAAR: Escape from Authority: The Perspectives of Erich Fromm   TB/1155

Sociology

JACQUES BARZUN: Race: A Study in Superstition. Revised Edition   TB/1172

BERNARD BERELSON, Ed.: The Behavioral Sciences Today   TB/1127

ABRAHAM CAHAN: The Rise of David Levinsky: A documentary novel of social mobility in early twentieth century America. Intro. by John Higham   TB/1028

THOMAS C. COCHRAN: The Inner Revolution: Essays on the Social Sciences in History   TB/1140

ALLISON DAVIS & JOHN DOLLARD: Children of Bondage: The Personality Development of Negro Youth in the Urban South ¶   TB/3049

ST. CLAIR DRAKE & HORACE R. CAYTON: Black Metropolis: A Study of Negro Life in a Northern City. △ Revised and Enlarged. Intro. by Everett C. Hughes
Vol. I TB/1086; Vol. II TB/1087

EMILE DURKHEIM et al.: Essays on Sociology and Philosophy: With Analyses of Durkheim's Life and Work. ¶ Edited by Kurt H. Wolff   TB/1151

LEON FESTINGER, HENRY W. RIECKEN & STANLEY SCHACHTER: When Prophecy Fails: A Social and Psychological Account of a Modern Group that Predicted the Destruction of the World ¶   TB/1132

ALVIN W. GOULDNER: Wildcat Strike: A Study in Worker-Management Relationships ¶   TB/1176

FRANCIS J. GRUND: Aristocracy in America: Social Class in the Formative Years of the New Nation △   TB/1001

KURT LEWIN: Field Theory in Social Science: Selected Theoretical Papers. ¶ △ Edited with a Foreword by Dorwin Cartwright   TB/1135

R. M. MAC IVER: Social Causation   TB/1153

ROBERT K. MERTON, LEONARD BROOM, LEONARD S. COTTRELL, JR., Editors: Sociology Today: Problems and Prospects ¶   Vol. I TB/1173; Vol. II TB/1174

ROBERTO MICHELS: First Lectures in Political Sociology. Edited by Alfred de Grazia ¶ °   TB/1224

BARRINGTON MOORE, JR.: Political Power and Social Theory: Seven Studies ¶   TB/1221

BARRINGTON MOORE, JR.: Soviet Politics—The Dilemma of Power: *The Role of Ideas in Social Change* ¶
TB/1222

TALCOTT PARSONS & EDWARD A. SHILS, Editors: Toward a General Theory of Action: *Theoretical Foundations for the Social Sciences* TB/1083

JOHN H. ROHRER & MUNRO S. EDMONDSON, Eds.: The Eighth Generation Grows Up: *Cultures and Personalities of New Orleans Negroes* ¶ TB/3050

ARNOLD ROSE: The Negro in America: *The Condensed Version of Gunnar Myrdal's An American Dilemma*
TB/3048

KURT SAMUELSSON: Religion and Economic Action: *A Critique of Max Weber's The Protestant Ethic and the Spirit of Capitalism.* ¶ ° *Trans. by E. G. French. Ed. with Intro. by D. C. Coleman* TB/1131

PHILIP SELZNICK: TVA and the Grass Roots: *A Study in the Sociology of Formal Organization* TB/1230

GEORG SIMMEL et al.: Essays on Sociology, Philosophy, and Aesthetics. ¶ *Edited by Kurt H. Wolff* TB/1234

HERBERT SIMON: The Shape of Automation: *For Men and Management* △ TB/1245

PITIRIM A. SOROKIN: Contemporary Sociological Theories: *Through the First Quarter of the 20th Century* TB/3046

MAURICE R. STEIN: The Eclipse of Community: *An Interpretation of American Studies* TB/1128

FERDINAND TÖNNIES: Community and Society: *Gemeinschaft und Gesellschaft. Translated and edited by Charles P. Loomis* TB/1116

W. LLOYD WARNER & Associates: Democracy in Jonesville: *A Study in Quality and Inequality* TB/1129

W. LLOYD WARNER: Social Class in America: *The Evaluation of Status* TB/1013

RELIGION

Ancient & Classical

J. H. BREASTED: Development of Religion and Thought in Ancient Egypt. *Intro. by John A. Wilson* TB/57

HENRI FRANKFORT: Ancient Egyptian Religion: *An Interpretation* TB/77

G. RACHEL LEVY: Religious Conceptions of the Stone Age and their Influence upon European Thought. △ *Illus. Introduction by Henri Frankfort* TB/106

MARTIN P. NILSSON: Greek Folk Religion. *Foreword by Arthur Darby Nock* TB/78

ALEXANDRE PIANKOFF: The Shrines of Tut-Ankh-Amon. △ *Edited by N. Rambova.* 117 *illus.* TB/2011

H. J. ROSE: Religion in Greece and Rome △ TB/55

Biblical Thought & Literature

W. F. ALBRIGHT: The Biblical Period from Abraham to Ezra TB/102

C. K. BARRETT, Ed.: The New Testament Background: *Selected Documents* △ TB/86

C. H. DODD: The Authority of the Bible △ TB/43

M. S. ENSLIN: Christian Beginnings △ TB/5

M. S. ENSLIN: The Literature of the Christian Movement △ TB/6

JOHN GRAY: Archaeology and the Old Testament World. △ *Illus.* TB/127

JAMES MUILENBURG: The Way of Israel: *Biblical Faith and Ethics* △ TB/133

H. H. ROWLEY: The Growth of the Old Testament △
TB/107

D. WINTON THOMAS, Ed.: Documents from Old Testament Times △ TB/85

The Judaic Tradition

LEO BAECK: Judaism and Christianity. *Trans. with Intro. by Walter Kaufmann* JP/23

SALO W. BARON: Modern Nationalism and Religion
JP/18

MARTIN BUBER: Eclipse of God: *Studies in the Relation Between Religion and Philosophy* △ TB/12

MARTIN BUBER: For the Sake of Heaven TB/801

MARTIN BUBER: The Knowledge of Man: *Selected Essays.* △ *Edited with an Introduction by Maurice Friedman. Translated by Maurice Friedman and Ronald Gregor Smith* TB/135

MARTIN BUBER: Moses: *The Revelation and the Covenant* △ TB/27

MARTIN BUBER: The Origin and Meaning of Hasidism △
TB/835

MARTIN BUBER: Pointing the Way. △ *Introduction by Maurice S. Friedman* TB/103

MARTIN BUBER: The Prophetic Faith TB/73

MARTIN BUBER: Two Types of Faith: *the interpenetration of Judaism and Christianity* ° △ TB/75

ERNST LUDWIG EHRLICH: A Concise History of Israel: *From the Earliest Times to the Destruction of the Temple in A.D.* 70 ° △ TB/128

MAURICE S. FRIEDMAN: Martin Buber: *The Life of Dialogue* △ TB/64

GENESIS: *The NJV Translation* TB/836

SOLOMON GRAYZEL: A History of the Contemporary Jews
TB/816

WILL HERBERG: Judaism and Modern Man TB/810

ARTHUR HERTZBERG: The Zionist Idea TB/817

ABRAHAM J. HESCHEL: God in Search of Man: *A Philosophy of Judaism* JP/7

ISAAC HUSIK: A History of Medieval Jewish Philosophy
JP/3

FLAVIUS JOSEPHUS: The Great Roman-Jewish War, *with The Life of Josephus. Introduction by William R. Farmer* TB/74

JACOB R. MARCUS: The Jew in the Medieval World TB/814

MAX L. MARGOLIS & ALEXANDER MARX: A History of the Jewish People TB/806

T. J. MEEK: Hebrew Origins TB/69

C. G. MONTEFIORE & H. LOEWE, Eds.: A Rabbinic Anthology. JP/32

JAMES PARKES: The Conflict of the Church and the Synagogue: *The Jews and Early Christianity* JP/21

PHILO, SAADYA GAON, & JEHUDA HALEVI: Three Jewish Philosophers. *Ed. by Hans Lewey, Alexander Altmann, & Isaak Heinemann* TB/813

CECIL ROTH: A History of the Marranos TB/812

CECIL ROTH: The Jews in the Renaissance. *Illus.* TB/834

HERMAN L. STRACK: Introduction to the Talmud and Midrash TB/808

JOSHUA TRACHTENBERG: The Devil and the Jews: *The Medieval Conception of the Jew and its Relation to Modern Anti-Semitism* JP/22

Christianity: General

ROLAND H. BAINTON: Christendom: *A Short History of Christianity and its Impact on Western Civilization.* △ *Illus.* Vol. I TB/131; Vol. II TB/132

Christianity: Origins & Early Development

AUGUSTINE: An Augustine Synthesis. △ *Edited by Erich Przywara* TB/335

ADOLF DEISSMANN: Paul: *A Study in Social and Religious History* TB/15

EDWARD GIBBON: The Triumph of Christendom in the Roman Empire (*Chaps. XV-XX of "Decline and Fall," J. B. Bury edition*). § △ *Illus.* TB/46

MAURICE GOGUEL: Jesus and the Origins of Christianity. ° △ *Introduction by C. Leslie Mitton*
Volume I: *Prologemena to the Life of Jesus* TB/65
Volume II: *The Life of Jesus* TB/66

EDGAR J. GOODSPEED: A Life of Jesus TB/1

ROBERT M. GRANT: Gnosticism and Early Christianity. △ *Revised Edition* TB/136

ADOLF HARNACK: The Mission and Expansion of Christianity in the First Three Centuries. *Introduction by Jaroslav Pelikan* TB/92

R. K. HARRISON: The Dead Sea Scrolls : *An Introduction* ᵒ ᐃ TB/84

EDWIN HATCH: The Influence of Greek Ideas on Christianity. § ᐃ *Introduction and Bibliography by Frederick C. Grant* TB/18

ARTHUR DARBY NOCK: Early Gentile Christianity and Its Hellenistic Background TB/111

ARTHUR DARBY NOCK: St. Paul ᵒ ᐃ TB/104

ORIGEN: On First Principles. ᐃ *Edited by G. W. Butterworth. Introduction by Henri de Lubac* TB/310

JAMES PARKES: The Conflict of the Church and the Synagogue: *The Jews and Early Christianity* JP/21

SULPICIUS SEVERUS et al.: The Western Fathers: *Being the Lives of Martin of Tours, Ambrose, Augustine of Hippo, Honoratus of Arles and Germanus of Auxerre.* ᐃ *Edited and translated by F. R. Hoare* TB/309

F. VAN DER MEER: Augustine the Bishop: *Church and Society at the Dawn of the Middle Ages* ᐃ TB/304

JOHANNES WEISS: Earliest Christianity: *A History of the Period A.D. 30-150. Introduction and Bibliography by Frederick C. Grant* Volume I TB/53
Volume II TB/54

Christianity: The Middle Ages and The Reformation

JOHN CALVIN & JACOPO SADOLETO: A Reformation Debate. *Edited by John C. Olin* TB/1239

G. CONSTANT: The Reformation in England: *The English Schism, Henry VIII, 1509-1547* ᐃ TB/314

CHRISTOPHER DAWSON, Ed.: Mission to Asia: *Narratives and Letters of the Franciscan Missionaries in Mongolia and China in the 13th and 14th Centuries* ᐃ TB/315

JOHANNES ECKHART: Meister Eckhart: *A Modern Translation by R. B. Blakney* TB/8

DESIDERIUS ERASMUS: Christian Humanism and the Reformation: *Selected Writings. Edited and translated by John C. Olin* TB/1166

ÉTIENNE GILSON: Dante and Philosophy ᐃ TB/1089

WILLIAM HALLER: The Rise of Puritanism ᐃ TB/22

HAJO HOLBORN: Ulrich von Hutten and the German Reformation TB/1238

JOHAN HUIZINGA: Erasmus and the Age of Reformation. ᐃ *Illus.* TB/19

A. C. MC GIFFERT: Protestant Thought Before Kant.ᐃ *Preface by Jaroslav Pelikan* TB/93

JOHN T. MC NEILL: Makers of the Christian Tradition: *From Alfred the Great to Schleiermacher* ᐃ TB/121

G. MOLLAT: The Popes at Avignon, 1305-1378 ᐃ TB/308

GORDON RUPP: Luther's Progress to the Diet of Worms ᵒ ᐃ TB/120

Christianity: The Protestant Tradition

KARL BARTH: Church Dogmatics: *A Selection* ᐃ TB/95

KARL BARTH: Dogmatics in Outline ᐃ TB/56

KARL BARTH: The Word of God and the Word of Man TB/13

RUDOLF BULTMANN et al: Translating Theology into the Modern Age: *Historical, Systematic and Pastoral Reflections on Theology and the Church in the Contemporary Situation. Volume 2 of Journal for Theology and the Church, edited by Robert W. Funk in association with Gerhard Ebeling* TB/252

WINTHROP HUDSON: The Great Tradition of the American Churches TB/98

SOREN KIERKEGAARD: Edifying Discourses. *Edited with an Introduction by Paul Holmer* TB/32

SOREN KIERKEGAARD: The Journals of Kierkegaard. ᵒ ᐃ *Ed. with Intro. by Alexander Dru* TB/52

SOREN KIERKEGAARD : The Point of View for My Work as an Author: *A Report to History.* § *Preface by Benjamin Nelson* TB/88

SOREN KIERKEGAARD: The Present Age. § ᐃ *Translated and edited by Alexander Dru. Introduction by Walter Kaufmann* TB/94

SOREN KIERKEGAARD: Purity of Heart ᐃ TB/4

SOREN KIERKEGAARD: Repetition: *An Essay in Experimental Psychology.* ᐃ *Translated with Introduction & Notes by Walter Lowrie* TB/117

SOREN KIERKEGAARD: Works of Love: *Some Christian Reflections in the Form of Discourses* ᐃ TB/122

WALTER LOWRIE: Kierkegaard: *A Life* Vol. I TB/89
Vol. II TB/90

JOHN MACQUARRIE: The Scope of Demythologizing: *Bultmann and his Critics* ᐃ TB/134

PERRY MILLER & T. H. JOHNSON, Editors: The Puritans: *A Sourcebook of Their Writings* Vol. I TB/1093
Vol. II TB/1094

JAMES M. ROBINSON et al.: The Bultmann School of Biblical Interpretation: New Directions? *Volume 1 of Journal for Theology and the Church, edited by Robert W. Funk in association with Gerhard Ebeling* TB/251

F. SCHLEIERMACHER: The Christian Faith. ᐃ *Introduction by Richard R. Niebuhr* Vol. I TB/108
Vol. II TB/109

F. SCHLEIERMACHER: On Religion: *Speeches to Its Cultured Despisers. Intro. by Rudolf Otto* TB/36

PAUL TILLICH: Dynamics of Faith ᐃ TB/42

EVELYN UNDERHILL: Worship ᐃ TB/10

G. VAN DER LEEUW: Religion in Essence and Manifestation: *A Study in Phenomenology.* ᐃ *Appendices by Hans H. Penner* Vol. I TB/100; Vol. II TB/101

Christianity: The Roman and Eastern Traditions

DOM CUTHBERT BUTLER: Western Mysticism: *The Teaching of Augustine, Gregory and Bernard on Contemplation and the Contemplative Life* § ᵒ ᐃ TB/312

A. ROBERT CAPONIGRI, Ed.: Modern Catholic Thinkers I: *God and Man* ᐃ TB/306

A. ROBERT CAPONIGRI, Ed.: Modern Catholic Thinkers II: *The Church and the Political Order*ᐃ TB/307

THOMAS CORBISHLEY, S.J.: Roman Catholicism ᐃ TB/112

CHRISTOPHER DAWSON: The Historic Reality of Christian Culture TB/305

G. P. FEDOTOV: The Russian Religious Mind: *Kievan Christianity, the 10th to the 13th centuries* TB/70

G. P. FEDOTOV, Ed.: A Treasury of Russian Spirituality TB/303

ÉTIENNE GILSON: The Spirit of Thomism TB/313

DAVID KNOWLES: The English Mystical Tradition ᐃ TB/302

GABRIEL MARCEL: Being and Having: *An Existential Diary.* ᐃ *Introduction by James Collins* TB/310

GABRIEL MARCEL: Homo Viator: *Introduction to a Metaphysic of Hope* TB/397

GUSTAVE WEIGEL, S. J.: Catholic Theology in Dialogue TB/301

Oriental Religions: Far Eastern, Near Eastern

TOR ANDRAE: Mohammed: *The Man and His Faith* ᐃ TB/62

EDWARD CONZE: Buddhism: *Its Essence and Development.* ᵒ ᐃ *Foreword by Arthur Waley* TB/58

EDWARD CONZE et al., Editors: Buddhist Texts Through the Ages ᐃ TB/113

ANANDA COOMARASWAMY: Buddha and the Gospel of Buddhism. ᐃ *Illus.* TB/119

H. G. CREEL: Confucius and the Chinese Way TB/63

FRANKLIN EDGERTON, Trans. & Ed.: The Bhagavad Gita TB/115

SWAMI NIKHILANANDA, Trans. & Ed.: The Upanishads: *A One-Volume Abridgment* △   TB/114
HELLMUT WILHELM: Change: *Eight Lectures on the I Ching* △   TB/2019

Philosophy of Religion

NICOLAS BERDYAEV: The Beginning and the End § △ TB/14
NICOLAS BERDYAEV: Christian Existentialism: *A Berdyaev Synthesis.* △ *Ed. by Donald A. Lowrie* TB/130
NICOLAS BERDYAEV: The Destiny of Man △   TB/61
RUDOLF BULTMANN: History and Eschatology: *The Presence of Eternity* ⁰   TB/91
RUDOLF BULTMANN AND FIVE CRITICS: Kerygma and Myth: *A Theological Debate* △   TB/80
RUDOLF BULTMANN and KARL KUNDSIN: Form Criticism: *Two Essays on New Testament Research.* △ *Translated by Frederick C. Grant* TB/96
MIRCEA ELIADE: The Sacred and the Profane TB/81
LUDWIG FEUERBACH: The Essence of Christianity. § *Introduction by Karl Barth. Foreword by H. Richard Niebuhr* TB/11
ÉTIENNE GILSON: The Spirit of Thomism TB/313
ADOLF HARNACK: What is Christianity? § △ *Introduction by Rudolf Bultmann* TB/17
FRIEDRICH HEGEL: On Christianity: *Early Theological Writings. Ed. by R. Kroner and T. M. Knox* TB/79
KARL HEIM: Christian Faith and Natural Science △ TB/16
IMMANUEL KANT: Religion Within the Limits of Reason Alone. § *Intro. by T. M. Greene & J. Silber* TB/67
K. E. KIRK: The Vision of God: *The Christian Doctrine of the Summum Bonum* § △ TB/137
JOHN MACQUARRIE: An Existentialist Theology: *A Comparison of Heidegger and Bultmann.* ⁰ △ *Preface by Rudolf Bultmann* TB/125
PAUL RAMSEY, Ed.: Faith and Ethics: *The Theology of H. Richard Niebuhr* TB/129
PIERRE TEILHARD DE CHARDIN: The Divine Milieu ⁰ △   TB/384
PIERRE TEILHARD DE CHARDIN: The Phenomenon of Man ⁰ △   TB/383

Religion, Culture & Society

JOSEPH L. BLAU, Ed.: Cornerstones of Religious Freedom in America: *Selected Basic Documents, Court Decisions and Public Statements. Revised and Enlarged Edition* TB/118
C. C. GILLISPIE: Genesis and Geology: *The Decades before Darwin* § TB/51
KYLE HASELDEN: The Racial Problem in Christian Perspective TB/116
WALTER KAUFMANN, Ed.: Religion from Tolstoy to Camus: *Basic Writings on Religious Truth and Morals. Enlarged Edition* TB/123
JOHN T. MC NEILL: A History of the Cure of Souls TB/126
KENNETH S. MURDOCK: Literature and Theology in Colonial New England TB/99
H. RICHARD NIEBUHR: Christ and Culture △ TB/3
H. RICHARD NIEBUHR: The Kingdom of God in America TB/49
R. B. PERRY: Puritanism and Democracy TB/1138
PAUL PFUETZE: Self, Society, Existence: *Human Nature and Dialogue in the Thought of George Herbert Mead and Martin Buber* TB/1059
WALTER RAUSCHENBUSCH: Christianity and the Social Crisis. ‡ *Edited by Robert D. Cross* TB/3059
KURT SAMUELSSON: Religion and Economic Action: *A Critique of Max Weber's The Protestant Ethic and the Spirit of Capitalism.*¶ ⁰ △ *Trans. by E. G. French. Ed. with Intro. by D. C. Coleman* TB/1131
TIMOTHY L. SMITH: Revivalism and Social Reform: *American Protestantism on the Eve of the Civil War* △ TB/1229
ERNST TROELTSCH: The Social Teaching of the Christian Churches ⁰ △   Vol. I TB/71; Vol. II TB/72

NATURAL SCIENCES AND MATHEMATICS

Biological Sciences

CHARLOTTE AUERBACH: The Science of Genetics Σ △   TB/568
MARSTON BATES: The Natural History of Mosquitoes. *Illus.* TB/578
A. BELLAIRS: Reptiles: *Life History, Evolution, and Structure.* △ *Illus.* TB/520
LUDWIG VON BERTALANFFY: Modern Theories of Development: *An Introduction to Theoretical Biology* TB/554
LUDWIG VON BERTALANFFY: Problems of Life: *An Evaluation of Modern Biological and Scientific Thought* △   TB/521
HAROLD F. BLUM: Time's Arrow and Evolution TB/555
JOHN TYLER BONNER: The Ideas of Biology. Σ △ *Illus.* TB/570
A. J. CAIN: Animal Species and their Evolution. △ *Illus.* TB/519
WALTER B. CANNON: Bodily Changes in Pain, Hunger, Fear and Rage. *Illus.* TB/562
W. E. LE GROS CLARK: The Antecedents of Man: *An Introduction to Evolution of the Primates.* ⁰ △ *Illus.* TB/559
W. H. DOWDESWELL: Animal Ecology. △ *Illus.* TB/543
W. H. DOWDESWELL: The Mechanism of Evolution. △ *Illus.* TB/527
R. W. GERARD: Unresting Cells. *Illus.* TB/541
DAVID LACK: Darwin's Finches. △ *Illus.* TB/544
ADOLF PORTMANN: Animals as Social Beings. ⁰ △ *Illus.* TB/572
O. W. RICHARDS: The Social Insects. △ *Illus.* TB/542
P. M. SHEPPARD: Natural Selection and Heredity. △ *Illus.* TB/528
EDMUND W. SINNOTT: Cell and Psyche: *The Biology of Purpose* TB/546
C. H. WADDINGTON: How Animals Develop. △ *Illus.* TB/553
C. H. WADDINGTON: The Nature of Life: *The Main Problems and Trends in Modern Biology* △ TB/580

Chemistry

J. R. PARTINGTON: A Short History of Chemistry. △ *Illus.* TB/522

Communication Theory

J. R. PIERCE: Symbols, Signals and Noise: *The Nature and Process of Communication* △ TB/574

Geography

R. E. COKER: This Great and Wide Sea: *An Introduction to Oceanography and Marine Biology. Illus.* TB/551
F. K. HARE: The Restless Atmosphere △ TB/560

History of Science

MARIE BOAS: The Scientific Renaissance, 1450-1630 ⁰ △   TB/583
W. DAMPIER, Ed.: Readings in the Literature of Science. *Illus.* TB/512
A. HUNTER DUPREE: Science in the Federal Government: *A History of Policies and Activities to 1940* △ TB/573
ALEXANDER KOYRÉ: From the Closed World to the Infinite Universe: *Copernicus, Kepler, Galileo, Newton, etc.* △   TB/31
A. G. VAN MELSEN: From Atomos to Atom: *A History of the Concept Atom* TB/517
O. NEUGEBAUER: The Exact Sciences in Antiquity TB/552
HANS THIRRING: Energy for Man: *From Windmills to Nuclear Power* △ TB/556
LANCELOT LAW WHYTE: Essay on Atomism: *From Democritus to 1960* △ TB/565

Mathematics

E. W. BETH: The Foundations of Mathematics: *A Study in the Philosophy of Science* △   TB/581
H. DAVENPORT: The Higher Arithmetic: *An Introduction to the Theory of Numbers* △   TB/526
H. G. FORDER: Geometry: *An Introduction* △   TB/548
S. KÖRNER: The Philosophy of Mathematics: *An Introduction* △   TB/547
D. E. LITTLEWOOD: Skeleton Key of Mathematics: *A Simple Account of Complex Algebraic Problems* △   TB/525
GEORGE E. OWEN: Fundamentals of Scientific Mathematics   TB/569
WILLARD VAN ORMAN QUINE: Mathematical Logic   TB/558
O. G. SUTTON: Mathematics in Action. ○ △ *Foreword by James R. Newman. Illus.*   TB/518
FREDERICK WAISMANN: Introduction to Mathematical Thinking. *Foreword by Karl Menger*   TB/511

Philosophy of Science

R. B. BRAITHWAITE: Scientific Explanation   TB/515
J. BRONOWSKI: Science and Human Values. △ *Revised and Enlarged Edition*   TB/505
ALBERT EINSTEIN et al.: Albert Einstein: Philosopher-Scientist. *Edited by Paul A. Schilpp*   Vol. I   TB/502
    Vol. II   TB/503
WERNER HEISENBERG: Physics and Philosophy: *The Revolution in Modern Science* △   TB/549
JOHN MAYNARD KEYNES: A Treatise on Probability. ○ △ *Introduction by N. R. Hanson*   TB/557
KARL R. POPPER: Logic of Scientific Discovery △   TB/576

STEPHEN TOULMIN: Foresight and Understanding: *An Enquiry into the Aims of Science.* △ *Foreword by Jacques Barzun*   TB/564
STEPHEN TOULMIN: The Philosophy of Science: *An Introduction* △   TB/513
G. J. WHITROW: The Natural Philosophy of Time ○ △   TB/563

Physics and Cosmology

JOHN E. ALLEN: Aerodynamics: *A Space Age Survey* △   TB/582
STEPHEN TOULMIN & JUNE GOODFIELD: The Fabric of the Heavens: *The Development of Astronomy and Dynamics.* △ *Illus.*   TB/579
DAVID BOHM: Causality and Chance in Modern Physics. △ *Foreword by Louis de Broglie*   TB/536
P. W. BRIDGMAN: Nature of Thermodynamics   TB/537
P. W. BRIDGMAN: A Sophisticate's Primer of Relativity △   TB/575
A. C. CROMBIE, Ed.: Turning Point in Physics   TB/535
C. V. DURELL: Readable Relativity. △ *Foreword by Freeman J. Dyson*   TB/530
ARTHUR EDDINGTON: Space, Time and Gravitation: *An Outline of the General Relativity Theory*   TB/510
GEORGE GAMOW: Biography of Physics Σ △   TB/567
MAX JAMMER: Concepts of Force: *A Study in the Foundation of Dynamics*   TB/550
MAX JAMMER: Concepts of Mass *in Classical and Modern Physics*   TB/571
MAX JAMMER: Concepts of Space: *The History of Theories of Space in Physics. Foreword by Albert Einstein*   TB/533
G. J. WHITROW: The Structure and Evolution of the Universe: *An Introduction to Cosmology.* △ *Illus.*   TB/504

Code to Torchbook Libraries:

TB/1+   : The Cloister Library
TB/301+   : The Cathedral Library
TB/501+   : The Science Library
TB/801+   : The Temple Library
TB/1001+   : The Academy Library
TB/2001+   : The Bollingen Library
TB/3001+   : The University Library
JP/1+   : The Jewish Publication Society Series